PHYSICS LIBRARY

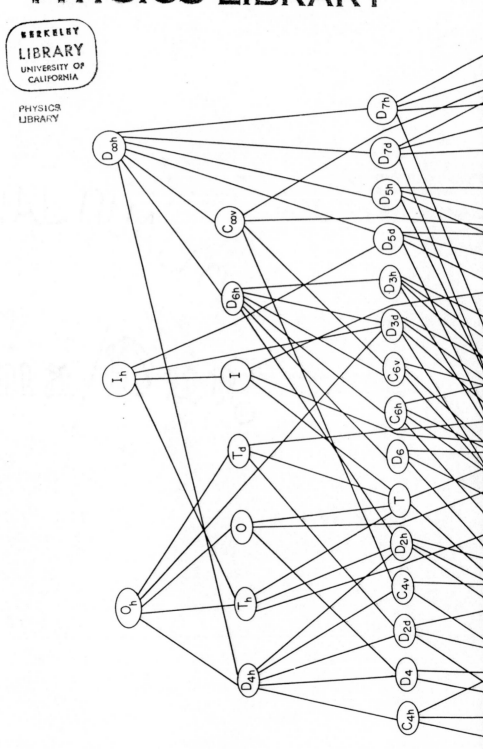

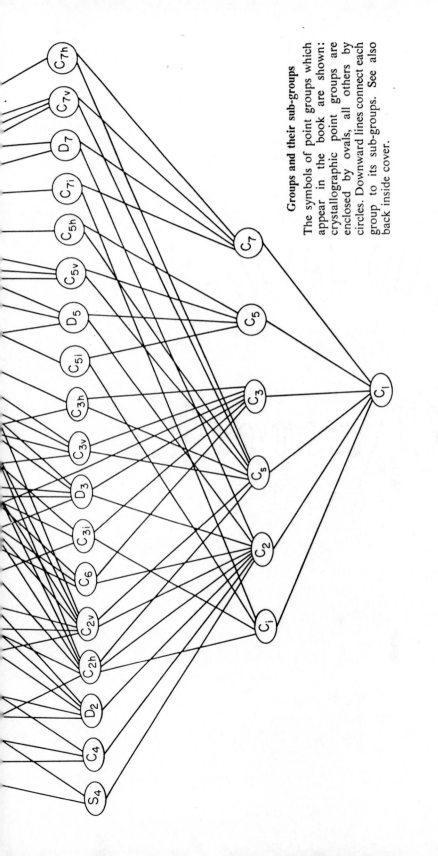

Groups and their sub-groups

The symbols of point groups which appear in the book are shown: crystallographic point groups are enclosed by ovals, all others by circles. Downward lines connect each group to its sub-groups. See also back inside cover.

The Jahn–Teller Effect
in
Molecules and Crystals

WILEY MONOGRAPHS IN CHEMICAL PHYSICS

Editor

John B. Birks, *Reader in Physics, University of Manchester*

Photophysics of Aromatic Molecules:
J. B. Birks, Department of Physics, University of Manchester

Atomic & Molecular Radiation Physics: *L. G. Christophorou, Oak Ridge National Laboratory, Tennessee*

The Jahn–Teller Effect in Molecules and Crystals: *R. Englman, Soreq Nuclear Research Centre, Yavne*

The Jahn–Teller Effect
in
Molecules and Crystals

R. Englman

Soreq Nuclear Research Centre,
Yavne

WILEY – INTERSCIENCE

a division of John Wiley & Sons Ltd

London   New York   Sydney   Toronto

Library of Congress catalog card number
77–37113

ISBN 0 471 24168 7

Set in 10/12 pt. Monotype Times, printed by letterpress, and bound
in Great Britain at The Pitman Press, Bath

An Historical Note

In the year 1934 both Landau and I were in the Institute of Niels Bohr at Copenhagen. We had many discussions. I told Landau of the work of one of my students, R. Renner, on degenerate electronic states in the linear CO_2 molecule. I explained that in this case an intricate coupling between the splitting of electron states and nuclear vibrations will arise which will modify the applicability of the Born–Oppenheimer approximation to these states. Landau objected.

He said that I should be very careful. In a degenerate electronic state the symmetry on which the degeneracy is based (in this case the fact that the three atoms are colinear in equilibrium) will in general be destroyed. I managed to convince Landau that his doubts were unfounded. (This may have been the only case in which I have won an argument with Landau.)

A year later in London I asked myself the question whether another exception to Landau's postulated statement might exist. It was clear that the electronic degeneracy may destroy the symmetry on which it is based. But how often will this necessarily occur? The question did not appear simple.

I proceeded to discuss the problem with H. A. Jahn who, as I, was a refugee from a German university. We went through all possible symmetries and found that linear molecules constitute the only exception. In all other cases Landau's suspicion was verified.

One problem was not solved. The proof of the so-called Jahn–Teller effect was obtained by the thoroughly inelegant method of enumerating every symmetry and discussing each symmetry one by one. No general proof is, to my knowledge, available even today.

This is the reason why the effect should carry the name of Landau. He suspected the effect, and no one has given a proof that a mathematician would enjoy. Jahn and I merely did a bit of spade work.

July 1971   EDWARD TELLER

Preface

For many years the implications of the Jahn–Teller effect were insufficiently understood. This is the probable explanation for the long period of relative stagnation which followed the formulation of the theorem of Jahn and Teller (1937). However, during the last five years or so our understanding of the Effect has greatly developed, and it has now become possible to write a book from the vantage-point if not of knowing, then at least of recognizing what was lacking before.

In this recent period there have been some outstandingly important pieces of work that were responsible for these developments. These introduced new concepts, removed some misconceptions and brought about several changes of emphasis. Alongside these modernizing developments there has been a massive growth of the literature, and the subject has become complex, fragmented and particular, thus getting a long way away from the one-sentence simplicity which characterized the original formulation some thirty-five years ago. In this volume I have made it my business to present all or most aspects of the modern Jahn–Teller effect. In addition, being intent on a unifying review of the subject I attempted to summarize the present status of the subject in a nutshell. The result is the introductory chapter.

Although aware of the growth and diversification of the subject I nevertheless believe that the theoretical turning point which dates with the paper of Jahn and Teller warrants the consideration of the subject of non-perturbational coupling as a whole, radiating as it were from a common origin. Sure enough, this conception of a variety of physical phenomena pivoting on a theoretical idea is not unassailable. Nor is it free from the writer's bias and in vain could I hope for a general acceptance of this essentially theoretician's point of view.

If it is suggested that kinship between various topics by a common parentage provides justification for this book, another pragmatic reason does so even more. I feel that there is bound to be a significant practical advantage in the collection into one volume of phenomena happening in very different systems which are subject to the same law and are describable by the same formalism. It may be expected that many ideas and methods which are the tools of trade in one field could be put to good use in another branch if only they were known. The fruits of cross-fertilization are difficult to define in advance, but a practice which has proven advantageous in so many walks of life can be safely recommended in this field as well.

It has been my intention all along to describe *how* the Jahn–Teller effect operates in various systems and not to describe the *systems* in which the Jahn–Teller effect operates. Still, a brief introduction of the various physical systems was thought unavoidable, if only in order to define the symbols, quantities and concepts which are used elsewhere in the text. The result was a roving chapter with no natural place in the book, to which frequent references are made in the other chapters and which would perhaps best be read parallel to the main text. I should emphasize that it was not meant to provide a self-contained source of instruction. For want of a better place it was laid to rest in the seventh chapter. Another didactic feature in the lay-out of the book is the setting in small print of material which has not yet stood the test of time or of work by the author which, partly, has not yet been published.

A word on the book's background will perhaps be allowed. This book was not written inside that enviable synthetic commodity: the hermetically sealed ivory tower. On the contrary it was mainly written or prepared in a tumultuous period and during rather hectic involvement in all sorts of non-scientific affairs. I am not trying to make excuses for the shortcomings of this work. Rather I would like to communicate the heavy feeling of writing a book (albeit on a scientific, impersonal topic) for a public which is rift in its human and moral values, as the scientific community nowadays indeed is. Being committed on some of these issues, by choice or by the force of circumstances, I was strongly conscious of addressing a divided audience. I knew that whichever way I would write this book I could not change this; still the lack of identification with my public (which identification must be one of the true sources of gratification in scientific writing) hovered over me depressingly.

Rehovoth, *November 1971*   R. ENGLMAN

Acknowledgements

At various stages in the preparation of this book, I have turned to numerous experts and have been helped by them through discussions or correspondence, encouragement or criticism. At times their well-informed, portentous silence was the best stimulus. My thanks are due to S. Alexander, J. Amran-Sussman, J. H. Freed, B. (Binyamin) Halperin, F. S. Ham, M. Karplus, B. Kirtman, A. S. Novick, A. (Amiram) Ron, P. S. Rudman, K. W. H. Stevens and M. Weger. Special gratitude is owed to M. D. Sturge for his instructive comments and incisive criticisms on the whole MS. With so many virtuoso soloists playing in the orchestra the reader might well have expected a flawless performance. If he still discerns a discordant tone or a false note, who is he to blame but the conductor.

In the hard work of typing I have had the devoted assistance of Val F. Guthrie, Tzila Grossman, Yvette Matatyah and Leon Soncino.

It is one of the established uses of Prefaces to present to the reader (no matter what his views will be on the scientific merits of the book) a favourable picture of the author as a person. It is almost as common to show him as a temporary defaulter in his roles as husband and father. The burden of this situation was borne by my wife Hasia and my children Matanyahu, Klilah and Shua, by the little ones speechlessly and by the rest helplessly. I ask their forgiveness, apologetic but unrepentant.

'The honour of a person derives from the constancy of his place' goes a Hebrew saying, and I apply this to my home Institute, who for many years have provided me with the facilities to undertake this project. Still, towards the end I have moved about a bit and I am indebted to K. W. H. Stevens, Professor of Theoretical Physics in the University of Nottingham for enabling me to carve off time from my working hours, during a Sabbatical Year in his Department, to complete this book. I also thank my hosts at a Zahal base for letting me do the same in my spare hours as the Chaplain's Chauffeur.

I acknowledge the source of figure 7.18. It is: Alefeld, B.: Die Messung der Gitterparameteränderung von Strontiumtitanat am Phasenübergang bei 108 °K. *Z. Physik*, **222,** 155–164 (1969), Berlin–Heidelberg–New York: Springer.

The rest of the borrowed figures are acknowledged in the customary place and manner. Manuscripts prior to publication were received from R. J. Elliott and his Oxford Collaborators, E. A. Hughes, M. C. M. O'Brien and E. Pytte.

R.E.

Glossary of Symbols

(with brief definitions and references to location)
(See also remarks on notation on page 4)

A	(amplitude of a vibronic function of a given representation in the vibronic eigenfunctions, equation (3.72)) p. 73
	(elastic energy due to shear strain) p. 152
a	(or a_γ^Γ, amplitude of the Γ, γ state in an electronic state-function) p. 219
	(half separation between parabolic potentials in the vertical direction, in units of a vibrational quantum) p. 230
	(linear dimension of a unit cell in a cubic lattice) p. 151
A_{ij}	(hyperfine coupling tensor, equation (3.44)) p. 44
α	$((\hbar\omega)/(4E_{\mathrm{JT}}))$ rotational energy in the q_θ, q_ϵ for $E \otimes \epsilon$ coupling, equation (3.22)) p. 33
	(amplitudes of harmonic oscillators localized in right-hand well for two intersecting potentials, equation (3.86)) p. 97
	(a numerical coefficient appearing in the distortive term of the energy) p. 167
B	(a constant of proportionality between the reorientation rate and T, equation (6.9)) p. 193
b	(half separation between intersecting parabolic potentials, in units of the zero point motion amplitude) p. 230
β	(resonance integral, equation (7.21)) p. 237
	(non-linear or anisotropic coupling parameter for $E \otimes \epsilon$, equation (3.20)) p. 32
	(amplitudes of harmonic oscillators localized in left-hand well for two intersecting potentials, equation (3.86)) p. 97
	(a numerical coefficient appearing in the distortive term of the energy, equation (5.5)) p. 167
β, β_1, β_2	(non totally symmetric, non-degenerate modes in tetragonal symmetry) p. 3
C_{ij}	(elastic constants) p. 160
c, c_l, c_t	(average, longitudinal and transverse sound velocities, equation (6.4)) p. 192
$\chi(q)$	(vibrational factor in wave-function, equation (2.3)) p. 11
D	(a geometric factor connecting the strain to the displacement of the local normal mode, equation (4.2)) p. 125

Δ	(or 10 Dq, cubic field splitting energy between e and t_2 orbitals) p. 214
	(the splitting of the first excited vibronic doublet into A_1 and A_2 for $E \otimes \varepsilon$; figure (3.9)) p. 38
	(the energy of the excited state in an Orbach process) p. 193
δ_1, δ_2	(splittings of a d^1, $^2T_2$-ion by the trigonal field and spin–orbit coupling) p. 141
ΔE	(energy difference between electronic states) p. 2
e	(amplitude of uniform strain) p. 160
E	(energy) p. 11
$\mathscr{E}, E$	(applied electric field) p. 141
ε	(doubly degenerate distortional mode) p. 3
ϵ	(a component of an E-doublet transforming as x^2-y^2 in O or T_d-symmetry) p. 214
$\varepsilon^0, \varepsilon^{00}$, etc.	(distortions in an ε-mode; figure 6.1, and Appendix VII) p. 197
E	(the ϵ-component of an electronic doublet) p. 20
E_F	(Fermi-energy for electrons) p. 162
E_G	(energy overlap in the bands of semi-metals) p. 169
e_ϵ, e_θ	(uniform strains in a crystal having ϵ or θ character at the site of the J-T centre) p. 125
E_{JT}	('the Jahn–Teller stabilization energy', the reduction in the energy of the system due to a vibronic coupling linear in a vibrational mode (equation 1.2, table 1.1)) p. 3
E_θ	(the θ-component of an electronic doublet transforming as $(2z^2 - x^2 - y^2)/\sqrt{3}$ in O- or T_d-symmetry) p. 20
ϕ	(the angular variable in the polar representation of a doubly degenerate mode, equation (3.6)) p. 25
	(the orientation angle for the magnetic field in the (110)-plane) p. 137
$\phi(\mathbf{r}), \phi(\mathbf{r}, q)$	(electronic factor in wave function, equations (2.2) and (2.5 −BO)) p. 11
$G(x)$	(second order quenching factor for $T \otimes \varepsilon$, equation (3.64)) p. 61
$G(\gamma)$	(coefficient of operator having symmetry γ, which describes the effect of an external force or stress, equations (3.34) and (3.58)) p. 42
Γ	(an irreducible representation of a molecular symmetry group) p. 4
	(for odd-electron systems: reducible but non-degenerate representation of the double group)
	(one third of the tunnelling splitting between the ground doublet states and the nearby singlet, equation (3.28)) p. 37
	(the inverse of the life time of an excited state) p. 202
$[\Gamma]$	(dimension of the representation Γ) p. 12
γ	(a fourfold vibrational mode in the icosahedral groups, shown inside back cover)
	(a component of a general irreducible representation of a molecular symmetry group)
γ_ϵ	(or γ_θ or γ, the strain energy due to a strain e_ϵ or e_θ or $\sqrt{e_\theta^2 + e_\epsilon^2}$, equation (4.1)) pp. 125, 132

g_{ij} (g-tensor, for $E \otimes \varepsilon$; equation (3.41)) p. 44
Γ_q (an irreducible representation for a vibrational state) p. 4
γ_q (a component of Γ_q) p. 4
H (Hamiltonian) p. 2
H   (magnetic field)
H_λ^Λ (derivative of H with respect to normal coordinate q_λ^Λ) p. 219
η (a component of a t_2- or T_2-triplet transforming as zx in O or T_d-symmetry) p. 214
     (a fivefold vibrational mode in the icosahedral groups, shown inside back cover).
I, **I**   (invariant or totally symmetric operator or matrix)
$I(\omega)$ (intensity or line-shape function for optical transitions) p. 49
J (total angular momentum) p. 51
j (a label specifying different vibronic states of the same symmetry, $j = 0, 1, 2, \ldots$ for states of increasing energy, equations (3.72), (7.29)) p. 73
     ($= l \pm \frac{1}{2}$, the rotational quantum number of the vibrational–electronic motion for linearly coupled $E \otimes \varepsilon$, equation (3.9)) p. 25
J' (quasi-angular momentum) p. 221
j' (quasi-angular momentum in aromatic hydrocarbons) p. 236
$\hat{J}_z$ (operator for the z-component of total angular momentum) p. 245
K (coefficient of the quadratic vibronic term for $E \otimes \varepsilon$, equation (3.4)) p. 20
     (constant representing the radial extension of an atomic wave function) p. 159
     (contact hyperfine coupling parameter) p. 45
     (eigenvalue of the component of the total angular momentum along the symmetry axis of a molecule) p. 51
     (expansion parameter in the Born–Oppenheimer approximation, of the order of the fourth root of the ratio of the electronic and ionic masses) p. 9
     (time reversal operator) p. 273
K   (reciprocal lattice vector) p. 150
k (Boltzmann constant)
k   (wave-vector for electronic states) p. 150
K_0 (complex-conjugation operator) p. 274
ξ (a component of a t_2- or T_2-triplet transforming as yz in O- or T_d-symmetry) p. 214
ξ (vibrational modes transforming as (x^2, y^2, z^2), equation (5.7)) p. 171
k_F (Fermi momentum) p. 157
$K(\Gamma)$ (quenching or reduction factor for operators of Γ symmetry, equations (3.61), (3.74), (3.79)) p. 60
L (the reduced matrix element for linear coupling) p. 2
l (the rotational quantum number of a two dimensional harmonic oscillator, equation (3.8)) p. 25
Λ (an irreducible representation for vibrational states) p. 4
     (rotational quantum number for electrons in a linear molecule, equation (7.30)) p. 248

λ (a component of an irreducible representation of a vibrational state or mode) p. 4

(numerical measure of co-operative interactions between distorted octahedra, equation (5.11)) p. 174

M (ionic mass or effective mass of vibrator) p. 4

(reduction factor for the vibrational angular momentum operator, figure (3.9)) p. 38

m (electronic mass) p. 9

$\mathbf{m}$ (vibrational angular momentum of τ_2-mode) p. 79

N (coefficient of the third order term in $E \otimes \varepsilon$, equation (3.4)) p. 20

(number of octahedral sites in a spinel-type crystal) p. 174

$3N$ (number of vanadium atoms in a V_3Si crystal of given volume) p. 160

n (vibrational quantum number, equations (3.68) and (3.69)) p. 70

ν (frequency of jumps between potential wells, equation (6.1)) p. 191

(measure of the non-linear coupling strength in $E \otimes \varepsilon$, equation (3.25)) p. 34

(number relating the J-T stabilization energy E_{JT} to the reduced matrix element L and $\hbar\omega$, equation (1.2) and table 1.1) pp. 3–4

N_e (number of occupied states in the upper split band of the x or y chains in tetragonal V_3Si) p. 161

$n(E_F)$ (density of electronic states at the Fermi energy) p. 162

ω (characteristic or relevant vibrational frequency) p. 2

(frequency of light in transition) p. 49

ω_D (Debye frequency)

P (reduction or quenching factor, defined in equation (3.29) or (3.38) for the ground vibronic doublet in $E \otimes \varepsilon$) p. 37

(momentum conjugate to q) p. 231

π (doubly degenerate mode in the continuous axial groups, shown inside back cover)

Ψ or $\psi\,(\mathbf{r}, q)$ (vibronic wave-function) pp. 11–12

Q (a local normal vibrational co-ordinate) p. 4

(a uniform distortion) p. 157

q (dimensionless normal co-ordinate) p. 4

(radial variable in the polar representation of doubly degenerate modes, equation (3.6)) p. 25

(reduction or quenching factor defined in equation (3.30) or (3.37) for the ground vibronic doublet in $E \otimes \varepsilon$) p. 37

$\{Q\}$ (set of normal co-ordinates) p. 11

Q^0 (value of the local vibrational co-ordinate at the point where the electronic states are degenerate) p. 12

q^0 (equilibrium position under stress, equation (4.1)) p. 125

q_0 (value of the radius at the bottom of the potential trough in the linear $E \otimes \varepsilon$ problem, equation (3.15)) p. 28

R (a group operation) p. 149

$\mathbf{r}$ (co-ordinate for electrons) p. 4

r, r' (reduction or quenching factor defined in equation (3.31) or (3.39–40) for $E \otimes \varepsilon$ as a matrix element between a vibronic doublet and A_1 (or A_2, for r')) p. 38

ρ	(Dirac 4×4 matrices, equation (3.77)) p. 81
S	(ratio of the J-T stabilization energy E_{JT} to the vibrational quantum $\hbar\omega$, equation (1.2)) p. 3
S'	($= L^2/(\hbar\omega\Delta E)$, ratio of the reduced linear coupling matrix element squared to the vibrational quantum times the electronic energy difference, equation (1.1)) p. 2
s	(label specifying different symmetry adapted harmonics, belonging to the same representation, equation (3.70)) p. 72
	(order parameter for tetragonal-cubic transition in spinels) p. 174
σ	(matrices, 2×2: equation (3.35), 4×4: equation (3.77)) p. 42
T	(temperature) p. 49
t	(time) p. 192
τ	(reorientation time between equivalent distortions, equations (6.6–8)) p. 193
τ_2	(triply degenerate vibrational mode) p. 3
τ_2^1, τ_2^2, etc.	(distortions in a τ_2-mode, figure 6.1, Appendix VII) p. 197
T_c	(semiconductor–conductor or superconductor–conductor transition temperature) p. 168
T_D	(orthorhombic–tetragonal transition temperature) p. 180
T_e	(electronic kinetic energy) p. 12
T_m	(martensitic tetragonal–cubic transition temperature) p. 159
T_N	(nuclear kinetic energy) p. 12
T_t	(tetragonal–cubic transition temperature) pp. 176, 182
θ	(a component of an E-doublet transforming as $(2z^2-x^2-y^2)/\sqrt{3}$ in O- or T_d-symmetry) p. 214
	(electronic angle in a linear molecule, equation (7.30)) p. 248
	(orientation angle of the strain-energy components γ_θ, γ_ϵ, equation (4.7)) p. 132
U	(energy at the bottom of the potential wells, equations (6.2–3)) p. 192
	(unitary transformation matrix for contragradience) p. 274
V	(potential for ion–electron interaction) p. 11
v	(radial or principal quantum number of a two-dimensional harmonic oscillator, equation (3.8)) p. 25
	(a trigonal field splitting parameter) p. 141
V°, V^+, V^-	(neutral, positively and negatively charged vacancy) p. 225
V_0	(half the band gap in insulating oxides at $T = 0$) p. 256
$V_0, V_1, \ldots$	(Fourier components of V) p. 151
$V_{11}, V_{12}, \ldots$	(interaction energies between neighbouring distorted octahedra in spinels) p. 175
W	($\frac{1}{2}$, $\frac{1}{3}$ or $\frac{1}{4}$ of the energy separation between states participating in a P–J–T–E) p. 90
	(energy) p. 11
	(off-diagonal matrix element linking two electronic states which cross in a configurational diagram) p. 97
	($= \hbar^2 k_F^2/2m$, width of the occupied part of the conduction band) p. 165
w	(half-width of the d-electron energy band in V_3Si) p. 161
	(number of potential wells made inequivalent by strain) p. 192
z	(concentration of J-T ions at B-sites in spinels) p. 173

ζ (a component of a t_2- or T_2-triplet transforming as xy in O- or T_d-symmetry) p. 214
                     (Coriolis coupling parameter) p. 246

ζ_d (one-electron spin–orbit coupling coefficient) p. 221

ζ_v (Coriolis coupling parameter in the vibronic state v, equation (7.29)) pp. 52, 247

Contents

Contents

1 *Introduction*

Outline

The Jahn–Teller and pseudo-Jahn–Teller effects (J-T-E and P-J-T-E) are introduced as those limiting cases when the correction to the Born–Oppenheimer approximation becomes comparable to the vibrational energy. The correction must then be included in the solution exactly and not as a perturbation. Characteristic of a strong J-T-E are marked changes in the electronic properties (I), great propensity to suffer distortion (II) and an inherent stability to maintain the distortion (III).

Molecules and solids consist of heavy and light constituents, atoms or ions and electrons. Much of our understanding of the behaviour of these systems is based on the separation of the motion of the heavy and light components. Because of this separation we can interpret with relative ease physical effects which are connected either with the ionic motion (e.g. sound propagation, heat conduction) or with electronic properties (e.g. magnetism). The theories which enable the separations to be formulated mathematically in a systematic way, are known as the Born–Oppenheimer or adiabatic approximations. They are described in various texts[1,2] and will be treated by us in Chapter 2.

At the same time there are a number of physical phenomena which depend for their existence on the interplay between the two types of motion, the ionic and the electronic. Examples of these are the relaxation of electronic spins in a lattice, the high temperature electrical resistance of metals, superconductivity, ultrasonic attenuation, etc., in *extended* systems and some less spectacular but nevertheless important phenomena in *localized* systems. By localized systems we mean entities composed of an atom and a few of its close neighbours in a solid, such that some physical event is confined to these entities. Small or medium-sized molecules are of course also localized systems. Then we know that phenomena like ligand field transitions in paramagnetic complexes,[3] or singlet → singlet absorptions in some hydrocarbons[4,5] very frequently take place thanks to the coupling in localized systems between the vibrations of the ions and the trajectory of the electrons.[6] This coupling and the effects following from it are often called vibronic (vibrational-electronic). This term is more commonly used for localized systems than in the solid state proper, where, although the vibronic coupling is vital for the existence of the phenomenon, there are a number of other effects which steal the limelight from the vibronic coupling. Another term which is frequently used for vibronic coupling is 'the break-down of the Born–Oppenheimer approximation'. This owes its origin to the circumstance that there are terms in the Hamiltonian H of the system which are neglected in that approximation but which re-enter in vibronic effects. (We again refer to References 1 or 2 and to the discussion in Chapter 2). Frequently it is possible to explain the experimental data by regarding the vibronic coupling as small. One includes then the lowest order correction to the Born–Oppenheimer treatment, namely the term in the Hamiltonian which is linear in the vibrational coordinates, and treats this term perturbationally. In second order perturbation theory we should expect an energy lowering of the order of $L^2/\Delta E$, where ΔE is a characteristic energy difference between two states which are admixed by the linear term. As regards L, this is simply meant to be a measure of the strength of the linear coupling, having the dimensions of energy. Its definition is given in equation (1.3), which is written in a sufficiently general form to enable us to employ the same letter L in all cases of linear vibronic coupling treated in this work. (Unfortunately, this wide generality is achieved at the expense of leaning on the rather erudite subject of tensorial sets.) In words, L is the reduced matrix element of the linear coupling term, in which the vibrational coordinates have the amplitude of the zero point motion. To obtain a pure number which is representative of the magnitude of the vibronic coupling we may divide $L^2/\Delta E$ by an energy associated with the motion of either the heavy particles, the ions, or the light particles, the electrons. Since, however, our object is to find out when the perturbation approach breaks down, it is more pertinent to divide by the smaller of these quantities, by $\hbar\omega$ say, which is the energy quantum of a vibration involved in the coupling.

The pure number ratio

$$S' = \frac{L^2}{\hbar\omega\Delta E} \tag{1.1}$$

is the most important of all the quantities appearing in this work and we should do well to examine under what conditions it is small, medium or large and to decide how to treat each case.

Characteristically, the magnitudes of L and $\hbar\omega$ are similar, but there may be in exceptional cases fluctuations by a few orders of magnitude either way. As regards ΔE three cases arise:

1. When ΔE is an energy difference between electronic states, and this is large compared to either L or $\hbar\omega$. Then the perturbational approach is justified, since even the lowest, second-order perturbation alters the energy by only a small fraction of the whole. This case falls altogether outside the range of this work.

2. When ΔE is an electronic energy difference which is however, in a rather uncharacteristic way, of the order of or smaller than L. Then the ratio in equation (1.1) is of the order of unity or larger. It is elementary to conclude that perturbation theory has to be abandoned and an exact solution of at least part of the vibronic problem is necessary. This case, especially when referring to localized systems, is frequently called the pseudo-Jahn–Teller-effect (P-J-T-E) and is dependent, as we have seen, on the near-degeneracy of some electronic states. Many of our comments for the next case, the Jahn–Teller-effect (J-T-E), apply also here.

3. When L denotes the matrix elements between *degenerate* electronic states then the electronic part in ΔE is zero, by definition. Nevertheless, ΔE does not vanish but is one vibrational quantum since L arises from a term in the Hamiltonian which is linear in a vibrational coordinate. The ratio S' in equation (1.1) will now be made precise by specifying the vibrational quantum $\hbar\omega$ and also by appending a number v^{-1} which varies from case to case, as shown in Table 1.1. We have then instead of equation (1.1)

$$\frac{1}{v} \frac{L^2}{(\hbar\omega)^2} \equiv \frac{E_{\mathrm{JT}}}{\hbar\omega} \equiv S \tag{1.2}$$

Here E_{JT} is the Jahn–Teller stabilization energy, so named since it was shown in the fundamental work of Jahn and Teller[7] that for any molecular system whose electronic state is degenerate there is at least one linear term in the Hamiltonian such that L is nonzero and consequently $E_{\mathrm{JT}} \neq 0$. (Appendix 2.) The exceptions to this statement are Kramers degeneracies in odd-electron systems (treated in Appendix 3) and linear molecules (discussed in Chapter 3, §§3.7. 1–4, and in Chapter 7, §7.11).

From our earlier estimates of the relative magnitudes of L and $\hbar\omega$ we now see that $E_{\mathrm{JT}}/(\hbar\omega)$ can be small, of order unity or large. We shall be occupied with the last two cases, which are the regimes for the operation of the J-T-E. The case when $E_{JT}/\hbar\omega \gg 1$ is often called the strong coupling limit. (The use of the term 'the static case' for the description of this limit will be avoided in this work, for reasons to be given shortly. Instead, we shall use the term 'the static problem' for the *mathematical* problem which arises when we let the nuclear masses approach infinity in the

Table 1.1 The Jahn–Teller stabilization energy

$\Gamma \otimes \Lambda$	$E \otimes \beta$	$E \otimes \varepsilon$	$T \otimes \varepsilon$	$T \otimes \tau_2$	$G_{3/2} \otimes \varepsilon$	$G_{3/2} \otimes \tau_2$
ν:	4	8	6	9	16	24

The number ν in $S = \dfrac{E_{JT}}{\hbar\omega} = \nu^{-1} \dfrac{L^2}{(\hbar\omega)^2}$ is shown for the simple cases arising in the cubic, trigonal and tetragonal classes of point symmetry groups. $\hbar\omega$ is the quantum of energy of the vibrational mode whose coordinate q^Λ (normalized as below) belongs to the representation Λ, and L is the reduced matrix element

$$L = \langle \Gamma \| \frac{\partial H}{\partial q^\Lambda} \| \Gamma' \rangle \equiv L(\Lambda) \qquad (1.3)$$

between the electronic states having representations Γ and Γ'. (In the J-T-E, $\Gamma = \Gamma'$.)

Important remarks on the notation to be used in this work

1. The Schönflies symbols are used for groups and the Mulliken notation[9] for their single-valued representations. For double-valued representations the symbols and conventions of Herzberg[10] are used. Appendix 1 may be useful to follow the nomenclature.
2. Small roman letters (e.g. t_{2g}) are used for representations of one-electron states, large letters (e.g. $^3T_{2g}$) for many-electron or vibronic states and the corresponding Greek letter (τ_{2g}) for representation of the vibrational species or its overtones. The g ('gerade' for even), u ('ungerade' for odd) and 1, 2, ... subscripts are occasionally omitted for typographical convenience, when their presence does not add any information to the reader. Thus in the table above we have the 2, 3 and 4-fold representations (E, T and $G_{3/2}$) of electronic states and the 1, 2 and 3-fold vibrational modes (β, ε and τ_2).
3. For generic notation of a representation and its component we shall adopt: Γ, γ for electronic or vibronic states

   Λ, λ or Γ_q, λ_q for vibrational states.

4. The vibrational coordinates are designated by Q, eventually with appropriate suffixes. However, we shall prefer the use of dimensionless vibrational coordinates, whose symbol is q. $q = Q(M\omega/\hbar)^{\frac{1}{2}}$, where M is the effective mass of the vibrator and ω its frequency. $\mathbf{r}$ will be the symbol for electronic coordinates.
5. The electronic-vibrational coupling is symbolized by $\otimes$, e.g. $t_{2g} \otimes \varepsilon_g$, $^3T_{1u} \otimes \tau_{2g}$, or more briefly $T \otimes \tau_2$.

Hamiltonian.) The weak coupling case ($E_{JT}/\hbar\omega \ll 1$) will only be mentioned for the sake of completeness or for didactic reasons. To the best knowledge of the author, the results of the weak coupling case can always be obtained perturbationally (although the electronic manifold is degenerate), so that this case does not belong to the realm of the J-T-E which is essentially a non-perturbational phenomenon.

Having thus attempted to trace the logical origin of our subject and its relation to the wider class of vibronic phenomena, we shall now go on to jot down, again in general terms, some of its outstanding properties.

Of course the J-T-E appears in the experimental results (of optical absorption, spin resonance and relaxation, infrared and Raman spectra, etc.) in many more ways than we could afford to list, but when we are concerned with the inherent properties of J-T systems (and not with their experimental detection) we find that we can characterize the effect of vibronic coupling on electrically degenerate states by the following:

I. The electronic degeneracy is replaced by a vibronic degeneracy of the same type which possesses *qualitatively* (i.e. symmetry-wise) similar and *quantitatively* different electronic and vibrational properties from those of the vibronically uncoupled system.

II. Vibronically strongly coupled systems possess an extraordinarily great tendency to undergo distortion when stressed in some favourable direction. The stress may be applied externally or may be due to some internal irregularity in the lattice.

A few calculated curves depicting strain *versus* stress (i.e. effect *vs.* cause) are seen in Figure 1.1. The details of this figure (as well as the lengthy caption) will probably be clear only after reading sections §3.2 and 3.6.2 on which Figures 1.1(a) and 1.1(b) are based, respectively. At this stage it is only necessary to recognize that $\beta/\alpha \gg 1$ and $b \gg 1$ represent cases of strong vibronic coupling. It is apparent from the figures that in these limits large, almost spontaneous distortions occur in strongly coupled systems even for small stresses. As explained in the caption of the figure there are a number of equivalent directions in any vibronic system along which nearly spontaneous distortions can take place. (A fluctuating distortion, which averages to zero, is present also in the absence of perturbation, but only when the system has been perturbed some time during its history is the distortion 'realized'.)

III. In the limit of strong coupling, the distorted state, which comes about owing to the stress as explained above, displays a considerable stability with respect to transitions between equivalent distorted configurations. The transitions may be of two kinds: real transitions (reorientations) caused by time dependent forces or strains, and virtual transitions (a term proper to quantum mechanics) which means the admixture into a state localized in one distorted configuration of a

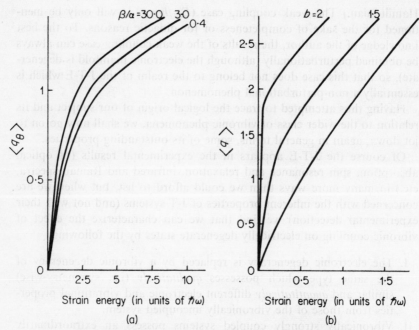

Figure 1.1 Strain–stress curves for Jahn–Teller systems.

The abscissa is the strain energy that exists in the absence of J-T coupling, the ordinate is the thermal average of the distortion (in units of the zero point motion amplitude) when the J-T centre is installed. Figure 1.1(a) is from calculations (by B. Halperin) for a case symbolized by $E \otimes \varepsilon$, in a cubic system. The stress and the resulting distortion is along one of the three tetragonal axis. This case is discussed in §3.2.2, where the symbols for the anisotropy energy β and the rotational energy α are defined. [Equation (3.20) and (3.22)]. The larger β/α, the more localized the J-T-system and the greater its tendency to distort.

Figure 1.1(b) comes from calculations (by B. Barnett) for $E \otimes \beta$ treated in §3.6.2, where b and a are defined. The stress and the resulting distortion are parallel (or antiparallel) to the direction of the β mode. The parameter W of §3.6.2 is 1 and the temperature in the averages of both Figures 1.1(a) and (b) is about 10 °K. These curves are further discussed in §4.1.

state localized in a second configuration. The rates of both transitions are reduced by the J-T-E, as a result of which the distortion becomes (once it occurs) more or less static.

The adjective 'static' has been given in the past to the strong J-T-E. We understand now that a static situation *may* come about by the characteristics II and III. However, as a consequence of the terminology the strong

J-T-E has often been regarded as equivalent to or even synonymous with 'a distorted configuration'. In truth, however, a distorted situation amounts to a reduction of the symmetry of the systems and this cannot occur as a result of internal interaction but only owing to the presence of a low symmetry external field. In fact there are cases, e.g. Ni^{3+} in Al_2O_3, which properly qualify for the appellation 'strong J-T-E' ($E_{JT}/\hbar\omega > 5$) and in which a distortion is not stable above 15 °K, because the transition rates are not sufficiently reduced. (Parameters and references for many vibronically coupled systems are collected in Appendix IX). In the author's view the term 'static J-T-E' could often be replaced with advantage by 'vibronically stabilized distortion'.

We have not yet mentioned the higher order terms (quadratic and higher order in the vibrational coordinates) in the vibronic Hamiltonian. Now a little reflection shows, even without any formal calculation, that a large linear vibronic coupling leads to large amplitude displacements of the system from equilibrium. How are the higher order terms treated? The answer is that they are generally neglected as long as their effects are quantitative and are taken into account (as in the $E \otimes \varepsilon$ case) when they lead to qualitatively new results. This answer is admittedly not satisfactory, but it has not hitherto caused any evident discrepancy and there the matter rests.

The relation of the J-T-E to cases where the minima of the potential energy are known to be off-centrally situated may be commented on in general terms. Such cases can arise, e.g. in hydrogen bonding[11] or with some small substitutional impurities like Li^+ in KCl,[12-4] because of the packing of the ions and are not caused by the degeneracies of the electronic states. Here also the characteristics II and III are found, but the quantitative changes in the electronic properties (I, above) are not expected to be as marked as in the J-T-E.

In the discussion of this chapter the denominator ΔE in equation (1.1) was treated as a quantity which takes discrete values, whereas in many important examples of vibronic coupling in solids (e.g. in superconductivity) ΔE is a nearly continuous function of the wave-vector. These instances fall outside the scope of this work, except for the treatment of phase-transitions in semi-metals later in Chapter 5. The reason for this omission is that many-electron effects (which will hardly occupy our attention) are so important in these cases that the vibronic aspect is, historically at any rate, left in the background.

1.1 References

1. L. Pauling and E. B. Wilson, *Introduction to Quantum Mechanics* (New York: McGraw-Hill, 1935), §34.
2. J. M. Ziman, *Principles of the Theory of Solids* (Cambridge: University Press, 1965), §6.10.
3. C. J. Ballhausen, *Introduction to Ligand Field Theory* (New York: McGraw-Hill, 1962), Chapter 8.
4. D. P. Craig, *J. Chem. Soc.*, **1950**, 59 (1950).
5. A. D. Liehr, *Z. Naturforschg*, **13a**, 429 (1958); **16a**, 641 (1961).
6. G. Herzberg and E. Teller, *Z. Physikal. Chem.*, **B21**, 410 (1933).
7. H. A. Jahn and E. Teller, *Proc. Roy. Soc. A*, **161**, 220 (1937).
8. J. S. Griffith, *The Irreducible Tensor Method for Molecular Symmetry Groups* (London: Prentice Hall, 1962).
9. E. B. Wilson, Jr., J. C. Decius and P. C. Cross, *Molecular Vibrations* (New York: McGraw-Hill, 1955).
10. G. Herzberg, *Electronic Spectra and Electronic Structure of Polyatomic Molecules* (Princeton: Van Nostrand, 1966).
11. R. Blinc and D. Hadzi, *Mol. Phys.*, **1**, 391 (1958).
12. G. Lombardo and R. O. Pohl, *Phys. Rev. Letters*, **15**, 291 (1965).
13. H. S. Sack and M. C. Moriarty, *Solid State Commun.*, **3**, 93 (1965).
14. V. Narayanamurti and R. O. Pohl, *Rev. Mod. Phys.*, **42**, 20 (1971).

2 Born–Oppenheimer and adiabatic approximations

Outline

Since the Jahn–Teller Hamiltonian H_{JT} is introduced in the following chapters as a correction to the 'Born–Oppenheimer approximation' we explain in this chapter why we prefer this to what we call (for the sake of distinction) the 'adiabatic approximation'.

When the dynamic nature of the ionic coordinates is taken into account, or in other words, when the kinetic energy of the ions is not neglected, one has a three or more body problem on one's hands, which one would then try to solve by means of the perturbation theory outlined (for the electronically non-degenerate case) in, e.g., Born and Huang's book.[1] Electronic degeneracy invalidates these schemes. This can be seen in two ways. One, the perturbation expansion parameter K^2 (K being of the order of the fourth root of the ratio of the electronic and ionic masses m and M) is inversely proportional to the energy separation of electronic states from the perturbed energy. This separation shrinks to zero in the J-T-E and the expansion parameter increases beyond any limit. Two, when the system is degenerate, it does not remain in the neighbourhood of zero distortion, but rather expatiates to a number of minima which lie at a distance of the order $q = L/(\hbar\omega)$. (For help with the notation we refer to Table 1.1.) We shall see that this distance may be beyond the region of validity of the 'effective potential' for the ionic motion.

There are two approximation schemes for the treatment of electron–ionic motions in the non-degenerate case, which part company in the zero-order states and which differ essentially in their physical contents. The first (chronologically) is what we shall call the Born–Oppenheimer approximation,[2] in which the basic electronic states are referred to the equilibrium positions of the ions. (This scheme also bears the discouraging appellation: 'the crude adiabatic approximation'). These points are frequently geometrically privileged points: they are the so-called 'reference configuration' of molecules[3] or the ideal periodic lattice points of the crystal. The applications of group theory (the use of representations, selection rules, etc.) are based on this approximation method and so are also most analytical and numerical calculations of electronic wave-functions. Since the true origin of forces are the instantaneous positions of the nuclei and not their equilibrium positions, a correction has to be added to that *static* potential which is based on the equilibrium positions. We shall call this correction the 'static interaction' between the electronic and ionic motions. As shown in detail in Reference 2, the electron and ionic motions may be decoupled approximately, so as to yield a 'self-consistent (or effective) potential' for the ionic motion which is correct to the order of $K^2 \times$ Electronic energies, that is to say, correct to the order of $Q^2(\partial^2/\partial r^2)H$ (r and Q stand for the totality of electronic and normal vibrational coordinates, respectively, and H is the total Hamiltonian). The terms, of order Q^3 or higher, which are beyond the validity of the effective potential, are inessential for many purposes, provided that the nuclear trajectories are shorter than the electronic ones. This is true in non-degenerate cases and also appears to be the situation in degenerate cases. However, temperature dependent, anharmonic effects cannot be properly based on this potential.

The second scheme, known as the adiabatic approximation,[4] starts with zero-order electronic wave-functions which are 'in principle' solution of the Schrödinger-equation containing a potential whose sources are the ions at their actual positions. What we have previously called the static interaction is now fully taken into account. There remains, however, the dynamic interaction which represents the effect of the recoil of the electronic motion on the nuclear coordinates. This approximation is clearly more accurate than the Born–Oppenheimer scheme and has some advantages over the latter. Thus, an effective potential can now be defined for nuclear motions, with the electrons in a particular adiabatic state, which is now correct to the order K^4. On the other hand, only very rarely, and then only in simple cases, were adiabatic wave-functions actually constructed[5-8] or utilized.[9] Mostly, calculations which are supposedly based on the adiabatic approximation are in fact carried out in the Born–Oppenheimer scheme. Moreover, the fact that, unlike the latter, the adiabatic approximation is not really a systematic perturbation scheme but rather something like a happy first guess, has led to some misunderstandings in the literature. Some inconsistencies are exposed and corrected in Reference 10. However, a remaining difficulty in the scheme as regards the J-T-E is that the zero order electronic functions are non-analytical near the point of degeneracy; in fact, their first derivatives with respect to vibrational coordinates tend to infinity at this point. This is a rather unexpected result since the physically meaningful solutions of the differential equations of Physics are generally well behaved. We shall shortly see the reasons for this misdemeanour—but before that we need to do some formal study.

The Hamiltonian for the electron-ion system can be written as

$$H = T_{el} + T_N + V(\{Q\}) + V(\mathbf{r}, \{Q\}) \qquad (2.1)$$

Here $\mathbf{r}$ denotes the electronic coordinates and $\{Q\}$, or Q for brevity, the totality of normal vibrational coordinates. T_{el} and T_N are the electronic and nuclear kinetic energies, respectively. The other two terms are the potentials, of which the second includes the electron-ion interaction. We shall now give a simplified version of the Born–Oppenheimer approximation, sufficient for our purposes. We start with the non-degenerate case. The basic pair of equations is:

$$[T_{el} + V(\mathbf{r}, Q)]\phi_e(\mathbf{r}, Q) = W_e(Q)\phi_e(\mathbf{r}, Q) \qquad (2.2)$$

and

$$[T_N + W_e(Q) + V(Q)]\chi_n(Q) = E_e\chi_n(Q) \qquad (2.3)$$

These equations define the electronic wave-functions $\phi_e(\mathbf{r}, Q)$ which contain the nuclear coordinates Q parametrically and the vibrational (ionic) wave-functions $\chi_n(Q)$ which arise from the effective potential $W_e + V$ appropriate to each electronic state. The Born–Oppenheimer wave-functions are however not these, but those arising from the first line H_{0e} of the Hamiltonian rewritten as below, with the second line ΔH_e regarded as perturbation.

$$\begin{aligned} H = H_{0e} + \Delta H_e \\ = T_{el} + T_N + V(\mathbf{r}, Q^0) + V(Q) + W_e(Q) - W_e(Q^0) \\ + V(\mathbf{r}, Q) - V(\mathbf{r}, Q^0) - W_e(Q) + W_e(Q^0) \qquad (2.4\text{–BO}) \end{aligned}$$

Q^0 is the position of stability of the system, where $\partial(W_e(Q) + V(Q))/\partial Q = 0$. This position depends on the electronic state, which is denoted by e, and so does the perturbation ΔH_e. This procedure is somewhat unusual, but is quite harmless as long as ϕ_e is non-degenerate, as here assumed. The zero-order solutions are

$$\psi_{e,n} = \phi_e(\mathbf{r}, Q^0)\chi_n(Q) \qquad (2.5\text{–BO})$$

The correction to this is obtained by straightforward perturbation theory, leading to

$$\delta\psi_{e,n} = \sum_{e',n'} \psi_{e'n'} \langle e'n' | \Delta H_e | en \rangle / (E_{e'n'} - E_{en}),$$

which is of order $K\psi_{e,n}$ in terms of the expansion parameter $K = O(m/M)^{1/4}$ introduced earlier. This may be verified by expanding ΔH_e around the equilibrium position Q^0, retaining only the lowest order term, of the type $(Q - Q^0)\partial(\Delta H_e)/\partial Q$, and noting that $|(Q - Q^0)|/|\mathbf{r}|$ is of order K. Similarly, the correction to the zero-order energy E_{en} arising from ΔH_e is given by standard perturbation theory. The inclusion in the second line of equation (2.4–BO) of the W-terms ensures that there are no first order corrections. This may be verified by calculating the expectation value of the second line in that equation. The second order perturbational corrections add terms of order $K^2 E_{en}$ to the original energy. This concludes our discussion of the non-degenerate case.

In situations, where the electronic state index e refers to two or more states which are degenerate at $Q = Q^0$, there is no unique way of splitting the Hamiltonian H into a zero-order part H_0 and a perturbation ΔH_e, in consequence of the variety of ways terms like $W_e(Q) - W_e(Q^0)$ with different e's may be added and subtracted off. On the other hand, we cannot afford to leave out these terms from H_{0e}, since these represent the electronic contributions to the position of stability and it is quite likely from physical considerations that in the absence of these terms the system would not attain stability at all. The following partition of H seems to be the best:

$$H = H_0 + \Delta H$$

where

$$H_0 = T_{el} + T_N + V(\mathbf{r}, Q^0) + V(Q^0) + \sum_e (W_e(Q) - W_e(Q^0))/[\Gamma]$$

$$(2.6\text{–BO})$$

$$\Delta H = V(\mathbf{r}, Q) - V(\mathbf{r}, Q^0) - \sum_e (W_e(Q) - W_e(Q^0))/[\Gamma]$$

$$(2.7\text{–BO})$$

Here the summations run over all degenerate states (whose number equals the dimensionality of the Γ-representation $\equiv[\Gamma]$) and the reference configuration is again given by the vanishing, at Q^0, of the derivatives of H_0 with respect to the nuclear coordinates.

The zero-order solutions are now superpositions of the degenerate set. They take the form

$$\psi_{\{e\},n} = \sum_{e=e_1 \ldots e_{[]}} \phi_e(\mathbf{r}, Q^0)\chi_{e,n}(Q) \qquad (2.8\text{–BO})$$

where the electronic parts ϕ_e are solutions of equation (2.2). The $[\Gamma]$-fold degenerate eigenvalue is $W(Q^0)$. The set of equations, whose eigenfunctions are the set $\chi_{en}(Q)$, is

$$[T_N + \Sigma W_e(Q)/[\Gamma] + V(Q) + \langle e|\Delta H|e\rangle]\chi_{en}(Q)$$
$$+ \sum_{e' \neq e} \langle e|\Delta H|e'\rangle\chi_{e'n}(Q) = E_n\chi_{en}(Q) \qquad (2.9\text{–BO})$$

The matrix elements $\langle e|\Delta H|e'\rangle$ are of course functions of Q. ΔH is shown in equation (2.7–BO).

To derive this result one multiplies $H\psi_{\{e\},n}$ by the functions ϕ_e in succession and integrates over the electronic coordinates. On diagonalizing the matrix $\langle e|\Delta H|e'\rangle$, a set $([\Gamma]$ in number) of functions of Q are found, which, when added to $\Sigma W_e(Q)/[\Gamma] + V(Q)$, determine a set of potentials in which the ionic motion proceeds.

The diagonalization of the set of equations, equation (2.9–BO) yields functions which are defined for all values of the nuclear coordinates, and solutions χ (analytical or numerical) can therefore be found throughout the range of Q. We must not forget, however, that the validity of the potential in the Born–Oppenheimer scheme stretches only as far as the quadratic terms. Higher order terms in the potentials are only qualitatively correct, although corrections to the wave-function and to the energy can be found up to any desired order, either for the degenerate or for the non-degenerate case, by admixing the excited states e'.

Turning now to the adiabatic approximation,[4,11-3] we again start with the non-degenerate case. The Hamiltonian, which consists of the same terms as in equation (2.1), is now partitioned according to the following two lines

$$H = T_{el} + T_N + V(Q) + V(\mathbf{r}, Q) + \langle e|T_N|e\rangle$$
$$- \langle e|T_N|e\rangle \qquad (2.4\text{–AD})$$

The Dirac-bracket is a function of Q, but is not a differential operator.

The zero order wave-function are

$$\psi_{e,n} = \phi_e(\mathbf{r}, Q)\chi_n(Q) \qquad (2.5\text{–AD})$$

The factor functions ϕ and χ satisfy two equations similar to those in the Born–Oppenheimer case: The first is equation (2.2) and the second

$$[T_N + W_e(Q) + V(Q) + \langle e|T_N|e\rangle]\chi_n(Q) = E_{e,n}\chi_n(Q), \qquad (2.10)$$

which is similar to equation (2.3), but note the enlargement of the effective potential for the ionic motion by the addition of the last term on the left hand side. It is shown in Reference 1 that the effective potential is correct up to, and including, the fourth order in the nucleus displacements

$(Q - Q^0)$ or in the expansion parameter K. To check the correctness of the approximation scheme, we evaluate the matrix elements of H in equation (2.4–AD). When this Hamiltonian is applied to $\psi_{e,n}$, equation (2.5–AD), we get, in addition to $E_{en}\psi_{en}$,

$$- \sum_i \frac{\hbar^2}{M_i}\left(\frac{\partial}{\partial Q_i}\phi_e(\mathbf{r}, Q)\right)\left(\frac{\partial}{\partial Q_i}\chi_n(Q)\right) + [\chi_n T_n \phi_e(\mathbf{r}, Q)$$
$$- \langle e|T_n|e\rangle\chi_n\phi_e(\mathbf{r}, Q)] \quad (2.11)$$

These terms represent the recoil effects mentioned earlier.

In the first term, partial derivatives of all the normal coordinates Q_i are to be taken. M_i is the effective mass corresponding to Q_i. The diagonal matrix element of this term, obtained by multiplying by the complex conjugate of $\psi_{e,n}$ in equation (2.5–AD) and integrating over the electronic and nuclear coordinates, vanishes since the non-degenerate wave-function can be taken as real. If so, then

$$\int \phi_e(\mathbf{r}, Q)\frac{\partial}{\partial Q_i}\phi_e(\mathbf{r}, Q)\, d\mathbf{r} = \tfrac{1}{2}\frac{\partial}{\partial Q_i}\int\phi_e^*(\mathbf{r}, Q)\phi_e(\mathbf{r}, Q)\, d\mathbf{r} = 0,$$

as stated. Non-diagonal matrix elements, obtained through multiplication by $\phi_{e'}^*\chi_{n'}^*$ and integration, survive only if e' denotes a state different from e. Their leading contribution to the energy is in the absence of degeneracy given by second order perturbation theory. This gives quantities of the order of $K^6 E_{en}$ and these we neglect.

The term in square brackets in equation (2.11) yields vanishing matrix elements both with $\phi_e^*\chi_n^*$ and $\phi_e^*\chi_{n'}^*$. In fact, the partitioning of the Hamiltonian was chosen with this end in view.

In the adiabatic approximation for degenerate states the zero order state are superpositions of the form

$$\psi_{\{e\},n} = \sum_{e=e_1 \ldots e_{[]}} \phi_e(\mathbf{r}, Q)\chi_{e,n}(Q) \quad (2.8\text{–AD})$$

The suggested partition of the Hamiltonian is for each state e as in equation (2.4–AD). The electronic factors satisfy equation (2.2) as before. The set of coupled differential equations for the vibrational factors are now derived in the way explained after equation (2.9–BO). One obtains

$$[T_N + W_e(Q) + V(Q) + \langle e|T_N|e\rangle]\chi_{en}(Q)$$
$$+ \sum_{e'=e_1 \ldots e_{[]}} \langle e|T_N|e'\rangle - \sum_i\frac{\hbar^2}{M_i}\left(\langle e|\frac{\partial}{\partial Q_i}|e'\rangle\right)\frac{\partial}{\partial Q_i}\chi_{e'n}(Q)$$
$$= E_n\chi_{en}(Q) \quad (2.9\text{–AD})$$

The non-vanishing terms which couple together e and e′ in the second line of the previous equation are the same as those discussed in the non-degenerate case. However, now it is no longer possible to use second order perturbations theory. Comparing the above set of equations with those of the Born–Oppenheimer case, equation (2.9–BO), we see that because of the dynamic type of interaction some matrix elements contain operators, rather than functions, of Q. Consequently, if we managed formally to diagonalize the matrix, these matrix elements would contribute to the kinetic energy rather than to the potential. This is a rather unsavoury situation, especially in view of the difficulties in the interpretations in simple terms of the higher than second-order derivatives with respect to the coordinates. However, the worst is yet to come: This is that the off-diagonal matrix elements diverge near the point of degeneracy, i.e. as $Q \to Q^0$. The equation defining ϕ_e shows this immediately, since

$$\frac{\partial \phi_{e_1}}{\partial Q_i} = -(T_{e1} + V(\mathbf{r}, Q) - W_{e_1}(Q))^{-1} \frac{\partial}{\partial Q_i} [V(\mathbf{r}, Q) - W_e(Q)]\phi_{e_1},$$

from which

$$\langle e_2 | \frac{\partial}{\partial Q_i} | e_1 \rangle = -[W_{e_2}(Q) - W_{e_1}(Q)]^{-1} \langle e_2 | \frac{\partial V(\mathbf{r}, Q)}{\partial Q_i} | e_1 \rangle \quad (2.12)$$

The energies W_{e_2} and W_{e_1} are not necessarily equal at a general point Q in configuration space. However, their values coincide as $Q \to Q^0$, so that the matrix elements of the first derivative, and also of higher derivatives, diverge (since $\langle e_2 | \partial V / \partial Q_i | e_1 \rangle \neq 0$, in general).

Let us investigate this phenomenon by reference to a doubly degenerate state coupled to a single vibrational mode Q. (One must beware,[12] however, of jumping to general conclusions on the basis of the properties of a single normal coordinate). Let us trace the wave-functions belonging to the two intersecting lines ACA′ and BCB′ as the normal coordinate Q is varied (Figure 2.1(a)). Since non-degenerate eigen-functions and eigen-values are analytical functions of the parameters of their defining equations, in this case of Q, the eigen-functions at the points A and B will 'follow' their respective energy lines. This is true up to the vicinity of the point of interaction C (where $Q = Q^0$). Here an ambiguity arises as to which wave function does one end up with on crossing the point $Q = Q^0$, the one belonging to CB′ or to CA′, having started with AC (say)?

Now, by looking at Figure 2.1(a) we should say that the wave function along CA′ is the analytical continuation of that along AC (and the function along CB′ the analytical continuation of that along BC) and no anomalies are expected at $Q = Q^0$. However, we have seen from equation (2.9–AD) and (2.12) that $\partial \phi / \partial Q$ is *singular* at $Q = Q^0$. We must conclude from this that the functions ϕ defined by the differential equation proceed along ACB′ and BCA′ as Q traverses Q^0 (and are therefore not analytical functions of the parameter $Q - Q^0$).

This apparently arbitrary behaviour of the differential equation can best be explained by adding a small increment ϵH_1 to the Hamiltonian, which increment is such that the degeneracy at Q^0 is lifted. With the removal of degeneracy, there is no further ambiguity left in the problem and it is clear that *one wave-function must be assigned to the energy surface $AC_1B′$ and another to $BC_2A′$* (Figure 2.1(b)). The degenerate case is to be regarded as the limit, as $\epsilon \to 0$, of the problem without degeneracy.

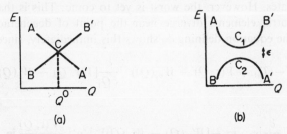

Figure 2.1 Energy *vs.* vibrational coordinate Q.
(a) When the two states are degenerate at $Q = Q^0$.
(b) When the two states are *near*-degenerate at $Q = Q^0$.

There appear to be two very good reasons for looking at the degenerate problem from this angle; a physical reason and a mathematical one. Physically, there is always expected to be a small external perturbation which resolves the degeneracy. From the modern mathematical point of view the singular case of degeneracy should not be considered in isolation but rather as a particular case of the wider set of problems in which it is embedded, namely the differential equation without degeneracy.[14]

There is actually another way for eliminating the singularity in the derivative across Q^0.[12] (This reference comments on an earlier work.)[15] This is artificially to join up, by a singular transformation, the solution AC (or BC) with that along CA′ (or CB′). The result will be a pair of functions analytic across C.

Let us illustrate this in a very simple case. Suppose that the degenerate states $\phi_a(\mathbf{r})$, $\phi_b(\mathbf{r})$ are (for simplicity's sake) independent of the ionic coordinate. Let the two states be coupled by the following J-T matrix

$$\begin{bmatrix} 0 & \alpha(Q - Q^0) \\ \alpha(Q - Q^0) & 0 \end{bmatrix}$$

The solutions of this are

$$\phi_{\pm}(\mathbf{r}, Q) = \frac{1}{\sqrt{2}}\left[\phi_a \pm \frac{Q - Q^0}{|Q - Q^0|}\phi_b\right] \tag{2.13}$$

$$E = \pm\alpha|Q - Q^0|$$

The derivative of $\phi_{\pm}$ is seen to be singular at $Q = Q^0$. Let us now derive a new pair of functions by applying ro $\phi_{\pm}$ a transformation as in the equation below:

$$\begin{pmatrix}\phi_A \\ \phi_B\end{pmatrix} = \tfrac{1}{2}\begin{pmatrix}1 - (Q - Q^0)/|Q - Q^0| & 1 + (Q - Q^0)/|Q - Q^0| \\ 1 + (Q - Q^0)/|Q - Q^0| & 1 - (Q - Q^0)/|Q - Q^0|\end{pmatrix}\begin{pmatrix}\phi_+ \\ \phi_-\end{pmatrix} \tag{2.14}$$

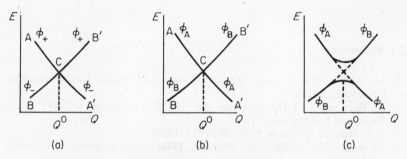

Figure 2.2 Designation of wave functions on the two branches of the energy *vs.* Q curve.

(a) Using the states of equation (2.13). (Note the break across Q^0.)

(b) Using the states of equation (2.14). ϕ_A and ϕ_B are continuous along the lines AA' and BB'.

(c) As in (b) but with only near-degeneracy near $Q = Q^0$.

The pair ϕ_A and ϕ_B are analytic functions of Q near $Q = Q^0$, and their derivatives are everywhere finite. Since the proper assignment of the function across the singularity only is necessary, the same transformation which is applied to Born–Oppenheimer functions near the singularity will also remove the singularity from the adiabatic wave-function.[12] The drawback of this and similar transformations has already been indicated. It is that across the singularity an electronic state jumps from one potential surface to another. (Figure 2.2(c) shows, what this means if there is a small perturbing term ϵH_1 present, which removes the degeneracy.) This can only be avoided at the expense of having singular derivatives at $Q = Q^0$, so that an entirely satisfactory procedure does not appear to be possible within the adiabatic approximation. These difficulties are absent in the Born–Oppenheimer scheme, which is in fact the customary starting point for the J-T-E, and *which will also be adopted in this work.* [Following the suggestion of Reference 13, the J-T-Hamiltonian of the Born–Oppenheimer approximation can perhaps be improved by adding the 'adiabatic' term $\langle e|T_N|e \rangle$, equation (2.10), to the diagonal potential.]

Having started the vibronic calculation in the Born–Oppenheimer scheme, we shall nevertheless find that frequently the best approximate wave-functions are of the adiabatic type, like those appearing in equation (2.8–AD) with χ determined by equation (2.9–AD) and ϕ_e being linear combinations of the degenerate Born–Oppenheimer states. This occurs especially for strong vibronic coupling (cf. equation (3.12) later), where the system has a number of equivalent minima, whose positions are sufficiently far from the point $Q = Q^0$, so that the singularity at this point is harmless.

2.1 References

1. M. Born and K. Huang, *Dynamical Theory of Crystal Lattices* (Oxford: Clarendon, 1951).
2. M. Born and R. J. Oppenheimer, *Ann. Physik (Leipzig)* **89**, 457 (1927); also Reference 1, Appendix VII.
3. J. T. Hougen, *J. Chem. Phys.*, **37**, 1433 (1962).
4. M. Born, *Gött. Nachr. Math. Phys. Kl.*, **1931**, 1; also Reference 1, Appendix VIII.
5. L. C. Snyder, *J. Chem. Phys.*, **33**, 619 (1960).
6. L. L. Lohr, Jr. and W. N. Lipscomb, *Inorg. Chem.*, **2**, 911 (1965).
7. L. L. Lohr, Jr., *Inorg. Chem.*, **6**, 1890 (1967).
8. W. Kolos and L. Volniewicz, *J. Chem. Phys.*, **41**, 3674 (1964).
9. R. N. Porter, R. M. Stevens and M. Karplus, *J. Chem. Phys.*, **49**, 5163 (1968).
10. R. Englman, *Phys. Rev.*, **129**, 551 (1963).
11. W. D. Hobey and A. D. McLachlan, *J. Chem. Phys.*, **33**, 1695 (1960).
12. A. D. McLachlan, *Mol. Phys.*, **4**, 417 (1961).
13. M. Gouterman, *J. Chem. Phys.*, **42**, 351 (1965).
14. L. Finkelstein (Private communications, 1968).
15. H. M. McConnell and A. D. McLachlan, *J. Chem. Phys.*, **34**, 1 (1961).

3 Ideal vibronic systems

Outline

This chapter is the backbone of this book. It contains the solutions to the mathematical J-T and pseudo J-T problems in those simple, idealized systems for which the solutions are known. In more complex cases the nature of the solution, that is to say, the characteristic behaviour of the system is indicated without rigour, proof or detail. As far as possible within the limitation of space and knowledge, experimental counterparts of the theoretical results are added at each stage. The arrangement of the various cases is such that simple cases precede complicated ones. The chapter ends with the Renner–Teller effect for linear molecules.

For an understanding of the physical background, the derivation of the Hamiltonians and much of the terminology, frequent cross-references to the later Chapter 7 are suggested.

3.1 E ⊗ β

Mathematically the simplest case of the Jahn–Teller effect is when a doubly degenerate state becomes coupled by a single mode. We shall see this occurring in the cyclooctatetraene molecular ion (§7.8.5), which is a particular case of molecules having a four-fold axis of symmetry,[1] and in the resonant excitation of double molecules or dimers (§7.7). It is only fair to point out at the outset that the experimental importance of the E ⊗ β case is rather limited and its right of inclusion here is really earned by its mathematical simplicity, which helps to understand the more complex cases of the pseudo J-T-E (§3.6.1). In fact, the strictly degenerate case, characterized by the Hamiltonian (equation (7.14) with $W = 0$ and $q_- \equiv q$)

$$H'(q) = -\tfrac{1}{2}\hbar\omega\left(\frac{\partial^2}{\partial q^2} - q^2\right) - \frac{1}{\sqrt{2}}Lq\sigma_\theta, \qquad (3.1)$$

$$\left[\sigma_\theta = \begin{pmatrix} -1 & 0 \\ 0 & 1 \end{pmatrix}\right]$$

is so simple as to be trivially soluble by the two-fold set of vibronic wave functions

$$\psi_n' = \begin{pmatrix} \chi_n(q + L/\hbar\omega\sqrt{2}) \\ \chi_n(q - L/\hbar\omega\sqrt{2}) \end{pmatrix} \equiv \begin{pmatrix} \chi_n^{\mathrm{L}} \\ \chi_n^{\mathrm{R}} \end{pmatrix} \tag{3.2}$$

and

$$\psi_n'' = \begin{pmatrix} \chi_n(q + L/\hbar\omega\sqrt{2}) \\ -\chi_n(q - L/\hbar\omega\sqrt{2}) \end{pmatrix} \equiv \begin{pmatrix} \chi_n^{\mathrm{L}} \\ \chi_n^{\mathrm{R}} \end{pmatrix} \tag{3.3}$$

in which χ_n^{L} and χ_n^{R} stand for the nth linear harmonic oscillator wavefunction localized in the left and right displaced harmonic well, respectively. In the absence of off-diagonal terms, the energies are $(n + \frac{1}{2})\hbar\omega$, two-fold degenerate.

3.2 E $\otimes$ ε

3.2.1 The four stages A, B, C, D of the vibronic problem

The matrix Hamiltonian in the function space of the electronic doublet $|E_\theta\rangle$ and $|E_\epsilon\rangle$ is

$$H = \frac{1}{2}\hbar\omega\left[\frac{\partial^2}{\partial q_\theta^2} + \frac{\partial^2}{\partial q_\epsilon^2} + q_\theta^2 + q_\epsilon^2\right]\mathbf{I} + \frac{1}{2}L\begin{pmatrix} -q_\theta & q_\epsilon \\ q_\epsilon & q_\theta \end{pmatrix}$$
$$+ \frac{1}{4}K\begin{pmatrix} q_\theta^2 - q_\epsilon^2 & 2q_\theta q_\epsilon \\ 2q_\theta q_\epsilon & q_\epsilon^2 - q_\theta^2 \end{pmatrix} + \frac{\sqrt{2}}{4}N(3q_\theta q_\epsilon^2 - q_\theta^3)\mathbf{I} \tag{3.4}$$

The linear part, the first line in equation (3.4), will be derived for various physical systems in Chapter 7 (§§7.1, 7.8.1). The expression for non-linear coupling, the second line in equation (3.4) is derived in Appendix IV. The coefficients L (linear coupling), K (quadratic) and N (cubic) have the dimensions of energy and are given in terms of the electronic potential $V(\mathbf{r}; q_\theta, q_\epsilon)$ by

$$L = -\langle E_\theta|\frac{\partial V}{\partial q_\theta}|E_\theta\rangle + \langle E_\epsilon|\frac{\partial V}{\partial q_\theta}|E_\epsilon\rangle \equiv \langle E||\frac{\partial V}{\partial q^\epsilon}||E\rangle$$

$$K = -\langle E_\theta|\frac{\partial^2 V}{\partial q_\theta^2}|E_\theta\rangle - \langle E_\epsilon|\frac{\partial^2 V}{\partial q_\theta^2}|E_\epsilon\rangle$$

$$N = -\frac{2}{\sqrt{6}}\left[\langle E_\theta|\frac{\partial^3 V}{\partial q_\theta^3}|E_\theta\rangle + \langle E_\epsilon|\frac{\partial^3 V}{\partial q_\theta^3}|E_\epsilon\rangle\right]$$

The choice for the definitions of K and N is rationalized in Appendix IV. The definition of the vibrational coordinates q_θ, q_ϵ and other conventions were given after Table 1.1 in the introductory chapter.

Before attempting a quantitative solution of the Hamiltonian in equation (3.4) we shall give a qualitative description from no less than three angles. For the sake of presentation we recognize four stages of the problem: (A) No coupling, (B) Weak linear coupling, (C) Strong linear coupling and (D) Higher order coupling.

Let us first describe what will be the nature of some characteristic, low-energy vibronic state in each of these stages. In stage A the electronic and nuclear motions proceed independently of each other; i.e. either electronic state is compatible with a particular vibrational wave-function. At stage B the correlation between the electronic and vibrational motions begins to be felt: The system still vibrates in essentially the same manner as at the uncoupled stage A, however, now the vibratory motion causes periodic virtual excitation of the partner electronic state, as shown in Figure 3.1.

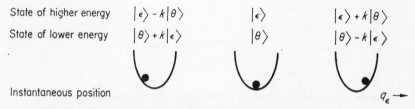

State of higher energy $|\epsilon\rangle - k|\theta\rangle$ $|\epsilon\rangle$ $|\epsilon\rangle + k|\theta\rangle$

State of lower energy $|\theta\rangle + k|\epsilon\rangle$ $|\theta\rangle$ $|\theta\rangle - k|\epsilon\rangle$

Instantaneous position $q_\epsilon \longrightarrow$

Figure 3.1 Electronic states as functions of the vibrational coordinate q_ϵ (for $Lq_\theta > 0$) in the weak linear coupling stage. There is a similar dependence of the states on q_θ. The coefficient k has the same sign as the linear coupling constant L.

In the third stage C the nuclei rotate freely round the distorted configuration of the system, avoiding the undisturbed configurations and carrying along the electronic states with the distortions. In the following figure (Figure 3.2(a)) three successive phases of the distortion can be seen, together with the ground state which belongs to each phase. The configurations of the nuclei in physical space appropriate to each of the three phases, are illustrated for a triatomic molecule in Figure 3.2(b). For an octahedral or a tetrahedral molecule the corresponding configurations of the ligands in real space cannot be drawn as simply as in Figure 3.2(b). It is evident, however, from the definition of the angle ϕ ($=\cotan^{-1} q_\theta/q_\epsilon$) that there exists a simple correspondence, between the motions of the ligands in the modes q_θ, q_ϵ and the distortions in Figure 3.2.

Finally, at stage D with sizeable higher order coupling, the nuclear motions take place in the neighbourhoods of three potential minima. These minima are the positions denoted by Roman one, two or three in

Figure 3.2(b). Between the minima tunnelling takes place. An interesting feature of stage D is that the vibronic states are not all doubly degenerate as in the previous stage: There exists also non-degenerate states whose behaviour is in many ways similar to that of electronic (orbital) singlets. Likewise, the doubly degenerate states show qualitatively similar behaviour to that of the $|E_\theta\rangle$, $|E_\epsilon\rangle$ doublets in the Born–Oppenheimer approximation. However, from a quantitative viewpoint, the magnitudes of many physical parameters, e.g. the g-factor, undergo significant adjustments from the uncoupled stage (§3.2.5).

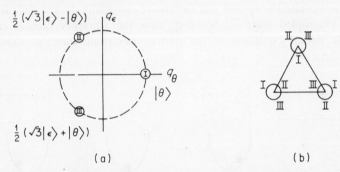

        (a)   (b)

Figure 3.2(a) Instantaneous distortions and wave-functions. Three phases of the distortion, characterized by the phase angle $\phi = \tan^{-1} q_\epsilon/q_\theta$, are shown together with the corresponding states. (For $L > 0$, $K = N = 0$ in equation (3.4).)

(b) Positions of the atoms of a triatomic molecule in the three phases of distortions shown in Figure 3.2(a). (After Reference 2.)

It is helpful to be able to visualize the positions of the minima in Figure 3.2(b) in real space. In a calculation for a hypothetical H_3-molecule the inward displacement of the atom on the C_2 axis at equilibrium is 0·5 Å (the stabilization energy, E_{JT} is 16,000 cm^{-1}, $\hbar\omega = 2050$ cm^{-1}).[3] This calculation is based on a semiempirical potential[4] from which the coefficients L, K and N of equation (3.4) are computed. In a first-principle calculation which utilizes 15 orbitals whose exponents are optimized, no stable minima are found in the q_θ, q_ϵ-plane but only dissociation channels.[4]

For a six- or four-coordinated complex whose ε-frequency (ω) is 200 cm^{-1}, the static displacement for the ligand in the z-direction is about $(E_{JT}/\hbar\omega) \times 10^{-1}$ Å. ($E_{JT}/\hbar\omega$ was defined in equation (1.2) and Table 1.1.)

The four stages A–D can also be described in an alternative way, by drawing the potential surfaces of the static problem. The surfaces are shown in the following figure (Figure 3.3).

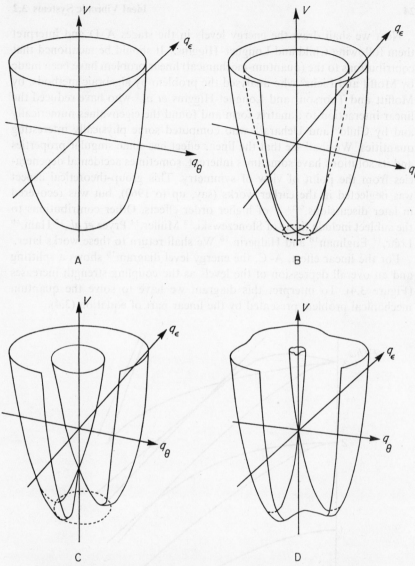

Figure 3.3 Potential surfaces in the q_θ, q_ϵ-plane for the four stages A–D of the text. The nature of the low-energy vibronic states is commented on briefly.

A. Two coinciding paraboloids of revolution.
B. Two potential surfaces slightly separated, except at $q_\theta = q_\epsilon = 0$ where they coincide. Low energy motion is in the proximity of this point.
C. Deep potential trough in lower potential. Low energy motion is around or near this trough. The point $q_\theta = q_\epsilon = 0$ is rarely touched.
D. Lower surface has three minima (as in Figure 3.2(b)) in which the low energy motion is concentrated.

Next we shall draw the energy levels in the stages A–D and interpret them following Child and Longuet–Higgins.[5] It should be mentioned that contributions to the (quantum-mechanical) linear problem have been made by Moffit and Liehr[6] who analysed the problem by algebraic methods, by Moffitt and Thorson[7] and Longuet-Higgins et al.[8] who have reduced the linear interaction to a matrix form and found the eigenvalues numerically and by Child[9] and Uehara[10] who computed some physically interesting quantities. We shall see that the linear effect has some singular properties and its solutions have sometimes inherent, sometimes accidental degeneracies from the point of view of symmetry. This group-theoretical aspect was neglected in the earlier works (say, up to 1960), but was recovered in later discussions,[5,11-2] of higher order effects. Other contributions to the subject include those of Slonczewski,[13] Müller,[14] Pryce et al.,[15] Ham,[16] Lohr,[17] Englman[18] and Halperin.[19] We shall return to these works later.

For the linear effect, A–C, the energy level diagram[10] shows a splitting and an overall depression of the levels as the coupling strength increases (Figure 3.4). To interpret this diagram we have to solve the quantum mechanical problem presented by the linear part of equation (3.4).

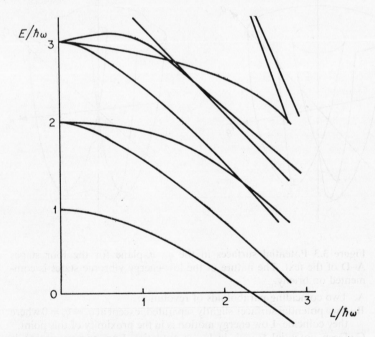

Figure 3.4 Vibronic energy levels (in units of $\hbar\omega$) as function of $L/\hbar\omega$. L is the linear coupling coefficient. [Equation (3.4).] After Uehara.[10] At $L/\hbar\omega = 1$ the order of levels is $j = \frac{1}{2}; \frac{3}{2}, \frac{1}{2}; \frac{5}{2}, \frac{1}{2}, \frac{3}{2}; \cdots$

The cylindrical symmetry of the potential surfaces in Figures 3.3A–3.3C suggests that we should rewrite the first line of the Hamiltonian, equation (3.4), in polar coordinates

$$H = \tfrac{1}{2}\hbar\omega\left[-\frac{\partial^2}{\partial q^2} - \frac{1}{q}\frac{\partial}{\partial q} - \frac{1}{q^2}\frac{\partial^2}{\partial \phi^2} + q^2\right] + \tfrac{1}{2}L\begin{pmatrix} 0 & q\,e^{-i\phi} \\ q\,e^{+i\phi} & 0 \end{pmatrix} \quad (3.5)$$

$$|-1\rangle \quad |+1\rangle$$

Here

$$\left.\begin{aligned} q_\theta &= q\cos\phi \\ q_\epsilon &= q\sin\phi \end{aligned}\right\} \tag{3.6}$$

and the complex representations[20]

$$|\mp 1\rangle = (\pm)i(2)^{-\frac{1}{2}}(|E_\theta\rangle \mp i|E_\epsilon\rangle) \tag{3.7}$$

have been adopted.

(Our representations differ from those in Reference 8 by the factors $(\pm i)$. The angular variable 'ϕ' of these authors is equal to $\pi - \phi$, in terms of our angle ϕ).

A: (When $L = 0$) the eigensolutions of equation (3.5) are of the form

$$\psi_{v,l}^{(\pm 1)} = e^{il\phi}\rho_{v,|l|}(q)|\pm 1\rangle \tag{3.8}$$

where the functions $\rho_{v,|l|}$ are the two-dimensional oscillator wave-functions defined in Reference 21 (pp. 110–11) and l takes the values $v, v - 2, \ldots -v$. The energy levels

$$E = (v + 1)\hbar\omega$$

are $(v + 1)$-fold degenerate. Evidently, v is the quantum number for radial oscillations. These energy levels, shown on the ordinate of Figure 3.4, are again depicted in the following coupling diagram (Figure 3.5), whose meaning is explained in the caption.

The new quantum number $j = l \pm \tfrac{1}{2}$, introduced in the figure will accompany us throughout the linear effect. Its values are the eigenvalues of the operator

$$\begin{aligned} \hat{j} &= \hbar^{-1}(q_\theta p_\epsilon - q_\epsilon p_\theta) - \tfrac{1}{2}\sigma_\theta \\ &= \frac{1}{i}\frac{\partial}{\partial\phi} + \tfrac{1}{2}\begin{pmatrix} 1 & 0 \\ 0 & -1 \end{pmatrix}, \end{aligned} \tag{3.9}$$

σ_θ being defined in the complex representation, equation (3.7).

B: *Weak linear coupling.* The off-diagonal matrix elements in equation (3.5) link the states of the following figure in the manner indicated by broken lines. Along the zig-zag of these lines $j = l \pm \tfrac{1}{2}$ is unchanged, and is evidently a good quantum number. It is not difficult to ascertain that $\hat{j}$ commutes with the linear Hamiltonian.

Two eigenfunctions of $\hat{j}$ having eigenvalues $\pm j$ are codegenerate eigenstates of the Hamiltonian and can be generated from each other by an operation which takes ϕ to $\pi - \phi$ and $|+1\rangle \rightarrow |-1\rangle$, $|-1\rangle \rightarrow -|+1\rangle$. This operation (which in the representations of equation (3.4) is $|E_\theta\rangle \rightarrow |E_\epsilon\rangle$, $|E_\epsilon\rangle \rightarrow |E_\theta\rangle$; $q_\theta \rightarrow -q_\theta$, $q_\epsilon \rightarrow q_\epsilon$) commutes with the linear Hamiltonian and anticommutes with $\hat{j}$.

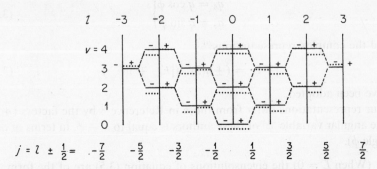

Figure 3.5 Coupling diagrams for weak linear coupling. The level heights for $L = 0$ (stage A) are shown by full horizontal lines, for $\hbar\omega > |L| \neq 0$ (B) by dotted horizontal lines. Broken lines connect the states coupled by the term in L; the signs $+$ and $-$ refer to the two states in equation (3.7).

The scheme of the perturbed energies of $|L| \ll \hbar\omega$ is shown on Figure 3.5 by dotted horizontal lines. Levels belonging to the same l but to different j are split apart by an amount $l \cdot L^2/(4\hbar\omega)$. There is also a uniform shift $L^2/(8\hbar\omega)$ downwards. Both the splitting and shift are apparent on Figure 3.4.

As noted earlier the linear coupling constant L used in this work is the reduced matrix element of the derivatives of the potential. L is related to equivalent symbols in the literature as follows:

$$L = 2k(\hbar\omega)^{8,22} = 2l\sqrt{(\hbar/M\omega)}^{\,7} = 2\sqrt{(2D)}\hbar\omega^{\,5,7}$$
$$= a'\hbar\omega^{\,10}$$

The quantum number $j = l$ (in the notation of Reference 8)

$$= \tfrac{1}{2}\Lambda.^{\,7}$$

Collecting the information contained in the Hamiltonian, equation (3.5), the coupling diagram (Figure 3.5) and the solutions of stage A, equation

(3.7), we observe that the solutions for linear coupling have the general form

$$F_{2l}(q)\, e^{il\phi}| + 1\rangle + F_{2l-1}(q)\, e^{i(l-1)\phi}| - 1\rangle \qquad (3.10)$$

In the low-energy solutions F_{2l} and F_{2l-1} will have the same sign if $L < 0$, and opposite signs if $L > 0$.

C: Strong linear coupling. The energy scheme for strong linear coupling is shown in the next figure (Figure 3.6). We shall now apply a unitary

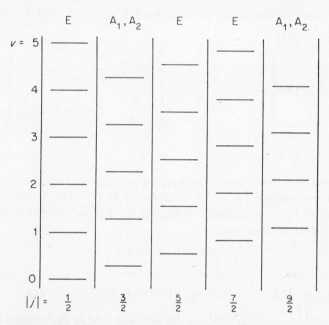

Figure 3.6 The energy scheme for strong linear coupling.

To obtain this figure we have first folded up Figure 3.5 about the central vertical line (noting that energy levels with j and $-j$ are degenerate). We have then omitted the numbers l, which are not even approximately good quantum numbers now, and introduced instead the group theoretical symbols A_1, A_2, E which will be explained in the text. v is the quantum number for radial oscillations, whose meaning will also emerge presently.

transformation S to the Hamiltonian, equation (3.5), with a view of separating the radial motions (whose quantum number is v) from the angular motion (quantum number j or group theoretical species A_1, A_2, E). This separation will be seen to be approximately correct for strong linear coupling.

The transformation matrix, which transforms the kets of equation (3.7) to those of equation (3.12) below, is

$$S^* = \frac{1}{\sqrt{2}} \begin{pmatrix} i\,e^{i\phi/2} & -i\,e^{-i\phi/2} \\ e^{i\phi/2} & e^{-i\phi/2} \end{pmatrix} \tag{3.11}$$

so that the new kets are expressible in terms of $|E_\theta\rangle$, $|E_\epsilon\rangle$ by

$$\left.\begin{aligned} |\psi_-\rangle &= \cos\phi/2|E_\theta\rangle - \sin\phi/2|E_\epsilon\rangle \\ |\psi_+\rangle &= \sin\phi/2|E_\theta\rangle + \cos\phi/2|E_\epsilon\rangle \end{aligned}\right\} \tag{3.12}$$

After the transformation the Hamiltonian, equation (3.5), becomes

$$H' = SHS^{-1}$$

$$= \tfrac{1}{2}\hbar\omega\left\{ -\frac{\partial^2}{\partial q^2} - \frac{1}{q}\frac{\partial}{\partial q} - \frac{1}{q^2}\begin{bmatrix} -\tfrac{1}{4} + \dfrac{\partial^2}{\partial\phi^2} & \dfrac{\partial}{\partial\phi} \\ -\dfrac{\partial}{\partial\phi} & -\tfrac{1}{4} + \dfrac{\partial^2}{\partial\phi^2} \end{bmatrix} + q^2 \right\}$$

$$+ \tfrac{1}{2}Lq\begin{bmatrix} -1 & 0 \\ 0 & 1 \end{bmatrix} \tag{3.13}$$

We are now using square brackets for matrices and eigenvectors in the new representation $|\psi_-\rangle$, $|\psi_+\rangle$, equation (3.12). The transformation, equation (3.11), is equivalent to that given by O'Brien[12] and Slonczewski.[13] The transformation matrix

$$\begin{pmatrix} \cos\phi/2 & -\sin\phi/2 \\ \sin\phi/2 & \cos\phi/2 \end{pmatrix}$$

of equation (3.12) is equivalent after multiplication by $\exp(-i\,\phi/2)$ to $\exp\left[-i\phi\begin{pmatrix} 1 & -i \\ i & 1 \end{pmatrix}/2\right]$, which is the operator used in Reference 23.

Referring back to the form of the solutions, equation (3.10), these are now rewritten in terms of the kets $|\psi_-\rangle$, $|\psi_+\rangle$ as

$$e^{i(l-\frac{1}{2})\phi}[(F_{2l} - F_{2l-1})|\psi_-\rangle + i(F_{2l} + F_{2l-1})|\psi_+\rangle] \tag{3.14}$$

In the static problem,[22] which arises when the kinetic energy is neglected or when the ionic masses tend to infinity, stability of the system occurs for

$$q = |L/(2\hbar\omega)| \equiv q_0 \tag{3.15}$$

This is the value of the radial coordinate at the bottom of the trough of the lower potential of Figure 3.3(c). The depth is known as the J-T-stabilization energy and its value is

$$E_{JT} = L^2/(8\hbar\omega)$$

independently of the angle ϕ. Group-theoretically, this result means that the dispositions with $\phi = 0$, $\pm 2\pi/3$ in Figure 3.2(a) and dispositions with a general value of ϕ have equal energy. (If the original point symmetry was O then the two types of dispositions are D_4 and D_2). This degeneracy is accidental, i.e. it arises from the insufficiency of the linear coupling, and will be removed in the next stage (D), when we introduce non-linear coupling.

The nature of the ground state depends on the sign of the linear coupling coefficient. Thus for $L > 0$ the vibronic ground state wave-function is proportional to $|\psi_-\rangle$ and for $L < 0$ to $|\psi_+\rangle$. This can be seen by making use of equation (3.4) and (3.12). We can, however, also start with equation (3.14) and note that for strong linear coupling the magnitudes of the radial factors $F_{2l}(q)$ and $F_{2l-1}(q)$ are for low values of l only weakly dependent on their subscript (cf. in the previous figure the weak dependence of the energy on j) so that $|F_{2l-1}| \sim |F_{2l}|$, while the relative signs of the radial factors are described in the sentence following equation (3.10). This determines the ground state for strong coupling.

It can be further shown that the actual form of the low-energy solutions of equation (3.13) is, for $L > 0$,

$$\psi_j = q^{-1/2}\chi_v(q - q_0)\chi_j(\phi)|\psi_-\rangle \tag{3.16}$$

with the energy

$$E = E_q + 2j^2(\hbar\omega)^3/L^2$$
$$= (v + 1/2)\hbar\omega - L^2/(8\hbar\omega) + 2j^2(\hbar\omega)^3/L^2 \tag{3.17}$$

Here χ_v is the v th oscillator wave-function, representing oscillations across the potential trough (which lies on a circle of radius $q_0 = |L/(2\hbar\omega)|$);

$$\chi_j(\phi) = e^{ij\phi} \tag{3.18}$$

where j is positive or negative, half integral. The requirement that j is half-integral is unusual for spinless particles in quantum-mechanics. In the present case it is due to the separation of the angular part of the state function ψ_j into two factors $\chi_j(\phi)$ and $|\psi_-\rangle$ in both of which the half-angle $\phi/2$ appears. The function ψ is however still single valued in the q, ϕ plane. The radial oscillations are described by the factor $q^{-1/2}\chi_v(q - q_0)$ in equation (3.16).

In addition to the low energy solutions ψ_j, there also exist high-energy solutions, associated with the upper of the two potential sheets in Figure 3.3C, which include the factor $|\psi_+\rangle$ rather than $|\psi_-\rangle$. This state has been interpreted[14] as representing a motion in which the electronic part is out of phase with the nuclear coordinates. The meaning of this statement can be understood by writing

$$|E_\theta\rangle \propto \sqrt{2} \cos \theta$$
$$|E_\epsilon\rangle \propto -\sqrt{2} \sin \theta$$

so that the states of equation (3.12) take the form $\cos(1/2\phi - \theta)$ and $\sin(1/2\phi - \theta)$ respectively. We see that the variation of the nuclear angle ϕ proceeds twice as fast as that of the electronic angle θ. Moreover 2θ and ϕ tend to keep in phase for $|\psi_-\rangle$ and to be out of phase for $|\psi_+\rangle$.

The solution in equation (3.16) has the form of an adiabatic product wave-function, (equation (2.8–AD)). This form of solution, or better, the absence of intermixing of the two states $|\psi_+\rangle$ and $|\psi_-\rangle$ is permissible provided that the off-diagonal matrix elements in the Hamiltonian, equation (3.13), which induce transitions between states belonging to two surfaces, are negligible compared to the separation of the potentials. This is the case when the coupling is strong, $|L|/\hbar\omega \gg 1$. Then the trough is deep and is sufficiently far from the origin, so that the singular nature of the off-diagonal matrix elements at $q = 0$ is innocuous. (The singularities at the points of intersection of the energy surfaces were discussed in Chapter 2.) The change in the energy of a low lying state on one surface due to the admixture of states on the other surface is easily shown to be of the order of $(\hbar\omega)^7/L^6$.

3.2.2 Slonczewski resonances

Details of the computed energy levels of Figure 3.4 were interpreted in an ingenious manner by Slonczewski and coworkers.[13–4, 24] The following figure (Figure 3.7) shows an inverse 'Dixon-plot' of some $E \otimes \epsilon$ energy levels for strong coupling and $j = 1/2$ and $3/2$. (A straight Dixon-plot,[25] in which differences in neighbouring energies are plotted against the mean energy, is shown in the section on the 'Renner–Teller effect'. §3.7.)

Since $(E_{n+1} - E_n)^{-1}$ is essentially the density of levels, the heights of the maxima in the figure have been associated[24] with lifetimes of some metastable states. The existence of these states was earlier suggested[13] as being due to the stabilization of the upper potential surface by the centrifugal term, namely

$$-\frac{\hbar\omega}{Lq^2}\frac{\partial^2}{\partial\phi^2}$$

in equation (3.13). The states are only metastable, because of the non-diagonal matrix-elements.

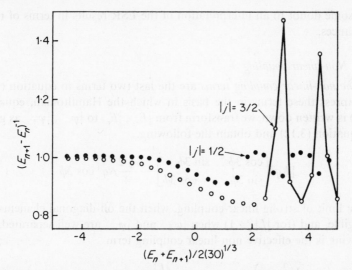

Figure 3.7 Inverse level separations *vs.* mean energy for $E_{\mathrm{JT}}/\hbar\omega = 15$. (From Reference 24.) The energies are in units of $\hbar\omega$.

The positions and force constants of the minima are determined from the following expression for the upper potential sheet of Figure 3.3C

$$\tfrac{1}{2}\hbar\omega\left(q^2 + \frac{j^2}{q^2}\right) + \tfrac{1}{2}Lq$$

The maxima seen in Figure 3.7 occur at the discrete energy levels of this potential. Approaching the problem from the angle of scattering theory Slonczewski and Moruzzi[24] calculated the times of crossing over to the lower potential of a particle entering on the upper potential, as function of its energy. The values of these staying times were shown to be comparable to the heights of the maxima of Figure 3.7. They are clearly only a few times ω^{-1}, so that the stability of the states is rather unfirm.

The theory is of interest because of its claim that it holds not only for a pair of quasi-molecular modes but also in a continuum of lattice vibrations. The condition for stability is that $L/k =$ (now meaning some average strength of the linear J-T-coupling coefficients) must be greater than the Deybe temperature of the lattice. Experimentally, the spin-lattice longitudinal relaxation of some E-state ions appeared to follow an exponential temperature dependence typical of an Orbach-mechanism with energy differences in the region of 500–2000 cm^{-1}.[26-30] While no alternative explanation (specifically, no alternative energy level existing in that region) is known at present, the short 'lifetimes' of the Slonczewski-resonances lend some doubt to an interpretation of the ESR results in terms of these resonances.

3.2.3 Non-linear coupling

D. The non-linear coupling terms are the last two terms in equation (3.4). To express these terms in the basis in which the Hamiltonian, equation (3.13) is written down, we transform from $|E_\theta\rangle$, $|E_\epsilon\rangle$ to $|\psi_-\rangle$, $|\psi_+\rangle$ as given by equation (3.12) and obtain the following.

$$\frac{1}{4}Kq^2 \begin{bmatrix} \cos 3\phi & \sin 3\phi \\ \sin 3\phi & -\cos 3\phi \end{bmatrix} - \frac{\sqrt{2}}{4}Nq^3 \cos 3\phi \qquad (3.19)$$

In the limit of strong linear-coupling, when the off-diagonal elements are negligible, and (for $|L| \gg 1$) when $|\psi_+\rangle$ and $|\psi_-\rangle$ are well separated, the following is the effective non-linear coupling term

$$\left(\pm \frac{1}{4}Kq^2 - \frac{\sqrt{2}}{4}Mq^3 \right) \cos 3\phi \simeq - \frac{|L|}{16(\hbar\omega)^3}(\tfrac{1}{2}\sqrt{2}ML^2 - \hbar\omega KL) \cos 3\phi$$

$$\equiv - \beta \cos 3\phi \qquad (3.20)$$

hereby defining β. This term will admix states, equation (3.16), which have quantum numbers j, with $j \pm 3$ states. Therefore only j modulo 3 will be good quantum numbers, not the individual j's. From their definition ($j = l \pm \frac{1}{2}$) we note that all j's must be half integers. A state belonging to $j = \frac{1}{2}$ modulo 3 will be degenerate, by time reversal, with one for which $j = -\frac{1}{2}$ modulo 3 $= \frac{5}{2}$ modulo 3, etc. We get in this way a set of vibronic doublets arising from combinations of states with $2j = \pm 1$, ∓ 5, ± 7, ∓ 11, . . . (where either the upper or the lower sign is to be taken throughout). There will also be states arising from admixtures among $2j = +3$, -3, $+9$, -9, . . ., which may be so combined as to form states which are either odd or even in ϕ. If one traces back the meaning of the angle ϕ to the symmetry operations of the octahedral group O, these states will be found to be odd or even under a $\pi/2$ rotation about a fourfold axis. These two vibronic states are A_2 or A_1 in O. (It may be seen that the vibronic factors, equation (3.12), do not form bases for irreducible representations in the octahedral group, since the j's are half integers. To overcome this inconvenience a larger group may be defined[31] in which the vibronic factors, equation (3.12), do form bases for irreducible representations. It may be noted in this connection that the two-valuedness of the vibronic factor has topological, rather than group-theoretical, roots. It was shown[32] that a similar situation will arise whenever two potential surfaces meet at one point in a two-dimensional space.)

The appropriate Schrödinger equation for these states is from equation (3.16) and (3.19) and under assumption of strong linear coupling ($q_0 = |L|/(2\hbar\omega) \gg 1$, $|q - q_0|q_0 \ll 1$),

$$\left[- \alpha \frac{\partial^2}{\partial\phi^2} - \beta \cos 3\phi + E_q \right] \chi(\phi) = E\chi(\phi) \tag{3.21}$$

Here

$$\alpha = \hbar\omega/2q_0^2 = (\hbar\omega)^2/(4E_{\text{JT}}) \tag{3.22}$$

E and E_q are as in (3.17) and $\chi(\phi)$ is the angular factor replacing in the wave function (3.16) the factor $\chi_j(\phi) = e^{ij\phi}$ which applies in the linear coupling case only. It follows from Floquet's theorem, introduced from a physicist's point of view in Reference 33, as applied to the periodic potential in the previous equation, that $\chi(\phi)$ can be written in the form

$$\chi(\phi) = \chi_{j'}(\phi) = e^{ij'\phi}u(\phi), \tag{3.23}$$

where $u(\phi)$ is periodic with a period of $2\pi/3$ and j' is half integer. The periodicity of $u(\phi)$ is connected to the circumstance that the x, y, z-axes of a cube can be interchanged by a rotation through the angles $\pm 2\pi/3$ about the body diagonal.

From the discussion in the last but one paragraph it is clear that j' can be taken as

$$\begin{aligned} j' &= \pm 1/2 \quad \text{for E-states} \\ &= 3/2 \quad \text{for A}_1 \text{ and A}_2\text{-states} \end{aligned} \right\} \tag{3.24}$$

More particularly, the partners of the E-representations have the angular factors

$$\text{Re } \chi_{(j'=\frac{1}{2})}(\phi) \quad \text{and} \quad \text{Im } \chi_{(j'=\frac{1}{2})}(\phi),$$

while the angular factor for the singlet is $\chi_{(j'=\frac{3}{2})}(\phi)$.

The eigenvalues $E - E_q$ belonging to the eigenfunctions $\chi(\phi)$ are shown in Figure 3.8 as functions of the non-linear coupling strength. The eigenvalues for linear coupling only, $2j^2(\hbar\omega)^3/L^2$, can be read off from the ordinate. In the figure $\beta > 0$ and all levels have the same radial quantum number v.

It is instructive to present explicit expressions for the three lowest-energy states from the point of view of Floquet's theorem in the strong non-linear coupling limit, since this case has never been fully treated in the literature[16,35-7] and has even given rise[11] to some confusion. We shall return to the larger manifold after the following essay (§3.2.4) on the three-state model.

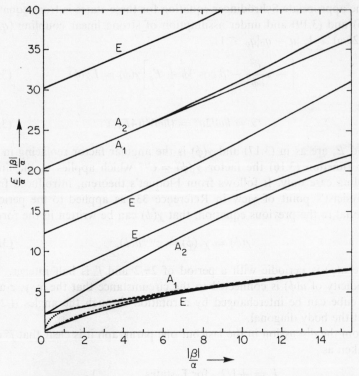

Figure 3.8 Vibronic energy levels *vs.* non-linear coupling strength. (After Reference 12.) The computed energies of levels (whose classifications in O-symmetry are shown) are plotted in units of α (the rotational energy) as function of $|\beta|/\alpha$, where β is defined in equation (3.20) of the text. In the lower part of the figure the energies obtained in the three state approximation (§3.2.4) are shown by broken lines in the region where this approximation is physically meaningful and by dotted lines where it is not. For the first and fourth cases appearing in Table 3.2 the subscripts 1, 2 on the singlets have to be interchanged in this figure.

3.2.4 *The three-state model*

In the strong coupling limit, when $|\beta/\alpha| \gg 1$, the system performs small oscillations about the angles at which the potential $-\beta \cos 3\phi$, equation (3.20), has minima. For octahedral $Cu^{2+}(d^9)$ and $Mn^{3+}(d^4)$ calculations indicate that $\beta > 0$.[22] Then the minima are at $\phi = 0$, $\pm 2\pi/3$. The frequency Ω of the angular oscillations about these minima is given by

$$\frac{1}{2}\frac{\Omega}{\omega}q_0^2 = \sqrt{\frac{9|\beta|}{8\alpha}} \equiv \nu \qquad (3.25)$$

In the limit of strong non-linear coupling ($|\beta/\alpha| \gg 1$) when the minima for the angular motion are deep, superpositions of three harmonic oscillator wave-functions, each of which is centred on a different minimum, should provide good approximate solutions for low lying states. The ground state doublet E and the lowest singlet A_1 or A_2 in Figure 3.8 can be obtained by taking the real or imaginary parts of the following functions (in which we drop the primes from the j's):

$$
\begin{aligned}
\chi_j(\phi) = \lim_{K \to \infty} \frac{N_{|j|}}{\sqrt{2K+1}} \, e^{ij\phi} \sum_{n=-K}^{K} & [e^{-\nu(\phi+2n\pi)^2} \, e^{-ij(\phi+2n\pi)} \\
& + e^{-\nu(\phi+2n\pi-2\pi/3)^2} \, e^{-ij(\phi+2n\pi-2\pi/3)} \\
& + e^{-\nu(\phi+2n\pi-4\pi/3)^2} \, e^{-ij(\phi+2n\pi-4\pi/3)}]
\end{aligned}
\tag{3.26}
$$

Here $j = \pm\frac{1}{2}, \frac{3}{2}$.

Higher lying states can be formed in a similar manner by writing higher order harmonic-oscillator wave-functions, instead of the zeroth order one (a Gaussian) as above, but the approximation soon becomes invalid except for extremely deep angular minima. The summation over n and the limit $K \to \infty$ are necessary in order to satisfy Floquet's theorem. One could perhaps dispense with these by regarding ϕ as the many valued angle $\phi = \phi' + 2n\pi$, where ϕ' is confined to a range of 2π and n is an integer (Reference 39, equation 2.4.6). However, then there is a danger that the overlaps between the various terms are not properly taken into account and the order of the levels $j = \pm\frac{1}{2}$ and $j = \frac{3}{2}$ comes out wrong.

The normalization factors are

$$
N_{\frac{1}{2}}^2 \sim \tfrac{1}{3}\sqrt{\frac{2\nu}{\pi}}(1+\gamma)^{-1}
$$

$$
N_{\frac{3}{2}}^2 \sim \tfrac{1}{3}\sqrt{\frac{2\nu}{\pi}}(1-2\gamma)^{-1}
$$

where

$$
\gamma = e^{-2\nu\pi^2/9}
\tag{3.27}
$$

The approximation sign signifies that in the normalization (as well as in all future expressions) we neglect overlaps between next nearest neighbour Gaussians in equation (3.26). The neglected quantities are smaller than γ by a factor $\exp(-\nu\pi^2/3)$, which is small [by equation (3.25)] for strong non-linear coupling.

Table 3.1 Numerical values of the tunnelling energy (in units of α) as a function of the anisotropic coupling strength $|\beta/\alpha|$. $[\alpha = (\hbar\omega)^2/(4E_{JT})]$. The values were obtained from numerical diagonalization of equation (3.21). (B. Halperin, private communication, 1969.)

$|\beta/\alpha|$	$3\Gamma/\alpha$
0·1	1·951
0·5	1·762
1·0	1·549
2·5	1·043
5·0	$5\cdot46 \times 10^{-1}$
10·0	$1\cdot8\ \times 10^{-1}$
20·0	$2\cdot5\ \times 10^{-2}$
50·0	$9\cdot7\ \times 10^{-5}$

Table 3.2 Specification of stable states in relation to the signs of L, equation (3.4), and of β, equation (3.20).

In an octahedral complex the case denoted by an asterisk is usual for one d electron and the one marked by a circle is common for the many-electron state containing one hole (Cu^{2+}, Mn^{3+}, Cr^{2+}). In calculations β has the tendency to be negative in an ionic point charge model and positive if covalency is taken into considerations.

	Conditions					In stable distortion		
	L	β	c/a	q_θ	$\cos\phi$	Ground state	Ground vibronic state, equation (3.12)	First excited singlet
	+	−	<1	−	−1	$|E_\epsilon\rangle$	$|\psi_-\rangle$	A_2
*	+	+	>1	+	+1	$|E_\theta\rangle$	$|\psi_-\rangle$	A_1
	−	−	<1	−	−1	$|E_\theta\rangle$	$|\psi_+\rangle$	A_1
°	−	+	>1	+	+1	$|E_\epsilon\rangle$	$|\psi_+\rangle$	A_2

[The exact expression for the normalization is rather formidable:

$$N^2_{|j|} = \frac{1}{3}\sqrt{\frac{2v}{\pi}}\left\{\sum_{n=-\infty}^{\infty} e^{-8n^2\pi^2 v} - e^{-2(2n+1)^2\pi^2 v} + 2\cos(2\pi j/3)(e^{-2(2n-\frac{1}{3})^2\pi^2 v}\right.$$

$$\left. - e^{-2(2n+\frac{2}{3})^2\pi^2 v})\right\}^{-1}]$$

The eigenenergies of equation (3.21) are found to be

$$E_E/\alpha = \langle\pm\tfrac{1}{2}|H|\pm\tfrac{1}{2}\rangle/\alpha$$

$$= \frac{E_q}{\alpha} + v - \frac{4v^2\pi^2}{9}\cdot\frac{\gamma}{1+\gamma} + \frac{8v^2}{9}e^{-9/8v}\cdot\frac{1-\gamma}{1+\gamma}$$

for the doublet, and for the next singlet

$$E_A/\alpha = \langle\tfrac{3}{2}|H|\tfrac{3}{2}\rangle/\alpha$$

$$= \frac{E_q}{\alpha} + v + \frac{8v^2\pi^2}{9}\cdot\frac{\gamma}{1-2\gamma} + \frac{8v^2}{9}e^{-9/3v}\cdot\frac{1+2\gamma}{1-2\gamma}$$

E_q is shown in equation (3.17).

The difference between these gives the important quantity which is called the tunnelling energy, and is denoted by 3Γ.

$$3\Gamma \equiv E_A - E_E = 3|\beta|\gamma[\pi^2(1+\gamma)/2 - 2\exp(-9/8v)] \quad (3.28)$$

This expression is dominated by the first, positive term, showing that the doublet is lower. This is also apparent in Figure 3.8. At a time a contrary claim was made,[11] which was then discussed critically by Ham,[16] Lohr[17] and others.

The quantity 3Γ is shown graphically in Figure 3.9 and is tabulated in Table 3.1. It has to be remarked, that the quantities in the Figure and the Table were calculated exactly and not in the three-state model (for which we have in equation (3.28) an analytical form).

Further important quantities in the theory and bearing directly on experimental data in the g-factors of degenerate ions[35-6,51-4] and on the magnitude of Coriolis coupling in molecules[55] are the reduction factors p, q, r and r' introduced by Ham.[16] His definitions refer to the lowest-lying vibronic doublet (Ψ'_θ, Ψ'_ϵ) and the singlet Ψ'_{A_1} or Ψ'_{A_2} immediately above these.

$$p = i\langle\Psi'_\theta|\begin{pmatrix}0 & -i\\ i & 0\end{pmatrix}|\Psi'_\epsilon\rangle \quad (3.29)$$

$$q = -\langle\Psi'_\theta|\begin{pmatrix}-1 & 0\\ 0 & 1\end{pmatrix}|\Psi'_\theta\rangle \quad (3.30)$$

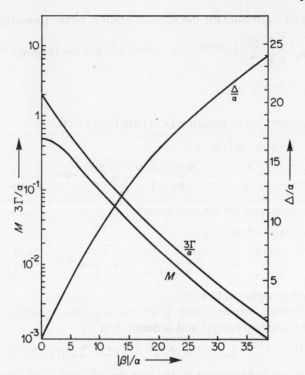

Figure 3.9 Some numerical results for non-linear coupling. (From Reference 34.)

(a) The energy separation 3Γ between the doublet and the nearest singlet, divided by α, *vs.* $|\beta|/\alpha$. (Left hand vertical scale.)

(b) Reduction factor for the vibrational angular momentum operator.

$$M = \langle \Psi_\theta | \partial/\partial\phi | \Psi_\epsilon \rangle$$

is shown *vs.* $|\beta|/\alpha$. (Left hand vertical scale.)

(c) The separation Δ of the energies of the singlets A_1 and A_2 divided by α, *vs.* $|\beta|/\alpha$. (Right hand vertical scale.)

and

$$\left. \begin{array}{l} r = \langle \Psi_{A_1} | \begin{pmatrix} -1 & 0 \\ 0 & 1 \end{pmatrix} | \Psi_\theta \rangle \\[2mm] r' = \langle \Psi_{A_2} | \begin{pmatrix} -1 & 0 \\ 0 & 1 \end{pmatrix} | \Psi_\epsilon \rangle \end{array} \right\} \tag{3.31}$$

(The integral q should not be confused with the vibrational amplitude defined earlier in equation (3.6).)

Thus p, q, r and r' will be functions of both the linear and the higher order couplings. In stage D and for extremely strong linear coupling when the adiabatic ansatz in equation (3.16) is found to be permissible (or, more accurately, when the off-diagonal matrix-elements in the Hamiltonian of equation (3.13) are negligible), p, q, r and r' can be redefined in terms of the vibrational factors χ in equation (3.16) as follows:

$$p = 0 \tag{3.29D}$$

$$q = \langle \chi_\theta | \cos \phi | \chi_\theta \rangle \tag{3.30D}$$

$$\left.\begin{aligned} r &= -\langle \chi_{A_1} | \cos \phi | \chi_\theta \rangle \\ r' &= -\langle \chi_{A_2} | \cos \phi | \chi_\epsilon \rangle \end{aligned}\right\} \tag{3.31D}$$

In this limit these integrals will only depend on one parameter $|\beta|/\alpha = 8\nu^2/9$, which represents the strength of non-linear coupling. Ham showed[16] that in the linear coupling limit ($\beta = 0$ or $\nu = 0$)

$$q = \tfrac{1}{2}(1 + p),$$
$$\tfrac{1}{2} < q \leqslant 1,$$

where the last equality holds in the absence of vibronic coupling. These two relations are generally true, i.e. are not dependent on the three-state model. However, they are valid only for the lowest doublet. On the other hand, in the regime of strong linear *and* non-linear couplings one finds that in the three-state model

$$p = 0$$

$$q = \tfrac{1}{2} e^{-1/(8\nu)} \frac{1 + 2\gamma}{1 + \gamma} \tag{3.32}$$

$$-r = r' = \frac{1}{\sqrt{2}} e^{-1/(8\nu)} \frac{1 - \gamma}{\sqrt{(1 + \gamma)(1 - 2\gamma)}} \tag{3.33}$$

[γ was defined in equation (3.27) and (3.25)].

In q, r and r' the dominant dependence ν is through the factor (which was first obtained by O'Brien[12]) preceding the fraction. We see therefore that in the non-linear coupling case ($\nu \neq 0$)

$$0 < q \leqslant \tfrac{1}{2}.$$

This inequality, which is based on the three-state approximation and equation (3.32), can be considerably strengthened when the matrix element q is evaluated from equation (3.30) after numerical diagonalization of the Schrödinger equation, equation (3.21). One finds[34,38,56] that

$$0 \cdot 484 < q \leqslant 0 \cdot 5$$

Computed values of q and r are shown in Figures 3.10 and 3.11. These are compared with the analytical expressions shown above in equation (3.32) and (3.33) which, we recall, are based on the three state approximation.

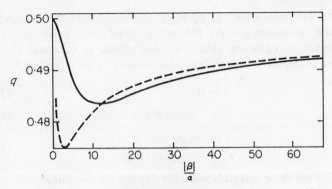

Figure 3.10 q vs. β/α

The diagonal reduction factor q is defined in equation (3.30). Exact, computed values are shown by full lines and the results of the three-state approximation by broken lines down to $|\beta|/\alpha = 1$ at which value the approximation is not expected to be valid any longer.

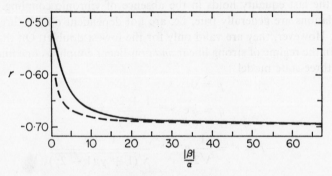

Figure 3.11 r vs. β/α

r is the off-diagonal reduction factor between the doublet and the nearest singlet. The exact values, as defined in equation (3.31), and those obtained in the three state approximation, are shown. (B. Halperin, unpublished, 1969.)

The physical importance and the general behaviour of p, q and r as functions of both linear and higher order couplings will be discussed later (§3.2.6).

The vibronic levels of Figure 3.8 were observed in the infrared-combination spectra of ReF_6[57-8] and TcF_6[59] and were interpreted[60-2] in terms of the J-T-E. Figure 3.12 shows part of the infrared spectrum of gaseous ReF_6 due to Claassen, Malm and Selig[58] in the region of 1470 cm^{-1}. This arises from the combination of the τ_{1u} mode (frequency $= 715 \text{ cm}^{-1}$) and ε_g-mode ($\omega \simeq 670 \text{ cm}^{-1}$), subject to a linear J-T-E, whose strength L

is of the order of the ε-frequency (670 cm^{-1}). Actually, the electronic state is the fourfold degenerate $G_{3/2}$, which is coupled to both ε_g and τ_{2g}, but for weak linear coupling the two couplings are independent (§3.4.1) and the energy levels for $G_{3/2} \otimes \varepsilon$ are in a simple two-to-one correspondence with those of E $\otimes$ ε. In Figure 3.12 a sharp line is apparent at 1470 cm^{-1} as well as an unresolved pair of lines near 1302 cm^{-1} about 42 cm^{-1} apart. These are, for increasing wave-numbers, the lowest E_g, A_{1g} and A_{2g} levels of Figure 3.8. The band at 1302 cm^{-1} seems to be rather stronger than the sharper band at 1470 cm^{-1}. This is in agreement with the theory[62-3] which accounts for the combination band intensities by a bilinear coupling term in the ε_g and τ_{1u}-coordinates.

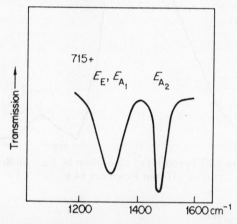

Figure 3.12 Portion of the infrared combination spectrum of ReF$_6$. (After Reference 58, 62). Above the transmission minima the vibronic energies + the τ_{1u} mode energy (715 cm^{-1}) are written.

The splitting due to tunnelling on the extreme right hand part of Figure 3.8 was observed in the excited configuration $4f^6 5d$ of Eu^{2+} in CaF$_2$ and SrF$_2$ by Kaplyanskii and Przhevuskii[64] (Figure 3.13). The two transitions shown by arrows were identified[65] as transitions to the lowest doublet and singlet in Figure 3.8 arising from the e(5d) orbital coupled to an ε-mode. The f-states are not coupled vibronically, neither in the excited nor in the ground state, because of their shielded nature. The doublet-singlet separations 3Γ are 6·5 cm^{-1} in SrF$_2$ and 15·3 cm^{-1} in CaF$_2$. The identification is based on the splitting of the 4011 Å doublet when stress is applied in a direction other than {111}, as will be explained in Chapter 4, and on the observation of asymmetric broadening of the cubic g-factor. In CaF$_2$ these techniques yield the value of 18 ± 4 cm^{-1} for 3Γ, in good agreement with the value (15·3) derived from optical spectroscopy.[64]

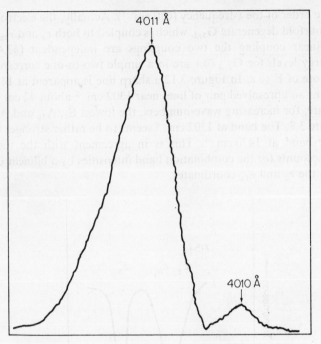

Figure 3.13 Spectrum of absorption by Eu^{2+} in SrF_2.
(From Reference 64.)

3.2.5 *The Ham effect*

The physical importance of the reduction or 'quenching' factors p, q, r and r', equation (3.29)–(3.31), was established by Ham.[16,39] His formulation of what is now called the 'Ham effect' proceeds along the following lines:

Any general form of interaction, due to any external physical cause, and acting on the electronic coordinates can be represented within the E electronic manifold as a Hamiltonian H (elect.) having the form

$$H(\text{elect.}) = G(A_1)\mathbf{I} - G(A_2)\sigma_2 + G(E_\theta)\sigma_\theta + G(E_\epsilon)\sigma_\epsilon \qquad (3.34)$$

The matrices

$$\mathbf{I} = \begin{pmatrix} 1 & 0 \\ 0 & 1 \end{pmatrix} \quad \sigma_\theta = \begin{pmatrix} -1 & 0 \\ 0 & 1 \end{pmatrix} \quad \sigma_\epsilon = \begin{pmatrix} 0 & 1 \\ 1 & 0 \end{pmatrix} \quad \text{and} \quad \sigma_2 = \begin{pmatrix} 0 & i \\ -i & 0 \end{pmatrix}$$

$$(3.35)$$

operate within the electronic doublet. (The three σ-matrices are defined so as to obey the angular momenta commutation rules.) The coefficients G

depend on the external perturbation but are independent of the vibrational coordinates. Thus an orthorhombic shear strain will enter through the coefficients $G(E_\theta)$ and $G(E_\epsilon)$, while the combination of a trigonal strain and a magnetic field will give rise to $G(A_2)$.

The same interaction operating within the nth vibronic doublet states $|\Psi_\theta'^n\}$, $|\Psi_\epsilon'^n\}$ will appear in the form

$$H(\text{vibr.}) = G(A_1)\mathbf{I}^{(n)} - p^{(n)}G(A_2)\sigma_2^{(n)} + q^{(n)}[G(E_\theta)\sigma_\theta^{(n)} + \\ + G(E_\epsilon)\sigma_\epsilon^{(n)}] \tag{3.36}$$

where the superfix (n) has been temporarily added to the matrices to indicate that the matrices refer to the nth vibronic doublet level. The reduction factors $p^{(n)}$, $q^{(n)}$ are defined thus:

$$q^{(n)} = -\{\Psi_\theta'^n|\sigma_\theta^{(n)}|\Psi_\theta'^n\} = \{\Psi_\epsilon'^n|\sigma_\theta^{(n)}|\Psi_\epsilon'^n\} \\ = \{\Psi_\epsilon'^n|\sigma_\epsilon^{(n)}|\Psi_\theta'^n\} \tag{3.37}$$

$$p^{(n)} = -i\{\Psi_\theta'^n|\sigma_2^{(n)}|\Psi_\epsilon'^n\} \tag{3.38}$$

$$\{\text{el } [\text{Hamil}(16)]\}$$

(Remember, $|\Psi_\theta'^n\}$ and $|\Psi_\epsilon'^n\}$ are vectors in the two-dimensional electronic space).

The most important case is clearly that when n denotes the vibronic ground state doublet, $n = g$. p, q without any superscript will refer to this case.

A physically important extension of equation (3.36) is achieved by noting, on the basis of group theory, that the quantities $G(E_\theta)$ and $G(E_\epsilon)$ occurring in the last two terms of equation (3.36) will appear in the matrix element of the interaction Hamiltonian between a component of a vibronic doublet state and a vibronic state of A_1 or A_2 symmetry. The matrix element between the θ or ϵ-component of the vibronic ground doublet and the nearest vibronic singlet of A_1 symmetry will be $G(E_\theta)$ or $G(E_\epsilon)$ multiplied by the factor

$$r = \{\Psi_{A_1}|\sigma_\theta|\Psi_\theta'^{(g)}\} = \{\Psi_{A_1}|\sigma_\epsilon|\Psi_\epsilon'^{(g)}\} \tag{3.39}$$

With a nearest vibronic singlet of A_2 symmetry the multiplying factor is

$$r' = -\{\Psi_{A_2}|\sigma_\epsilon|\Psi_\theta'^{(g)}\} = \{\Psi_{A_2}|\sigma_\theta|\Psi_\epsilon'^{(g)}\} \tag{3.40}$$

Before discussing the magnitudes of the reduction factors p, q, r and r' as function of the coupling coefficients L, K, N in equation (3.4), let us concentrate our thinking on a very important example where the formalism of equation (3.34) and (3.36) is useful, an electron spin-Hamiltonian. Consider for instance a Cu^{2+} ion in O-symmetry, so that the ground term is ^2E. In this case, one finds for the coefficients G in equation (3.34) the following expressions[66-7]

$$G(A_1) = \beta\tfrac{1}{2}\sum_{ij}S_i(g_{ij}^{\theta\theta} + g_{ij}^{\epsilon\epsilon})H_j$$

$$G(A_2) = 0$$

$$G(E_\theta) = \beta\tfrac{1}{2}\sum_{ij}S_i(g_{ij}^{\epsilon\epsilon} - g_{ij}^{\theta\theta})H_j$$

$$G(E_\epsilon) = \beta\sum_{ij}S_ig_{ij}^{\theta\epsilon}H_j$$

Here S_i is a component ($i = x, y, z$) of the electronic spin, $\mathbf{H}$ is the magnetic field and β is the Bohr magneton. Further

$$\left.\begin{array}{l}
g_{zz}^{\theta\theta} = 2\cdot0023 = g_\parallel^\theta \qquad\qquad g_{xx}^{\theta\theta} = g_{yy}^{\theta\theta} = 2\cdot0023 - 6\lambda/\Delta = g_\perp^\theta \\[4pt]
g_{zz}^{\epsilon\epsilon} = 2\cdot0023 - 8\lambda/\Delta = g_\parallel^\epsilon \quad g_{xx}^{\epsilon\epsilon} = g_{yy}^{\epsilon\epsilon} = 2\cdot0023 - 2\lambda/\Delta = g_\perp^\epsilon \\[4pt]
g_{xx}^{\theta\epsilon} = -g_{yy}^{\theta\epsilon} = -2\sqrt{3}\lambda/\Delta
\end{array}\right\} \quad (3.41)$$

λ is the spin-orbit coupling constant ($\lambda < 0$ in Cu^{2+}) and Δ is the energy separation of the electronic doublet from the $^2T_2$ term.

The expression for the interaction Hamiltonian in the lowest vibronic manifold is given above in equation (3.36) with $n = g$. This superscript will now be dropped for simplicity. The eigenvalues of the matrix H(vibr.), equation (3.36), are

$$G(A_1) \pm q\sqrt{\{[G(E_\theta)]^2 + [G(E_\epsilon)]^2\}}$$

from which the g-values are easily found to be

$$g_\pm = g_1 \pm qg_2\sqrt{[1 - 3(m^2n^2 + n^2l^2 + l^2m^2)]} \qquad (3.42)$$

Here

$$\left.\begin{array}{l}
g_1 \equiv \tfrac{1}{3}(g_\parallel^\epsilon + 2g_\perp^\epsilon) = \tfrac{1}{3}(g_\parallel^\theta + 2g_\perp^\theta) = 2\cdot0023 - 4\lambda/\Delta \\[4pt]
g_2 \equiv \tfrac{2}{3}(g_\parallel^\epsilon - g_\perp^\epsilon) = -\tfrac{2}{3}(g_\parallel^\theta - g_\perp^\theta) = -4\lambda/\Delta
\end{array}\right\} \quad (3.43)$$

and l, m, n are the direction cosines of the magnetic field. The result, equation (3.42),[16] exhibiting cubic symmetry accounts well for the spectra of the 'third type' found at low temperatures by Coffman[40] in $MgO:Cu^{2+}$ and by Höchli and Estle[35-6] in $SrF_2:Sc^{2+}$ and $CaF_2:Sc^{2+}$. The result with $q = 1$ in equation (3.42) is appropriate in the absence of vibronic coupling.[68]

To include also hyperfine interactions in equation (3.34) one must add

$$\tfrac{1}{2}\sum_{ij}I_i(A_{ij}^{\theta\theta} + A_{ij}^{\epsilon\epsilon})H_j \qquad (3.44)$$

to $G(A_1)$ and similarly for the other coefficients $G(E_\theta)$ and $G(E_\epsilon)$. Here I_i are the components of the nuclear spin and the A's are the components of the hyperfine coupling tensor. (Reference 69, sections 10 and 13(8).) In the result for the resonance energies $hv = (g_\pm)\beta H$, where $g_\pm$ is shown in equation (3.42), add $A_1 m$ to $g_1 \beta H$ and $A_2 m$ to $g_2 \beta H$. m is an eigenvalue of the nuclear spin $\mathbf{I}$ along $\mathbf{H}$, A_1 and A_2 are defined analogously to equation (3.43). Explicitly,

$$\left.\begin{aligned} A_1 &= \tfrac{1}{3}(A_{zz}^{\epsilon\epsilon} + 2A_{xx}^{\epsilon\epsilon}) \simeq -P(K + 4\lambda/\Delta) \\ A_2 &= -\tfrac{2}{3}(A_{zz}^{\epsilon\epsilon} - A_{xx}^{\epsilon\epsilon}) = -P[\tfrac{4}{7} + (4\lambda/\Delta) + \tfrac{3}{14}(4\lambda/\Delta)] \end{aligned}\right\} \quad (3.45)$$

Here $P = 2\gamma\beta\beta_N \overline{r^{-3}}$, whereas γ is the nuclear g-factor, β_N is the nuclear magneton, $\overline{}$ denotes the quantum mechanical expectation value and K (~ 0.3 in Cu^{2+}) is a number which characterizes the contact hyperfine interaction.

We shall return to the study of the spin-Hamiltonian and of the g-values after the inclusion (in §4.3) of low-symmetry fields in the Hamiltonian. Now we resume our study of the reduction factors and generalize the formulae of equation (3.32) and (3.33) and the inequality $0.484 < q \leqslant 0.5$ to cases when the linear coupling is not excessively strong. One of the purposes of the next section is to derive for q an interpolation formula which is valid when neither the purely linear coupling result

$$\beta = 0, \qquad \tfrac{1}{2} < q < 1$$

nor the extremely strong linear coupling formulae, (equation (3.32) and (3.33)) and $0.484 < q \leqslant 0.5$, are justified.

3.2.6 A guess for the E ⊗ ε vibronic ground-state

A number of properties of the vibronic ground state doublet arising from E ⊗ ε can be studied by manipulation of a guessed analytic form of these states. Not only leads this guess to the correct behaviour of the states in the limits of strong or weak coupling, but it also provides good physical insight into its behaviour. The method is applicable generally, to any case of linear J-T-E;[18] we shall here describe it by reference to E ⊗ ε using it to derive the reduction factors p and q. We shall find it possible to extend the original guess[18,67] to non-linear coupling. Since there does not exist any general solution for this case (not even a numerical solution in the q, ϕ-space; the only 'treatment' which is not restricted to a single potential surface is the perturbation theory of Reference 34), there may be some practical use in an analytical approximation.

The rationale for the suggested guess is that when the electronic energy-spacing is much smaller (as in a degenerate situation) than the vibrational level spacing, then it is reasonable to apply the adiabatic approximation in reverse and to suggest that the ions adjust themselves to the occupancy of particular electronic states in the sense, that we have now vibrational wave-functions whose equilibrium positions depend parametrically on the occupancy of the electronic states. This is also the physical meaning of the method[70] (similar in essence to the one described below) which is based on a unitary transformation and was applied to a P-J-T-E.

3

Starting with linear coupling, the first line in equation (3.4), we can try a solution of the form

$$\exp - \tfrac{1}{2}\left\{ \left[q_\theta + \tfrac{1}{2}\frac{L}{\hbar\omega}\sigma_\theta \right]^2 + \left[q_\epsilon + \tfrac{1}{2}\frac{L}{\hbar\omega}\sigma_\epsilon \right]^2 \right\}$$

$$= \exp\left[-\tfrac{1}{4}\frac{L^2}{(\hbar\omega)^2} \right] e^{-q^2/2}\left[\cosh\left(\frac{qL}{2\hbar\omega}\right)\mathbf{I} - \right.$$

$$\left. - \sinh\left(\frac{qL}{2\hbar\omega}\right)\begin{pmatrix} -\cos\phi & \sin\phi \\ \sin\phi & \cos\phi \end{pmatrix} \right]$$

A doublet-state may be obtained from this when we let it operate on any orthogonal combination of the state vectors. Choosing the combination $2^{-\frac{1}{2}}\binom{1}{i}$ and $2^{-\frac{1}{2}}\binom{1}{i}$ we obtain the two states Ψ and Ψ^*, where

$$\Psi = \frac{1}{\sqrt{2}}\exp\left[-\tfrac{1}{4}\frac{L^2}{(\hbar\omega)^2} \right] e^{-q^2/2}\begin{pmatrix} \cosh\,(qL/2\hbar\omega) + \sinh\,(qL/2\hbar\omega)\,e^{i\phi} \\ -i\,[\cosh\,(qL/2\hbar\omega) - \sinh\,(qL/2\hbar\omega)\,e^{i\phi}] \end{pmatrix}$$

(3.46)

and Ψ^* is its complex conjugate.

It may be noted that for strong linear coupling ($|L| \gg \hbar\omega$) and when also $L > 0$,

$$\Psi \sim \frac{1}{\sqrt{2}}\exp\left[-\tfrac{1}{4}\frac{L^2}{(\hbar\omega)^2} \right] e^{-q^2/2 + (qL/2\hbar\omega)}\begin{pmatrix} \cos\,\phi/2 \\ -\sin\,\phi/2 \end{pmatrix}e^{i\phi/2}$$

which agrees, apart from the use of the column vector notation, with the form of the solution $|\psi_-\rangle$ in equation (3.12) on the lower potential surface. The expectation value $\bar{E}$ of the linear part of the Hamiltonian, equation (3.4), is found to be

$$\bar{E} = 1 - \tfrac{1}{8}\frac{L^2}{(\hbar\omega)^2} - [\tfrac{1}{2}\left(\frac{L}{(\hbar\omega)}\right)\exp\left(\tfrac{1}{4}\frac{L^2}{(\hbar\omega)^2}\right)\mathrm{Erf}\left(\tfrac{1}{2}\frac{L}{\hbar\omega}\right)$$

$$- \int_0^{L/(\hbar\omega)} e^{t^2}\,\mathrm{Erf}\,(t)\,\mathrm{d}t]p(L/2\hbar\omega)$$

(3.47)

where

$$p(s) = [1 + 2s\,e^{s^2}\,\mathrm{Erf}\,(s)]^{-1}$$

(3.48)

and the error function Erf (s) is defined as

$$\mathrm{Erf}\,(s) = \int_0^s e^{-t^2}\,\mathrm{d}t$$

(3.49)

The expression (3.47) for $\bar{E}$ is compared with the computed eigenenergies of Longuet-Higgins et al.[8] in the following figure (Figure 3.14).

The reduction factors p and q, defined in equation (3.29) and (3.30) are related in our approximation by

$$q = \tfrac{1}{2}(1 + p),$$

p is given as a function of $\tfrac{1}{2}L/\hbar\omega$ by equation (3.48) above, with $s = |L|/(2\hbar\omega)$.

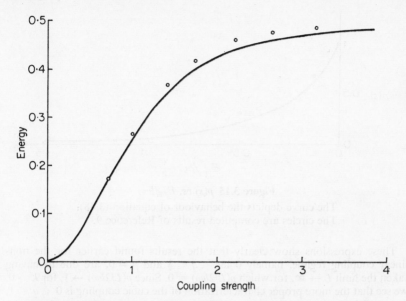

Figure 3.14 Comparison of approximate analytic energies, equation (3.47), (full line) and computed values of Longuet-Higgins *et al.*[8] (circles). $E/\hbar\omega - 1 + E_{JT}/\hbar\omega$ is plotted against $L/2\hbar\omega = \sqrt{(2E_{JT}/\hbar\omega)}$. (After References 18 and 67).

The following figure (Figure 3.15) compares p with computed values of Child and Longuet-Higgins.[5] For values of $s > 1.5$, when $p(s)$ as given by equation (3.48) decays too rapidly, the empirical formula[16] $\exp[-1.974s^{0.761}]$ may be used. Another result[34] $p = 0.5s^{-2}$ was derived by use of a closure procedure which is expected to be valid for $s \gg 1$.

A generalization[19] of the wave-function to non-linear coupling is achieved by adding the factors

$$\chi(\phi) = \frac{1}{\sqrt{3}}[e^{-\nu\phi^2/2} + e^{-\nu(\phi - 2\pi/3)^2/2} + e^{-\nu(\phi + 2\pi/3)^2/2}]$$

to the suggested solution Ψ shown above [equation (3.46)]. This correction factor is expected to be accurate for $\nu^2 = |9\beta/8\alpha| \gg 1$, but rather faulty for $\nu^2 < 1$. (In fact, $\chi(\phi)$ does not tend to 1 for $\nu^2 \to 0$. We shall remedy this below). Evaluating now the reduction factors, we find

$$q\left(\nu, \frac{L}{2\hbar\omega}\right) = \left(\frac{1}{2}\frac{1 + 2\gamma}{1 + \gamma} e^{-1/8\nu}\right)[1 + p(L/2\hbar\omega)] \qquad (3.50)$$

$$p\left(\nu, \frac{L}{2\hbar\omega}\right) = 1 . p(L/2\hbar\omega)$$

where $\gamma = e^{-\nu\pi^2/q}$ and $p(L/2\hbar\omega)$ was shown above, equation (3.48).

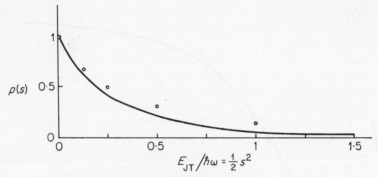

Figure 3.15 $p(s)$ vs. $E_{JT}/\hbar\omega$.
The curve depicts the behaviour of equation (3.48).
The circles are computed results of Reference 9.

These expressions show clearly that the results found earlier for the non-linear coupling regime, namely $0.484 < q \leqslant \frac{1}{2}$ and $p = 0$, are due to having taken the limit $L \to \infty$, for which $p(L/2\hbar\omega) = 0$. Since $p(L/2\hbar\omega) \to 1$, for $L \to 0$, we see that the more proper characterization of the cubic coupling is $0 < q \leqslant 1$, $0 < p \leqslant 1$.

To include in the discussion weak non-linear coupling, too, we assume that the factorization of p and q into one factor depending on linear coupling L and one involving the cubic coupling v is still possible at least approximately. We then replace the expression (equation (3.50)) for $q(v, \frac{1}{2}L/\hbar\omega)$ by

$$q\left(v, \frac{L}{2\hbar\omega}\right) = q(v)[1 + p(L/2\hbar\omega)]$$

where $q(v)$ is the quantity drawn by full lines in Figure 3.10 above. We see that $q(v)$ tends to $1/2$ as $v \to 0$, so that we recapture the result $q = \frac{1}{2}(1 + p)$ for purely linear coupling. For large v, $q(v)$ tends to the expression in round brackets in equation (3.50), as it should do.

Some general remarks on reduction factors from a broader perspective may be appropriate here.

In a vibronic state when the system resonates among different electronic states, the ionic displacement will adjust itself to the instantaneous electronic state. Thus here, because of the degeneracy of the electronic states, there occurs the reverse of the situation familiar from the adiabatic approximation when the electrons adjust to the ionic motion.[18] This situation (when called 'self-trapping'), is well known from the theory of strongly bound electrons in a lattice[71-2] or from the theory of strongly bound, Frenkel excitons.[73] The prime effect of the polarization of the lattice in these situations is the narrowing of the conduction band of the 'dressed' particle, due to the reduction of the overlap integrals by the vibrational factors. Exactly what happens in these cases as the electron hops among equivalent lattice-points, happens also in the J-T-E for the electron resonating between the degenerate states: the off-diagonal matrix elements are reduced. Reductions of some matrix elements were noted for E ⊗ ε [9,74] and for E ⊗ β [75]. The systematic formulation of the reduction of matrix elements was left, however, to Ham, who interpreted[16,66] through the J-T-E a number of previously unexplained results and provided a unified framework for an understanding how the J-T-E works in diverse situations. In addition to the theory of the reduction factors for E ⊗ ε already given in §3.2.5, we shall meet the 'Ham effect' again in §3.3.3 for T ⊗ ε, in §3.3.6 (T ⊗ τ_2) and in §3.4.2 [$G_{3/2}$ ⊗ ($\varepsilon + \tau_2$)].

3.2.7. Optical effects in E ⊗ ε

Here the semi-classical configuration diagram method, Appendix VI, delivers the following result for the intensity $I(\omega)$ of an A → E transition as function of the light frequency ω. ω_ε is the ε-mode frequency.

$$I(\omega) = \frac{|\hbar\omega - E_{el}|}{2\hbar\omega_\varepsilon E_{JT}} \tanh \frac{\hbar\omega_\varepsilon}{2kT} \exp\left[-(\hbar\omega - E_{el})^2 \middle/ \left(2\hbar\omega_\varepsilon E_{JT} \coth \frac{\hbar\omega_\varepsilon}{2kT}\right)\right]$$

Here E_{el} is the energy difference between the electronic states at the origin of configuration space. The above formula predicts a two-peaked band. The peak separation is $2\left[\hbar\omega_\varepsilon E_{JT} \coth \dfrac{\hbar\omega_\varepsilon}{2kT}\right]^{\frac{1}{2}}$, and the width goes at high temperatures as $\sqrt{T}$.

The quantum mechanical line shape was investigated analytically by Perlin[41] and Wagner.[42] The latter derived the following expression for the Fourier-transform of the line-shape

$$\mathscr{I}(t) = 1 + iL[C(t)]^{\frac{1}{2}}e^{-L^2C(t)/4} \, \mathrm{Erf}[\tfrac{1}{2}iL\sqrt{C(t)}] \tag{3.51}$$

where

$$C(t) = [1 + i\,\omega_\varepsilon t - e^{i\omega_\varepsilon t}]/(\hbar\omega_\varepsilon)^2$$

and the error function was defined in equation (3.49). The theoretical line shape for the transition A → E, computed by O'Brien (1971, private communication) based on Reference 8, is shown in Figure 3.16. This double-peaked structure may be compared with the electronic reflectance spectra of Cotton and Meyers[43] (Figure 3.17) showing the $^5T_{2g} \to \,^5E_g$ transition on Co^{3+} in some CoF_6 centres. Similar results for Fe^{2+} embedded in a number of cubic, trigonal and orthorhombic ionic crystals are due to Jones.[44] The maximum peak to peak separation in his spectra is, for FeF_2, 3500 cm^{-1}. The $^2T_{2g} \to \,^2E_g$ transitions of $Ti^{3+}(d^1)$ in Al_2O_3 [45] and in aqueous surroundings[46] show quite clearly two split bands separated by about 1850 cm^{-1}. On the other hand, the transition on Cu^{2+} in $CuSO_4$. $5H_2O$[47] appears to be better resolved into *three* Gaussian bands. These

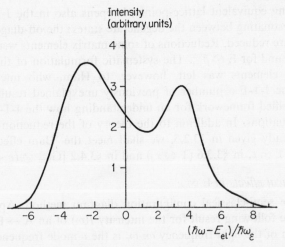

Figure 3.16 The line shape in an $A \to E$ transition computed for $E_{JT}/\hbar\omega_\varepsilon = 15$ and $0\,°K$. The $E \otimes \varepsilon$ vibronic states were obtained by numerical diagonalization,[8] and the transitions to them were broadened by Gaussian curves. (M. C. M. O'Brien, 1971 unpublished.) Note the dissimilar shapes of the two bands and see the remark at the end of Appendix VI.

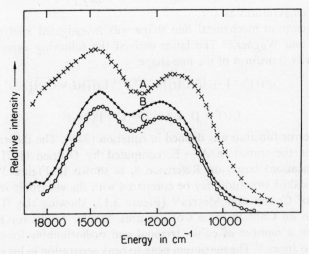

Figure 3.17 Electronic reflectance spectra (from Reference 43) in a transition to an $E \otimes \varepsilon$ system. Energy decreases from left to right.

A: Li_3CoF_6
B: K_2NaCoF_6
C: Na_3CoF_6

can be interpreted as caused by the splitting of the upper state, the $^2T_{1g}$ state in O_h symmetry, first into $^2A_{2g}$, ^2E_g in the 'truer' molecular symmetry group D_{4h}. Then a further reduction of symmetry due to E ⊗ ε coupling in the excited state doublet leads to a further splitting of ^2E_g. This interpretation is not unimpeachable, e.g. because of the small rhombic component in the crystal field.[48] On the other hand, an alternative explanation, also utilizing the J-T-E but in a different way: through an E ⊗ ε coupling in ground state (^2E_g in cubic symmetry) but no J-T-E in the final triplet, will not give a split band, since the E → A transition is not split.[8]

Several split bands appear in the spectra of Cu^{2+} in perovskites due to Reinen.[49] However, the J-T-E seen here is cooperative (discussed in §5.3) and, therefore, the spectra should be viewed, at any rate in the sense of the molecular field approximation, as the spectra in an externally distorted centre (Chapter 4). The low symmetry spectrum of Cr^{2+} (a rather rare d^4 configuration) in AgCl or AgBr is possibly due to a neighbouring charge compensating Ag-vacancy and not to the J-T-E as supposed.[50] The same is true for the spectra of Fe^{2+}.[50] The charge compensating vacancy may be present even at room temperatures, due to the vacancy being anchored to the divalent ion. It must be admitted, however, that the temperature dependence of the band splitting in the $T_{2g} \to E_g$ transitions on Ti^{3+} and Fe^{2+} in some silver halides[51] is not unlike the predicted (coth $\hbar\omega_\varepsilon/2kT)^{1/2}$ relation.

The double band separated by about 900 cm^{-1} in the spin–orbit allowed transition (involving a 4f → 5d one electron jump) of Sm^{2+} in CaF_2 and SrF_2 was correctly identified in terms of an E ⊗ ε coupling.[52]

The split zero phonon lines of the $^2T_2 \to {}^2E$ transition on Cu^{2+} in $ZnAl_2O_4$ [53] were interpreted (Reference 54, p. 193) as arising from a tunnelling splitting in E ⊗ ε.

3.2.8 Rotation–vibration interaction

Child[9] presented a treatment for symmetric top molecules occupying a state which is a rotational doublet (quantum numbers K and $-K$), as well as a vibronic doublet arising from a linear E ⊗ ε J-T-E. The Coriolis coupling term, equation (7.27), is essentially of the form

$$\frac{1}{I_\parallel}\hat{J}_z(\hat{J}_{ze} + \hat{J}_{z\varepsilon}) + \frac{1}{I_\perp}[\hat{J}_x(\hat{J}_{xe} + \hat{J}_{x\varepsilon}) + \hat{J}_y(\hat{J}_{ye} + \hat{J}_{y\varepsilon})]$$

(Here the operators $\hat{J}_e$, $\hat{J}_\varepsilon$ and $\hat{J}$ refer to the electronic, the vibrational and the total angular momenta, respectively. $I_\parallel$ and $I_\perp$ are the two principal moments of inertia.) Due to this term acting in the first order, the rotational level heights instead of going like K^2 are now proportional to

$(K \pm \zeta_{v,j})^2$ analogously to the vibrational–rotational interaction discussed in §7.9. The numbers $\zeta_{v,j}$ (v for vibronic) depend on the vibronic quantum number j which characterizes the states in the linear effect. For weak J-T coupling

$$\zeta_{v,j} \rightarrow \zeta_e - (j - \tfrac{1}{2})\zeta_\varepsilon$$

(ζ_e, ζ_ε are the electronic and vibrational Coriolis coefficients, respectively), while in the strong coupling limit the total vibronic angular momentum j enters as a whole,

$$\zeta_{v,j} \rightarrow -j\zeta_v$$

whereas the separate electronic or vibrational contributions are 'quenched' out. For a detailed dependence of $\zeta_{v,j}$ on j the original paper (Reference 9) or Herzberg's book (Reference 76, p. 64) may be consulted.

Since the spacings due to vibronic interaction (whenever this operates) are generally larger than rotational spacings, it is important to consider specifically those cases where the Coriolis term couples levels whose vibronic spacing is small or zero. Now $j = \pm\tfrac{1}{2}$ are a pair of equienergetic vibronic states. Since the Coriolis-term couples together states whose j values differ by one and for which $K + 2j$ remains the same, we expect an added strong *shift* in a $(K, |j| = \tfrac{1}{2})$-level due to the Coriolis-term acting in the second order upon a state $(K \pm 2, |j| = \tfrac{1}{2})$, as well as a first order Coriolis *splitting* of the degenerate $(K = 1, j = -\tfrac{1}{2})$ and $(K = -1, j = \tfrac{1}{2})$ levels. From consideration of l-type doubling in *vibronic* states, we realize that the vibrational–rotational term, equation (7.28),

$$A(\hat{J}_+^2 \hat{J}_{+\varepsilon} + \hat{J}_-^2 \hat{J}_{-\varepsilon})$$

exhibiting the trigonal symmetry of the molecule is still applicable; however, there will now be additional terms in the Hamiltonian of the same symmetry, e.g. the electron–rotational term which in the representation previously introduced, equation (3.7), can be written as

$$B\begin{pmatrix} 0 & \hat{J}_+^2 \\ \hat{J}_-^2 & 0 \end{pmatrix} \tag{3.52}$$

Here $\hat{J}_\pm$ are the total angular momentum raising and lowering operators and A and B are constants. These two terms do not exhaust all the possible terms of D_3-symmetry; they suffice however for our qualitative discussion, which will in fact be based only on the electron–rotational coupling. Other terms can be treated in almost equivalent ways. A more detailed and precise treatment is available in the paper of Child and Strauss,[55] who actually calculate A and B and show that they individually dominate the other terms of D_3-symmetry; however, together they tend to cancel each other.

The question that we now raise is under what circumstances will the electron–rotation term have non-vanishing matrix elements within an equienergetic manifold?

To answer this question we note that the operator of equation (3.52) obeys the following selection rules:

The change ΔK in the angular momentum quantum number K must satisfy:

$$\Delta K = \pm 2.$$

The change in the electronic angular momentum $= \pm 1$ (where upper and lower signs are taken consistently).

The change Δl in the vibrational angular momentum $= 0$.

We recall now that in the linear J-T-E the values of $j = l + \frac{1}{2}$ and $-l - \frac{1}{2}$ denote an equienergetic set. We conclude, therefore that $|j| = \frac{1}{2}$, $l = 0$ and $|K| = 1$ represent a set (and in fact the only one) to which our question above referred. Otherwise, e.g. for a vibrational quantum singly excited, $|l| = 1$, the operator matrix, equation (3.52), will act between non-degenerate sets belonging to different $|j|$'s. Then l-type doubling will be effective only in second order and will be much reduced. As we reach the stage C of §3.2.1, when the linear coupling becomes strong, l-type doubling will continue to be felt, but on a lesser scale due to the quenching of $\zeta_{v,j}$. This quenching will ultimately reduce $\zeta_{v,j}$ to half its value. Higher order coupling (stage D in §3.2.1) will restore the doubling to its original value.

The large l-type doubling of the order of 30 cm^{-1} observed in the planar configurations of ammonia[77] excited to a 3p(E″) Rydberg state is regarded as evidence that degenerate in-plane vibrations are not excited ($l = 0$).

3.3 T ⊗ (ε + τ₂)

3.3.1 T ⊗ (ε + τ₂)

The Jahn–Teller Hamiltonian, equation 3.53 below, can be written down from equation (7.8) with the aid of the V-coefficients.[20] For a set of degenerate T_2 orbitals ($|\xi\rangle$, $|\eta\rangle$, $|\zeta\rangle$)

$$H_{\mathrm{JT}} = \frac{L_\varepsilon}{2\sqrt{3}} \begin{pmatrix} q_\theta - \sqrt{3}q_\epsilon & 0 & 0 \\ 0 & q_\theta + \sqrt{3}q_\epsilon & 0 \\ 0 & 0 & -2q_\theta \end{pmatrix} - \frac{L_\tau}{\sqrt{6}} \begin{pmatrix} 0 & q_\zeta & q_\eta \\ q_\zeta & 0 & q_\xi \\ q_\eta & q_\xi & 0 \end{pmatrix}$$

(3.53)

The definition of the coefficients L_ε, L_τ is given in equation (1.3) and Appendix IV. Equation (3.53) is also appropriate for a T_1 orbital triplet. Both doubly (ε) and triply (τ_2) degenerate modes enter. (The symbolism used in this part is appropriate to the group O.) The ε-modes tend to stabilize tetragonal distortion, the τ_2-modes a trigonal distortion and one of the most interesting aspects of the situation is the competition of these two tendencies. This competition will now be studied. Afterwards, we describe the simpler cases of $T \otimes \varepsilon$ and $T \otimes \tau_2$ separately.

The stationary points for the static problem were found by Van Vleck[78] and by Öpik and Pryce[22] and are collected in Table 3.3.

Table 3.3 Stationary points of the potential surface for $T \otimes (\varepsilon + \tau_2)$

Nature of stationary points	Tetragonal	Intermediate	Trigonal
Symmetry	D_4	C_2	D_3
Number of points	3	6	4
Stabilization energy E_{JT}	$L_\varepsilon^2/(6\hbar\omega_\varepsilon)$	$\dfrac{1}{12}\left[\dfrac{L_\varepsilon^2}{2\hbar\omega_\varepsilon} + \dfrac{L_\tau^2}{\hbar\omega_\tau}\right]$	$L_\tau^2/(9\hbar\omega_\tau)$
Barrier height between minima	$L_\varepsilon^2/(8\hbar\omega_\varepsilon)$		$L_\tau^2/(36\hbar\omega_\tau)$
q_θ	$L_\varepsilon/(\sqrt{3}\hbar\omega_\varepsilon)$	$-\dfrac{L_\varepsilon}{2\hbar\omega_\varepsilon\sqrt{3}}$	$-\dfrac{L_\varepsilon}{2\hbar\omega_\varepsilon\sqrt{3}}$	0
q_ε	0	0	0
q_ξ	0	0	0	$\dfrac{\sqrt{2}L_\tau}{3\hbar\omega_\tau\sqrt{6}} \times m_1$
q_η	0	0	0	$\times m_2$
q_ζ	0	$\dfrac{L_\tau}{\hbar\omega_\tau\sqrt{6}} \times 1$	$\times(-1)$	$\times m_3$
Ground state	$\begin{pmatrix}0\\0\\1\end{pmatrix}$	$\dfrac{1}{\sqrt{2}}\begin{pmatrix}1\\1\\0\end{pmatrix}$	$\dfrac{1}{\sqrt{2}}\begin{pmatrix}1\\-1\\0\end{pmatrix}$	$\dfrac{1}{\sqrt{3}}\begin{pmatrix}m_1\\m_2\\m_3\end{pmatrix}$

Only one of the tetragonal points and two of the intermediate points have been listed explicitly, the others can be found by symmetry. For the trigonal points the set of integers (m_1, m_2, m_3) take the values $(1, 1, 1)$, $(-1, -1, 1)$, $(-1, 1, -1)$ and $(1, -1, -1)$. The conditions that the tetragonal and trigonal stationary points be the minima, and not saddle points, are that

$$L_\varepsilon^2/(6h\omega_\varepsilon) > L_\tau^2/(9h\omega_\tau) \quad \text{and} \quad L_\varepsilon^2/(6h\omega_\varepsilon) < L_\tau^2/(9h\omega_\tau) \quad (3.54)$$

respectively. As may be seen from the stabilization energies in the preceding Table, these are also the conditions for the stationary points to be absolute minima. On the other hand the points of C_2 symmetry can never be stable with respect to all displacements in the harmonic approximation. [With suitable anharmonic terms the intermediate points can also be stable.] The intermediate points are further discussed in §3.3.5.

When a low-symmetry field forces the system to be localized at a minimum of the potential, the triple electronic degeneracy will be commuted into a singlet and a doublet level (or three singlets in C_2). The states shown in Table 3.3 are the ground state singlets.

The vibrational modes also lose their degeneracy.[22] Thus at one tetragonal point listed in the table the force constant of q_ζ is $\frac{1}{2}M\omega_\tau^2$ as before, but that of q_ξ or q_η gets reduced to $\frac{1}{2}M\omega_\tau[\omega_\tau - (2/3)\omega_\varepsilon L_\tau^2/L_\varepsilon^2]$. However, the ε modes stay accidentally degenerate with the same vibrational frequency as in octahedral symmetry. At a trigonal point the vibrational mode along the trigonal axis gets split off from the 'rest'. Its frequency is as in O-symmetry. The 'rest' consists of the original ε-mode, which is not split by trigonal distortion and the remainder of the triplet-modes which form another vibrational doublet ε. The two doublets interact and the eigenfrequencies are given by the solution of the matrix (occurring twice)

$$\frac{1}{2}M\begin{bmatrix} \omega_\varepsilon(\omega_\varepsilon - \omega_\tau L_\varepsilon^2/L_\tau^2) & 3^{-\frac{1}{2}}(\omega_\varepsilon\omega_\tau)^{\frac{1}{2}}\omega_\tau L_\varepsilon/L_\tau \\ 3^{-\frac{1}{2}}(\omega_\varepsilon\omega_\tau)^{\frac{1}{2}}\omega_\tau L_\varepsilon/L_\tau & \frac{2}{3}\omega_\tau^2 \end{bmatrix}$$

It is evidently of interest to find out in any particular case which inequality in equation (3.54) applies. We now evaluate the ratio of the J-T stabilization energies $E_{JT}(\varepsilon)/E_{JT}(\tau)$ for some frequently studied systems.

When a paramagnetic ion with an unfilled d-shell is surrounded by six octahedrally arranged ligands constituting O_h-symmetry, then for the triplet states arising from the highest-spin atomic terms (the ones which lie lowest by Hund's rule) we find for the T_{2g} terms in the d^2, d^3, d^7 and d^8 configurations

$$\frac{E_{JT}(\varepsilon)}{E_{JT}(\tau)} = \frac{9}{4}\frac{\omega_\tau^2}{\omega_\varepsilon^2}\left[\frac{175}{162}\right]^2\left[\frac{\eta}{1 - \frac{5}{9}\eta}\right]^2 \quad (3.55)$$

Here $\eta \equiv \langle r^4 \rangle/(R^2 \langle r^2 \rangle)$ when a point charge model is chosen for the ligands. η is 3/2 times this in a dipole model. (R is a ligand-ion distance, $\langle r^2 \rangle$ and $\langle r^4 \rangle$ are moments of the 3d-electron). For the present purposes it is essential to recognize that η is characteristically between 0·2 and 0·4.

For other T-terms we must take cognizance of configuration interaction between t_2, and e-subshells. (e.g., a $^3T_{1g}$ state in d^2 is $\cos \alpha |t_2^2\rangle + \sin \alpha |et_2\rangle$). We have then

$$\frac{E_{JT}(\varepsilon)}{E_{JT}(\tau)} = \frac{9}{4}\frac{\omega_\tau^2}{\omega_\varepsilon^2}\left[\frac{1 - \frac{25}{27}\eta - t^2(2 - \frac{25}{108}\eta)}{1 - \frac{5}{9}\eta + 2t(1 + \frac{5}{12}\eta) + t^2(\frac{1}{2} - \frac{5}{8}\eta)}\right]^2 \qquad (3.56)$$

where t (which is actually $\tan \alpha$ in the case of d^2) ranges between $-\frac{1}{2}$ and 0 for the lower of the T_{1g}-terms in d^2, d^3, d^7 and d^8 and between 2 and ∞ for the upper T_{1g}-state. The weak crystal field case corresponds to $-\frac{1}{2}$ or 2. This was studied in Reference 78. For T_{2g} in d^1, d^4, d^6 or d^9 we put $t = 0$ in equation (3.56).

Let us now try to make some predictions. In a complex, $\omega_\tau^2/\omega_\varepsilon^2$ is presumably between 0·4 and 0·15 where the first number applies to covalently bonded complexes and the second number to ionic ones. In 4d, 5d hexafluorides the same ratio is close to 0·2.[62] Consequently, we would not expect an ε-stabilization for a T_{2g}-state in d^2, d^3, d^7 or d^8 in this model. On the other hand, Scott and Sturge[79] explained the $^3T_1 \rightarrow {}^3T_2$ zero-phonon lines of $V^{3+}(d^2)$ in Al_2O_3 by an ε-distortion in $^3T_2$. It is possible, however, that the relevant vibrational frequencies are characteristic of the corundum lattice and not of the complex; so that the ratio of the frequencies is different, perhaps closer to unity. Alternatively, the phonon-induced jumps between trigonal wells are too fast for the system to settle down in one of them. (Reorientation rates for different types of distortions are listed in Table 6.1.)

For T_{2g} in d^1, d^4, d^6 or d^9 we see from equation (3.56) ($t = 0$) that again only for highly covalent cases when $\omega_\tau^2/\omega_\varepsilon^2 > 0·4$ will the ratio $E_{JT}(\varepsilon)/E_{JT}(\tau)$ approach or exceed unity. For T_{1g} in d^1, d^4, d^6 or d^9 the contours of the square brackets in equation (3.56) are shown as function of η and t in Fig. 3.18. It is apparent that for a tetragonal distortion to be favoured

in the lower triplet, $(-\frac{1}{2} < t < 0)$, η must be small, t relatively large

in the upper triplet, $(2 < t < \infty)$, η must be large, t large.

As a rule the first set of criteria corresponds to low crystal field strength 10 Dq, the second to a large value of 10 Dq.

For a triplet state in a ps configuration Kamimura and Sugano[80] found

$$\frac{E_{JT}(\varepsilon)}{E_{JT}(\tau)} = \frac{9}{4}\frac{\omega_\tau^2}{\omega_\varepsilon^2}\left[\frac{1 - \frac{25}{6}\eta}{1 - \frac{5}{2}\eta}\right]^2$$

For KCl:Tl they estimate $\eta = 0·708$ so that the square brackets amount to 2·5 and a tetragonal distortion should result.

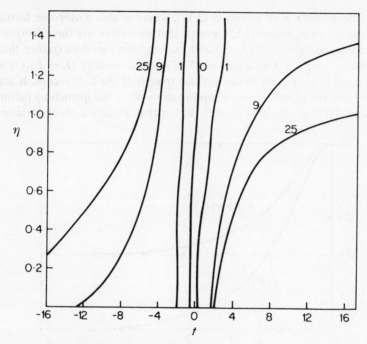

Figure 3.18 Contours of the ratio

$$\frac{4}{9} \frac{\omega_\varepsilon^2}{\omega_\tau^2} \frac{E_{\mathrm{JT}}(\varepsilon)}{E_{\mathrm{JT}}(\tau)}$$

in the η–t plane for octahedral complexes. η and t are defined following equation (3.55) and (3.56).

For tetrahedrally or cubically coordinated Jahn–Teller ions it is not possible to write out closed expressions as in the octahedral coordination, since the trigonal distortion may occur in various ways. Intuitively it is expected that (for a given pure strong-field configuration) trigonal distortions will be more effective (in the sense of $E_{\mathrm{JT}}(\varepsilon)/E_{\mathrm{JT}}(\tau)$ being smaller) in four or eight than in six-coordination. For the vibrations of a covalently bonded tetrahedron, e.g. diamond lacking a C-atom, the ratio $\omega_\tau^2/\omega_\varepsilon^2$ was estimated as $1/2$ (Lidiard and Stoneham 1967, unpublished).

O'Brien[81] has solved a Schrödinger equation for the vibronic problems, including a set of τ_2 and a set of ε-modes for the case when the two energies, $E_{\mathrm{JT}}(\tau)$ and $E_{\mathrm{JT}}(\varepsilon)$, and the two frequencies, ω_τ and ω_ε, are equal or nearly so. The vibronic ground state is a triplet, just as in §§3.3.2, 3.3.5 where one stabilization energy dominates the other. The triplet is formed by a threefold degenerate vibrational factor, comparable to the doubly degenerate χ_j of equation (3.18). There is also a vibronic factor, similar to ψ_- in equation (3.12), except that now there are three (and not two) electronic kets $|\xi\rangle$, $|\eta\rangle$, $|\zeta\rangle$ and two angular variables (rather than one), which map on a spherical shell of constant energy ($E = E_{JT}(\tau) = E_{JT}(\varepsilon)$) that is analogous to the circular trough of the $E \otimes \varepsilon$ case. It was found[81] that for equally strong coupling to ε and τ_2, the quenching factors $K(E)$ and $K(T_2)$ are equal (Figure 3.19) but that already a small deviation

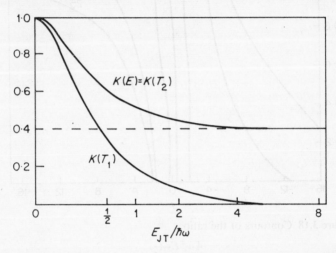

Figure 3.19 Reduction factors $K(\Gamma)$ for perturbations of Γ symmetry in the vibronic ground state of $T \otimes (\varepsilon + \tau_2)$, when the coupling to ε and τ_2 is equally strong. (M. C. M. O'Brien, *J. Phys. C. Solid State Phys.* (1971).) (By permission of The Institute of Physics and The Physical Society.)

from the equality of the energies $E_{JT}(\tau)$ and $E_{JT}(\varepsilon)$ will make one coupling, $T \otimes \varepsilon$ or $T \otimes \tau_2$, dominant. We turn now to these limiting cases.

3.3.2 $T \otimes \varepsilon$

If the coupling to the τ_2-modes is disregarded, i.e. if $L_\tau = 0$ in equation (3.53), one finds the following solution for the ε-modes only[66]

$$|\xi; v_\theta, v_\epsilon\rangle = |\xi\rangle\phi_{v\theta}(q_\theta + (3)^{-\frac{1}{2}}L/(2\hbar\omega))\phi_{v\epsilon}(q_\epsilon - L/(2\hbar\omega))$$

$$|\eta; v_\theta, v_\epsilon\rangle = |\eta\rangle\phi_{v\theta}(q_\theta + (3)^{-\frac{1}{2}}L/(2\hbar\omega))\phi_{v\epsilon}(q_\epsilon + L/(2\hbar\omega)) \qquad (3.57)$$

$$|\zeta; v_\theta, v_\epsilon\rangle = |\zeta\rangle\phi_{v\theta}(q_\theta - (3)^{-\frac{1}{2}}L/(\hbar\omega))\phi_{v\epsilon}(q_\epsilon)$$

The subscript ε has been omitted from L and ω. The ϕ's are wavefunctions for simple harmonic oscillators displaced from the origin by amounts that depend on the electronic state to which they are attached. Because of these displacements, the vibrational factors of different components in (3.57) are not orthonormal. Nevertheless, the vibronic kets still constitute an orthogonal set, since the electronic kets are orthogonal. On the other hand, the set of kets in equation (3.57) with a given v_θ, v_ϵ do not belong to an irreducible representation of the symmetry group, except that for $v_\theta = v_\epsilon = 0$ the set transforms as T_2. Nonetheless, the equienergetic manifold [of size $3(v_\theta + v_\epsilon + 1)$] which arises by taking in equation (3.57) all combinations of v_θ and v_ϵ such that $v_\theta + v_\epsilon$ is a constant, can be made to yield states which belong to the irreducible representations T_2 and T_1 of O. We denote these states by $|\psi_j^{\Gamma_v \gamma_v}\}$, where γ_v is one of the three components of $\Gamma_v = T_1$ or T_2 and $j = 0, 1, 2, \ldots$ is a further label. Levels with different Γ_v or j, and the same $v_\theta + v_\epsilon$ split up in higher approximation than equation (3.53).

The potential surfaces arising from the static problem will be shown later (Figure 3.24 below). It will be seen that they consist of three displaced paraboloids corresponding to the three displaced oscillators of equation (3.57). Because of the reduced overlap between oscillators in different paraboloids, we encounter 'the Ham effect'.

3.3.3 The Ham-effect for T $\otimes$ ε

As in the case of E $\otimes$ ε, here also the distortion of the normal modes accompanying each component of the electronic triplet is expected to have a marked effect on off-diagonal operators, such as the spin-orbit coupling or the trigonal field. Some experimental evidence for these effects will be presented later.

Following Ham,[66] we write out in the representation of the electronic triplet $|\xi\rangle$, $|\eta\rangle$, $|\zeta\rangle$ the matrix form of a Hamiltonian H(elect.) which operates on the electronic coordinates.

$$H(\text{elect.}) = G(A_1)\mathbf{I} + G(E_\theta)\mathscr{E}_\theta + G(E_\epsilon)\mathscr{E}_\epsilon + G(T_{2\xi})\tau_\xi$$
$$+ G(T_{2\eta})\tau_\eta + G(T_{2\zeta})\tau_\zeta$$
$$+ G(T_{1x})\mathscr{L}_x + G(T_{1y})\mathscr{L}_y + G(T_{1z})\mathscr{L}_z$$
$$\tag{3.58}$$

The matrix $\mathbf{I}$ is the 3×3 unit matrix.

$$\mathscr{E}_\theta = \begin{pmatrix} \frac{1}{2} & 0 & 0 \\ 0 & \frac{1}{2} & 0 \\ 0 & 0 & -1 \end{pmatrix} \qquad \mathscr{E}_\epsilon = \begin{pmatrix} -\dfrac{\sqrt{3}}{2} & 0 & 0 \\ 0 & \dfrac{\sqrt{3}}{2} & 0 \\ 0 & 0 & 0 \end{pmatrix}$$

These transform as the E-representation in the group O.

$$\tau_\xi = -\begin{pmatrix} 0 & 0 & 0 \\ 0 & 0 & 1 \\ 0 & 1 & 0 \end{pmatrix} \quad \tau_\eta = -\begin{pmatrix} 0 & 0 & 1 \\ 0 & 0 & 0 \\ 1 & 0 & 0 \end{pmatrix} \quad \tau_\zeta = -\begin{pmatrix} 0 & 1 & 0 \\ 1 & 0 & 0 \\ 0 & 0 & 0 \end{pmatrix}. \quad (3.59)$$

These transform as T_2.

$$\mathscr{L}_x = \begin{pmatrix} 0 & 0 & 0 \\ 0 & 0 & -i \\ 0 & i & 0 \end{pmatrix} \quad \mathscr{L}_y = \begin{pmatrix} 0 & 0 & i \\ 0 & 0 & 0 \\ -i & 0 & 0 \end{pmatrix} \quad \mathscr{L}_z = \begin{pmatrix} 0 & -i & 0 \\ i & 0 & 0 \\ 0 & 0 & 0 \end{pmatrix}$$

transform as T_1.

The coefficients G appearing in equation (3.58) are assumed to be independent of the coordinates (electronic or vibrational). They represent the effect of some field external to the system. The effect of the field can be represented within any vibronic triplet manifold $|\psi_j^{\Gamma_v \gamma_v}\}$ by a Hamiltonian H(vibr.) of the same form, but with the coefficients reduced as follows:

$$\begin{aligned} H(\text{vibr.}) = \ &1 \cdot G(A_1)I^{(v)} + 1 \cdot [G(E_\theta)\mathscr{E}_\theta^{(v)} + G(E_\epsilon)\mathscr{E}_\epsilon^{(v)}] \\ &+ K^{(v)}(T_2)[G(\xi)\tau_\xi^{(v)} + G(\eta)\tau_\eta^{(v)} + G(\zeta)\tau_\zeta^{(v)}] \\ &+ K^{(v)}(T_1)[G(x)\mathscr{L}_x^{(v)} + G(y)\mathscr{L}_y^{(v)} + G(z)\mathscr{L}_z^{(v)}] \end{aligned} \quad (3.60)$$

where v is an abbreviation for the labels Γ_v and j and

$$\begin{aligned} K^{(v)}(T_2) &= -\{\psi_j^{\Gamma_v \xi} | \tau_\zeta | \psi_j^{\Gamma_v \eta}\} \\ K^{(v)}(T_1) &= i\{\psi_j^{\Gamma_v \xi} | \mathscr{L}_z | \psi_j^{\Gamma_v \eta}\} \end{aligned} \quad (3.61)$$

In the ground state vibronic triplet, for which the superscript $v = 0$ (or g) is omitted,

$$K(T_2) = K(T_1) = \exp[-L^2/(2\hbar\omega)^2] = \exp(-3E_{\text{JT}}/2\hbar\omega) \quad (3.62)$$

and a marked reduction in the strength of off-diagonal electronic operators is found for strong vibronic coupling. Since electronic operators in a coordinated system can suffer a reduction from other causes too (thus the orbital magnetic moment,[82] the spin–orbit coupling constant[83] and other quantities[84-5] are reduced by covalency) it is a frequently occurring problem to separate the vibronic reduction from other factors.[53,86] We shall presently return to this problem.

In the excited vibronic states, because of increased overlap the quenching is less pronounced. Thus for levels at a height v above the ground vibronic level, the K's in equation (3.62) get enhanced by a factor of the order of $(3E_{\text{JT}}/2\hbar\omega)^{v/2}$, for strong vibronic coupling.

When the quenching is extreme, the electronic Hamiltonian, equation (3.58), has to be diagonalized within the degenerate vibronic states by use of Van Vleck perturbation theory (Reference 87, p. 158) and including second order corrections due to H (elect.). These corrections are important and were found to be larger than the first order corrections in many cases (References 79, 88–90).

A general comparison of the first and second order corrections was undertaken in Reference 91 for a system which is similar but not identical to the $T \otimes \varepsilon$ system of this section. The conclusions of Reference 91 are expected to hold qualitatively for $T \otimes \varepsilon$.

We return now to the theory of second order perturbation. The second order matrix element of H(elect.) between the components γ_v and γ'_v of the ground state $\Gamma_v = T_2, j = 0$ is

$$\{\psi_0^{T_2\gamma_v}|H|\psi_0^{T_2\gamma_v'}\} = -\sum_i \frac{\{\psi_0^{T_2\gamma_v}|H(\text{elect.})|i\}\{i|H(\text{elect.})|\psi_0^{T_2\gamma_v'}\}}{E_i} \quad (3.63)$$

in terms of the intermediate states $|i\}$ and their energies E_i, measured from the energy of the ground state. In that part of the sum in which $|i\}$ arises from the ground state electronic manifold we have for the intermediate states

$$|i\} = |\psi_j^{\Gamma_v'' \gamma_v''}\}, \qquad \Gamma_v'' = T_2 \quad \text{or} \quad T_1, \qquad \gamma_v' \neq \gamma_v'' \neq \gamma_v$$

The last set of conditions arises since only non-diagonal operators in H(elect.) connect T-states while diagonal operators $(I, \mathscr{E}_\theta, \mathscr{E}_\epsilon)$ give a vanishing contribution because of the orthogonality of vibrational factors. This partial sum can be simplified[66] to take the form, for $\gamma_v \neq \gamma'_v$,

$$(\hbar\omega)^{-1} e^{-3E_{JT}/\hbar\omega} G[3E_{JT}/(2\hbar\omega)] \sum_{\substack{\Gamma_v'', j \\ \gamma_v \neq \gamma_v'' \neq \gamma_v''}} \{\psi_0^{T_2\gamma_v}|H(\text{elect.})|\psi_j^{\Gamma_v'' \gamma_v''}\}$$

$$\times \{\psi_j^{\Gamma_v'' \gamma_v''}|H(\text{elect.})|\psi_0^{T_2\gamma_v'}\}$$

while for $\gamma_v = \gamma'_v$ one gets a similar expression, but in which the argument of G is $3E_{JT}/(\hbar\omega) = 3S$. Here

$$G(x) = \int_0^x (e^u - 1)u^{-1}\, du = \sum_{n=1}^\infty \frac{x^n}{n\, n!} \quad (3.64)$$

It will be realized that, although this is not made explicit, the summation in equation (3.63) is in effect only over the components of the electronic triplet. A graphical plot of the dependence of the matrix elements on $E_{JT}/\hbar\omega$ is seen in Figure 3.20. Note that for $x \to \infty$, $G(x) \to e^x/x$. Therefore, for a strong J-T-E, the off-diagonal elements, $\gamma_v \neq \gamma'_v$, of the second order perturbation also become quenched by the factor of equation (3.62). On the other hand, the diagonal elements, $\gamma_v = \gamma'_v$, turn out to be identical with those calculated for a purely electronic component which interacts

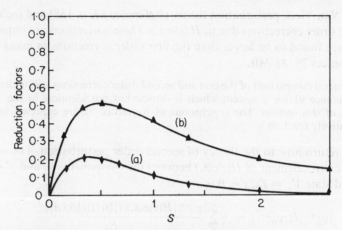

Figure 3.20 Reduction factors in second order perturbation theory
for $T \otimes \varepsilon$ vs. $E_{JT}/\hbar\omega = S$
(a) For off-diagonal matrix-elements: $e^{-3S}G(\frac{3}{2}S)$
(b) For diagonal matrix-elements: $e^{-3S}G(3S)$.
(From Reference 90.)

in the second order with the two other electronic components higher in
energy by the amount $3E_{JT}$. This result agrees with the naïve application of
the Franck–Condon principle to the situation shown in Figure 3.24 below.

The remaining terms in equation (3.63), those in which the intermediate
states $|i\rangle$ arise from different electronic manifolds, show analogous
behaviour. The non-diagonal terms $\gamma_v \neq \gamma'_v$ are again quenched by a
factor of the order of $\exp[-3E_{JT}/(2\hbar\omega)]$ while the diagonal terms again
agree with the Franck–Condon principle.

Summarizing the second order perturbation, we find that the off-
diagonal terms are again very small. However, the diagonal elements
may be large and if we recall that the coefficients G in equation (3.58) (not
to be confused with $G(x)$ in equation (3.64)) may in several cases be
variables or operators, e.g. of spin, we realize that these elements may in
fact make the dominant contribution, e.g. to a spin-Hamiltonian (Reference 39, §3.1).

In spite of their importance in the study of d^n-ions, we shall not discuss
spin-Hamiltonians, since they were exhaustively covered in Ham's
review[39] not long ago. Instead, we turn to those optical effects which are
modified by $T \otimes \varepsilon$ coupling.

3.3.4 Optical effects in $T \otimes \varepsilon$

We have already noted that the potential surfaces in the two dimen-
sional space of q_θ and q_ε are three separate paraboloids. Considering the

electrical dipole absorption pattern from a single quadratic potential surface associated with a singlet electronic state, we can easily see that this pattern will not be any more complicated than in a transition to one of the paraboloids. This follows since each component, x, y or z of the transition operator will induce transitions to only one of the paraboloids. Experimentally this is fairly well confirmed in the broad band spectra of a number of J-T centres, where there is good evidence that a $T \otimes \varepsilon$ coupling operates. The evidence appears in many cases in the shape of a marked reduction in the spin–orbit coefficient or of the effective trigonal field strength, or both. Such reduction is anticipated from the theory of the Ham-effect (§3.3.3). One finds examples of this in tetrahedrally coordinated Cu^{2+} in ZnO [53] and in ZnS,[92] Fe^{2+} in CdTe, ZnS and $MgAl_2O_4$ [93-3'] or in octahedrally coordinated Mn^{2+} (excited state $^4T_{1g}$) in $RbMnF_3$ [94] and V^{2+} in $KMgF_3$. [95] In the last work the spacings of four levels (exhibited in Figure 3.21) of the excited $^4T_{2g}$ term (split by spin–orbit coupling) and the intensities of Zeeman lines were fitted very well by the Ham theory assuming only the value 0·9 for the single parameter $E_{JT}/\hbar\omega$.

In References 88–9 the result expressed by equation (3.62) was refined and different reduction factors were calculated for the spin–orbit coupling coefficients and for the trigonal field splitting in $Al_2O_3 : V^{3+}$. In Reference 89 a P-J-T-problem in trigonal symmetry was actually treated; however, the same results could have been obtained within a J-T-problem in cubic symmetry by assuming[89] that the (perturbational) trigonal potential depended parametrically on the tetragonal distortion. It may be remarked that a similar dependence (though possibly a weaker one) could also be present in the spin–orbit coupling coefficient.

Returning to the optical spectra, it was shown[96] that theoretically the $A \to T$ band is expected to be unsplit and to have a Gaussian shape (Appendix VI). Returning to our original notation ω_ε for the frequency of the ε-modes and writing ω for the frequency of the absorbed light and E_{el} for the energy difference at the origin between the paraboloids for the ground state and for $T \otimes \varepsilon$, we obtain for the band intensity $I(\omega)$ the expression (normalized to unity)

$$I(\omega) = \left(2\pi E_{JT}\hbar\omega_\varepsilon \coth \frac{\hbar\omega_\varepsilon}{2kT}\right)^{-\frac{1}{2}} \exp\left[-(\hbar\omega - E_{el})^2 \Big/ \left(2E_{JT}\hbar\omega_\varepsilon \coth \frac{\hbar\omega_\varepsilon}{2kT}\right)\right]$$

This result is valid at least some distance away from the point of intersection of the three paraboloids. Near this point the electronic triplet states will be scrambled by spin–orbit coupling or some other mechanism, in a manner analogous to the scrambling occurring in intersystem crossing, to be discussed later in some detail in §7.6.

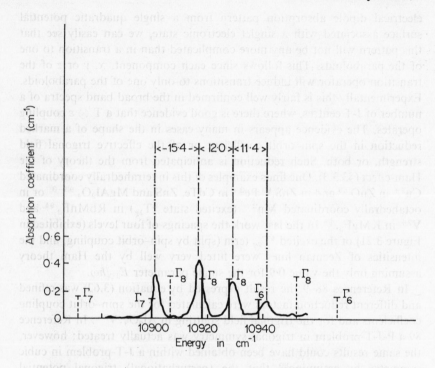

Figure 3.21 Spin–orbit split levels of the $^4T_2$ term in $KMgF_3$: V^{2+} as observed by zero-phonon-absorption at 2 °K. (From Reference 95.) The broken vertical lines show the spacings of the levels when calculated without vibronic coupling. The full vertical lines show the predicted lines including the Ham-effect with $S = 0.9$.

If the transition upon illumination is to one of the potential surfaces, i.e. by using polarized light, one also observes a polarized spectrum in emission. The spectrum of some heavy metal ions in alkali halides reported some time ago[97-8] was reinvestigated more recently for its dependence on the angle of incidence and on the wavelength of the exciting light by Fukuda *et al.*[99] A qualitative interpretation of the results in terms of a $T \otimes \varepsilon$ coupling is due to Toyozawa (unpublished, quoted in Reference 99). Earlier Öpik and Pryce[22] called attention to the relevance of the J-T-E to this situation.

The observations of Reference 99 relate to the A band (§7.3). It is rather remarkable at first sight that in emission this band should be dominated by $T \otimes \varepsilon$ coupling, since its absorption band shape[96] shows clear evidence of a $T \otimes \tau_2$ coupling (§3.3.7). The paradox can perhaps be resolved by supposing that it is the latter coupling which is dominant and should be seen in any fast process like light absorption. However, the electronic states in the four equivalent minima of T $\otimes$ τ_2 are not orthogonal and therefore the reorientation time (to be discussed later under relaxation) for transitions between these minima is expected to be quite short. For durations of time longer than the reorientation time, the average electronic states can be subject only to the T $\otimes$ ε coupling, where reorientation processes between the minima are much slower. It is not unreasonable to suppose that the time between the absorption and the re-emission enables the first relaxation to occur but not the second. That time was estimated as 2×10^{-8} sec at $20°K$.[100-1]

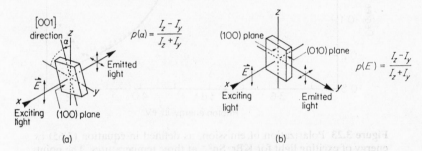

Figure 3.22 Schematic drawing of the polarization experiments of Fukudu *et al.*[99] (a) Azimuthal dependence, (b) Excitation spectra.

The experimental arrangement of Fukuda *et al.*[90] is shown in the figure above (Figure 3.22). The first experiment (a), testing the polarization p of the emitted light as function of the angle α between a cubic axis of the crystal and the direction of polarization of the exciting light, showed clearly maximal polarization at $\alpha = 0.90°$ and minimum, $p \sim 0$, at $\alpha \sim 45°$. This result eliminates the possibility of a static deformation due to T $\otimes$ τ_2 coupling, since in that case $\alpha = 45°$ would maintain the maximum polarization in emission. We are thus left with T $\otimes$ ε to account for the data of experiment (b), which measures, employing the geometry shown in the figure above, the polarization of the emitted light as function of the excitation frequency ω. We reproduce one of the experimental curves (KBr:Sn^{2+}) of the authors (Figure 3.23) and proceed to the theory.

Consider the potential surfaces (Figure 3.24) of the T $\otimes$ ε coupling in the q_θ, q_ε-plane. Anticipating the description of the normal modes of an octahedron (Table 7.2), we note that the three paraboloids represent distortions of the octahedral complex along x, y and z. The z-polarized exciting light will induce transition to some point on the z-paraboloid,

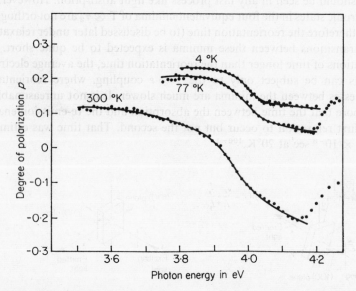

Figure 3.23 Polarization of emission, as defined in equation (3.65) *vs.* energy of exciting light for $KBr:Sn^{2+}$ at three temperatures. The points are the observations of Fukuda *et al.* (From Reference 99.)

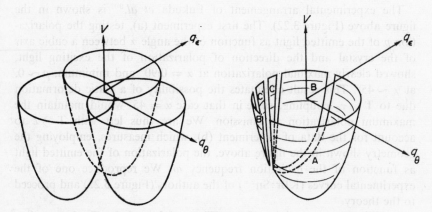

Figure 3.24 The potential surface of $T \otimes \varepsilon$ in the q_θ, q_ϵ plane, with the z-paraboloid dissected according to its relaxation behaviour (Table 3.4). (From Reference 90.)

from which the system will slide down (by all sorts of relaxation processes) with the intent to reach the minimum of the z-paraboloid. If, however, in the course of the downward motion the system encounters another paraboloid then the probabilities of staying on the first (f) and of moving over to the second (s) paraboloid are, respectively,

$$\Pi_f = \tfrac{1}{2}[1 + (1 + \eta)^{-1}] \sim 1/2$$

and

$$\Pi_s = \tfrac{1}{2}[1 - (1 + \eta)^{-1}] \sim 1/2$$

Here η is the ratio of probabilities of two simple events: of making an horizontal transition between the two surfaces and of moving down on the same surface. (It has been supposed that $\eta \gg 1$.)[102] There is also a chance, depending on the height and the position to which the excitation takes the system, for a further encounter with the third paraboloid. With some imagination we can dissect the z-paraboloid to five parts from which the probabilities to reach the minima of the x, y, z-paraboloids are as follows (Table 3.4)

Table 3.4 Probabilities Π_i for the system to reach the minimum of the paraboloid i, having started in a given part of the z-paraboloid ($i = x, y, z$).

Part	Π_x	Π_y	Π_z
A	0	0	1
B	Π_s	0	Π_f
B'	0	Π_s	Π_f
C	Π_s	$\Pi_s\Pi_f$	Π_f^2
C'	$\Pi_s\Pi_f$	Π_s	Π_f^2

With the probabilities of Table 3.4 we can calculate the degree of polarization p of the emission as function of the excitation frequency. The result is for $\eta \to \infty$

$$
\begin{aligned}
p(\omega) &= (I_z - I_x)/(I_z + I_x) \\
&= \frac{1 + 6\,\mathrm{Erf}\,(q)/\sqrt{\pi}}{3 + 2\,\mathrm{Erf}\,(q)/\sqrt{\pi}} \quad \text{for } q \leq 0 \\
&= \frac{1 - 3\,\mathrm{Erf}\,(q)/\sqrt{\pi}}{3 - \mathrm{Erf}\,(q)/\sqrt{\pi}} \quad \text{for } q \geq 0
\end{aligned}
\right\} \quad (3.65)
$$

where $\qquad q = q(\omega) = (\hbar\omega - E_{\mathrm{el}}) \Big/ \left[E_{\mathrm{JT}}\hbar\omega \coth\!\left(\frac{\hbar\omega}{2kT}\right) \right]^{\frac{1}{2}}$

I_x and I_z are the intensities of light polarized along x and z. The function Erf was defined in equation (3.49).

When the polarization is plotted versus frequency (Figure 3.25) the resemblance to the shape of the experimental curve above (Figure 3.23) is apparent. The change in sign of the room temperature curve is also explained. The observed value of p on the left hand side of the curve in

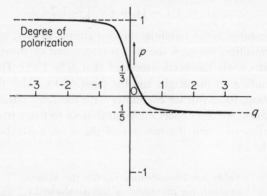

Figure 3.25 Degree of polarization p *vs.* q (Theory). [Both p and q are defined in equation (3.65). Roughly, q is proportional to the frequency of the exciting light *minus* $E_{el}/\hbar$.] (After Reference 99.)

Figure 3.23 is lower than that calculated because of thermal jumps and tunnelling between the minima. To include these, as well as a finite radiative decay, in the theory we write down rate equations for the time dependent probabilities $\Pi_i(t)$ of the system being at the bottom of the i-paraboloid at time t.

$$\Pi_i(t) = -\tau_r^{-1}\Pi_i(t) + (\Gamma + \nu\,e^{-3E_{JT}/4kT})[2\Pi_i(t) - \Pi_j(t) - \Pi_k(t)]$$
(3.66)
$$i \neq j \neq k$$

Here τ_r, Γ and $\nu\,e^{-3E_{JT}/4kT}$ are respectively the radiative lifetime, the tunnelling frequency and the barrier jumping rate. The equations (3.66) have to be solved subject to the initial conditions, which are those appearing in Table 3.4. This method is due to Kristoffel[103] who, however, considered the possibility of jumps only at the origin of the q_θ, q_ϵ-plane.

Some factors left out of the above considerations are the α_{1g} and τ_{2g} modes of the octahedron and the spin–orbit coupling, reduced no doubt by the Ham effect (§3.3.3) (although a $T \otimes \tau_2$ coupling was implied earlier by allowing inter-paraboloid jumps to occur).

It is useful to ask ourselves at this stage what is the rôle of vibronic coupling in the data and what result should we expect in its absence? Quite clearly, a negative polarization would not occur without the J-T-E. On the other hand it must be realized that for very short times the degree of polarization would be near unity after which it would fast decrease to zero. We see therefore that the J-T-E manifests itself in two apparently contrary tendencies: (a) it reduces the asymmetry (e.g. the polarization) of the situation, (b) it stabilizes in time, perpetuates the asymmetry. One should add that external causes for a preferred distortion along the cubic axes, e.g. a charge compensating vacancy, were also considered in Reference 99 and particularly in Reference 102. It was shown however, that the characteristic polarization *vs.* excitation frequency curves were due to the J-T-E even in cases (such as $KI:Sn^{2+}$) where an associated vacancy splits the x, y and z paraboloids sufficiently to make the emissions from inequivalent paraboloids resolvable.

Mention may be made at this point of a theoretical study of the polarization in magnetic dipole transitions in various J-T situations by Vekhter and Tsukerblat.[104]

Let us pause now to consider whether having solved the $T \otimes \varepsilon$ problem exactly as in equation (3.57) could we not go on to solve (exactly) the $T \otimes \tau_2$ problem, too. Suppose that we build up the J-T-Hamiltonian using the vibronic states, equation (3.57) with a given v_θ, v_ε, as the basis. Then we obtain a matrix identical to the second matrix in equation (3.53) except that its coefficient, let us denote this by L'_τ, will be reduced relative to the original coefficient of equation (3.53) by the overlap of vibrational factors, equation (3.61). For the vibrational ground state ($v_\theta = v_\varepsilon = 0$)

$$L'_\tau = L_\tau \exp\left[-3E_{JT}(\varepsilon)/2\hbar\omega_\varepsilon\right],$$

by equation (3.62). The remaining static problem for the τ_2-modes may now be solved, leading to the trigonal points of Table 3.3 with $L_\tau \rightarrow L'_\tau$. The $T \otimes \varepsilon$ problem is already solved, we recall. It would appear therefore that the intermediate type points of symmetry C_2, namely those at which the octahedron is distorted both tetragonally and trigonally, are of lower energy than a purely trigonal or tetragonal distortion, contrary to our earlier assertion. There is no contradiction, however, since as remarked earlier the static case holds true only in the limit of infinite ionic masses and in this limit the overlap, and therefore L'_τ, indeed vanishes. (Note, nevertheless, that for $E_{JT}(\varepsilon) \gg E_{JT}(\tau)$ the elimination of the ε-modes in the manner described is a very effective way to tackle the interwoven problem of the two types of modes.) Looking at the simple form of the eigenstates at the stationary points, last line of Table 3.3, one might conclude[105] that lower energy stationary points may be found, within the static J-T-E, by letting the eigenstates depend parametrically on the displacements. This is not so. The points given *are* the minima of the eigen-energies. The simplicity of the eigenstates is due to the symmetry of the problem.[106] The proof[105] that was given for the existence of minima of intermediate type was in error since it was based on a special vibrational factor (a product of T_{2g} and E_g-states) for each electronic ket. This is not justified. On the other hand, criticism[39,66,107] based on group theory is misplaced since, in the stationary states with which Bersuker and Vekhter[105] are concerned, the symmetry is broken. (The term 'tunnelling splitting' used by these authors may be misleading; nevertheless, their expression for the splitting clearly relates to the difference between potentials in a frozen-in situation.)

3.3.5 $T \otimes \tau_2$

With L_ε in equation (3.53) put equal to zero and $L_\tau \equiv L$, $\omega_\tau \equiv \omega$, we obtain the Hamiltonian for $T \otimes \tau_2$:

$$H = \tfrac{1}{2}\hbar\omega[p_\xi^2 + p_\eta^2 + p_\zeta^2 + q_\xi^2 + q_\eta^2 + q_\zeta^2] - \frac{L}{\sqrt{6}}\begin{pmatrix} 0 & q_\zeta & q_\eta \\ q_\zeta & 0 & q_\xi \\ q_\eta & q_\xi & 0 \end{pmatrix} \quad (3.67)$$

Let us start with two limiting cases of equation (3.67). For $|L| \ll \hbar\omega$ the eigen-energies of equation (3.67) follow the relation[7]

$$\frac{E}{\hbar\omega} = n + \tfrac{3}{2} + \frac{L^2}{24(\hbar\omega)}[l(l+1) - m(m+1) - 6] + O(L^4) \quad (3.68)$$

where $n = 0, 1, 2, \ldots$; $m = n, n - 2, \ldots 1$ or 0; $l = m + 1, m, m - 1$ when $m \geqslant 1$ and $l = 1$ when $m = 0$. The multiplicity of each level is $2l + 1$.

For large $|L|$ the eigenvalues are

$$\frac{E}{\hbar\omega} = -\frac{1}{9}\left(\frac{L}{\hbar\omega}\right)^2 + \sqrt{\tfrac{2}{3}}(n_1 + 1) + (n_2 + \tfrac{1}{2}) + O(L^{-2}) \quad (3.69)$$

where n_1 and $n_2 = 0, 1, 2, \ldots$. The level multiplicity is $4(n_1 + 1)$. In the ground state ($n_1 = n_2 = 0$) the last term in equation (3.69) has the value $0.0174(L/\hbar\omega)^{-2}$. In the presence of a strain along any one of the four body-diagonal axes of a cube, the system becomes vibronically stabilized (for $|L| \gg \hbar\omega$) in this direction. The character table of the reducible representations of the vibrations along and perpendicular to this direction is as shown in Table 3.5.

Table 3.5 Characters of vibrations parallel (∥) and perpendicular (⊥) to a trigonal axis.

Group Operations in O:		E	$8C_3$	$3C_2$	$6C_4$	$6C_2'$
Vibrations ∥	:	4	1	0	0	2
Vibrations ⊥	:	8	−1	0	0	0

For finite but large k the splittings occur as in Table 3.6.

Table 3.6 The splitting up of levels with a given n_1 (the principal quantum number for perpendicular vibrations) for finite L. m is the same as l in equation (3.8).

n_1	m	Representations	Multiplicity
0	0	$A_2 + T_1$	4
1	1	$E + T_1 + T_2$	8
2	0,2	$A_2 + E + 2T_1 + T_2$	12
3	1,3	$A_1 + A_2 + E + 2T_1 + 2T_2$	16

It is not easy to calculate the magnitude of the splitting even for $n_1 = 0$. The formulae

$$1 \cdot 2 E_{JT} e^{-1 \cdot 21 E_{JT}/(\hbar\omega)}, \quad E_{JT} = L^2/(9\hbar\omega)$$

give a fairly accurate representation of the computed splitting for strong coupling and $n_1 = 0$.

For intermediate values of the coupling strength computations[108] yield the eigen-energies shown in Figure 3.26. The degenerate electronic manifold is assumed to be T_2; for T_1 all the 1, 2 subscripts in the figure have to be interchanged. It is seen that the vibronic ground state has the symmetry of the electronic manifold for all values of the coupling strength L.

The $n = 1$ band of the infrared combination spectra of RuF_6 [109] was interpreted by means of the formula, equation (3.68), and the coupling strength in the molecule was determined.[62] The τ_{2g}-mode frequency was first established as 283 cm^{-1} by comparison with the rest of the series of $(4d)^n$ hexafluorides, then it was assumed that the bulk of the transition to the $n = 1$ band is to the level given in equation (3.68) by $l = m = 1$ (this level has the representation T_{1g} in the group O_h) and from the position of the absorption maximum the value $|L_\tau| = 270$ cm^{-1} was obtained. With the derived parameters one finds $E_{JT} \sim 30$ cm^{-1}. The spectra of another d^2 hexafluoride, OsF_6 also show signs of $T \otimes \tau_2$ coupling.

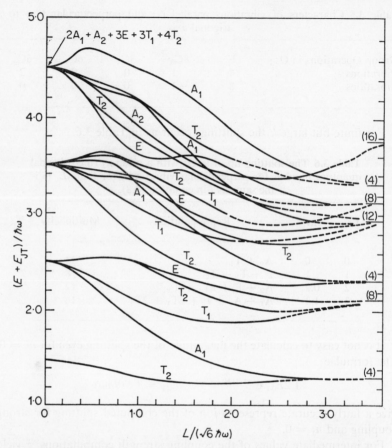

Figure 3.26 Vibronic energy levels as function of the coupling coefficient for an electronic triplet T_2 intereacting with τ_2-modes. On the extreme right the multiplicities of the levels for $L = \infty$ are shown in brackets. (After Reference 108.)

However, here the vibronic state arising from the electronic T_{2g} is not the ground state but lies some 10 cm^{-1} above the vibronic E_g arising from $E \otimes \varepsilon$. This blurs the spectrum. The theory of vibronic spectra of hexafluorides is given in great detail in the review of Weinstock and Goodman.[62]

Let us now return to the solution of equation (3.67). To construct the eigenfunctions of equation (3.67) in a systematic way we introduce in the first place the symmetry adapted vibrational eigenfunctions

$$|\Gamma_\tau, \gamma_\tau; n, m, s\rangle \tag{3.70}$$

(transforming as the γ_τ-component of the Γ_τ-representation) which arise from the solutions of the 3-dimensional oscillator. They were defined and given (up to the twelfth spherical harmonics) in Reference 110. In equation (3.70) n is the principal quantum of the oscillators, m is the associated rotational quantum number as in equation (3.68) and s is an additional label necessary to specify functions with the same Γ_τ, γ_τ, n and m. The *vibronic* functions belonging to the γ_v component of the Γ_v representation have the form

$$|\Gamma_v, \gamma_v; \Gamma_\tau; n, m, s\} = [\Gamma_v]^{1/2} \sum_{\gamma_\tau, \rho = \xi, \eta, \zeta} V\begin{pmatrix} \Gamma_\tau T_2 \Gamma_v \\ \gamma_\tau \rho \ \gamma_v \end{pmatrix} |\Gamma_\tau \gamma_\tau; n, m, s\rangle |\rho\rangle \quad (3.71)$$

The properties of the V-coefficients are described in Reference 20.

The vibronic eigenfunctions are the following linear combinations of the preceding:

$$|\psi_j^{\Gamma_v \gamma_v}\} = \sum_{\Gamma_\tau; n, m, s} A(\Gamma_v, j; \Gamma_\tau; n, m, s) |\Gamma_v \gamma_v; \Gamma_\tau; n, m, s\} \quad (3.72)$$

where $j (=0, 1, \ldots)$ labels states having the same representation Γ_v in order of increasing energy. In general, the coefficients A have to be obtained by numerical diagonalization of the Hamiltonian.

3.3.6 *Reduction factors for* $T \otimes \tau_2$

With the aid of the coefficients A in the preceding equation we can readily compute the reduction factors K introduced by Ham.[66] These can be defined for any electronic operator $O^{\Gamma\gamma}$ transforming as the γ-component of the irreducible representation Γ as follows:

$$\{\psi_j^{\Gamma_v \gamma_v} | O^{\Gamma\gamma} | \psi_{j'}^{\Gamma_v' \gamma_v}\} = K_{jj'}^{\Gamma_v \gamma \Gamma_v'}(\Gamma) \{\psi_j^{\Gamma_v \gamma_v} | O^{\Gamma\gamma} | \psi_{j'}^{\Gamma_v' \gamma_v}\}_{L=0} \quad (3.73)$$

provided the last factor is nonzero. The significance of these reduction factors is the same as of those for $T \otimes \varepsilon$ [equations (3.58), (3.60)], but now the definitions have been made more general than in either equation (3.61) or Reference 66. For the important case of the ground vibronic triplet

$$\Gamma_v = \Gamma_v' = T_2; j = j' = 0$$

$$K(\Gamma) = (-1)^\Gamma \cdot 3 \sum_{\Gamma_\tau, n, m, s} [A(T_2, 0; \Gamma_\tau; n, m, s)]^2 (-1)^{\Gamma_\tau} W\begin{pmatrix} T_2 T_2 \Gamma \\ T_2 T_2 \Gamma_\tau \end{pmatrix}$$

$$(3.74)$$

The W-coefficients of Reference 20 have been used. The reduction factors $K(E)$, $K(T_1)$ and $K(T_2)$ derived from equation (3.74) are shown in

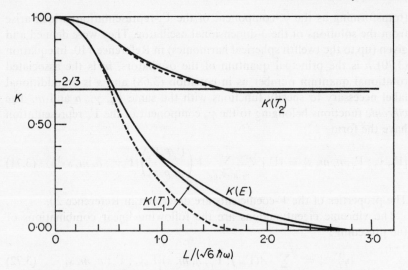

Figure 3.27 Reduction factors *vs.* coupling strength $[L_\tau/(\sqrt{6}\hbar\omega)]$.[108]
The broken curves represent the expressions given by Ham.[66]
(After Reference 108.)

Figure 3.27. On the same figure are shown for comparison, Ham's approximate analytical formulae:[66]

$$K(E) \sim K(T_1) \sim \exp\left[-9E_{\mathrm{JT}}/(4\hbar\omega)\right]$$
$$K(T_2) \sim \tfrac{1}{3}\{2 + \exp\left[-9E_{\mathrm{JT}}/(4\hbar\omega)\right]\}$$

A further reduction factor due to a *vibrational* operator will be introduced later in §3.3.8.

3.3.7 *Optical effects in* $T \otimes \tau_2$

The absorption line shape from a single hyperparaboloid of revolution to the potential surface of a $T \otimes \tau_2$ situation is derived by the finite-temperature configurational diagram method in the papers of Toyozawa and Inoue.[96] An approximate expression for the line shape is derived from the results of Appendix VI.

$$I(\omega) = \tfrac{1}{3}\left[\delta(q) + \frac{4}{\sqrt{\pi}}q^2\,\mathrm{e}^{-q^2}\right] \tag{3.75}$$

where $q = (\hbar\omega - E_{\mathrm{el}})/\sqrt{(3E_{\mathrm{JT}}kT_\tau)}, \quad kT_\tau = \tfrac{1}{2}\hbar\omega_\tau \coth \dfrac{\hbar\omega_\tau}{2kT}$

Here we have reinstated the notation ω_τ for the frequency of the τ-mode. E_{el} is the difference between the potentials at the origin, ω is the frequency of the absorbed light.

The total splitting is clearly $2\sqrt{(3E_{JT}kT_\tau)}$. The singularity in equation (3.75) at the origin is due to the branch point in the potential at that point, in contradistinction to the T ⊗ ε-case, when there is a simple intersection. The calculated curve has the characteristic spreadeagle shape shown by broken lines in the next figure (Figure 3.28). This shape is somewhat

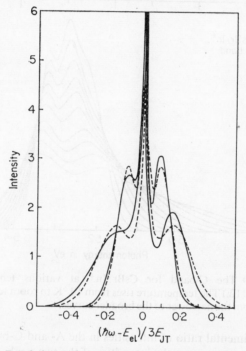

Figure 3.28 Band shapes for transition to a T ⊗ τ₂ system calculated in the semi-classical approximation. The normalized intensity function is plotted *vs.* $(\hbar\omega - E_{el})/3E_{JT}$. The broken curves arise from the calculation of Reference 96 including only linear T ⊗ τ₂ coupling. The full lines (Reference 111) include also a change in frequency (by a factor of 0.7) on going from the ground to the excited electronic state. (From Reference 112.) The broader (narrower) curves correspond to temperatures $9 \times 10^{-2}E_{JT}(3 \times 10^{-2}E_{JT})$.

changed when the symmetric broadening due to the symmetric mode is also included.

The variations of the A- and C-bands in absorption as function of temperature and other physical parameters were interpreted in Reference

96. (In emission the dominant coupling in these bands is $T \otimes \varepsilon$, §3.3.4.) The experimental C-band[112] (Figure 3.29) bears a strong resemblance to the theoretical curves shown previously. It also spreads out with increase of temperature, as predicted. In the A-band of Fukuda *et al.*[113] due to KCl: In, the two low-energy components merge into a single peak. Toyozawa and Inoue[96] were able to account for this by adding to the Hamiltonian a quadratic vibronic coupling term and adjusting the coefficient to agree

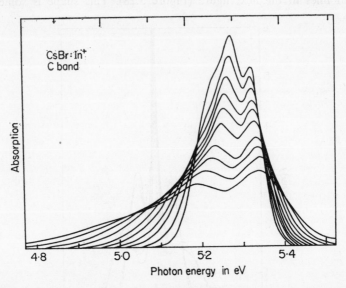

Figure 3.29 The C-band for CsBr:In$^+$ at various temperatures. (Reference 112.) The temperature rises from 10 °K to room temperature on going from the narrower to the broader curves.

with the experimental ratio of intensities in the A- and C-bands. Even the change with temperature of the intensities of the two peaks of the A-band was reproduced in this way. More elaborate calculations of the band shape in a transition to $T \otimes \tau_2$ are to be found in the works of Honma[111] and Cho.[114] These workers base their results on the semi-classical approximation (Appendix VI) which is valid for high temperature and strong coupling (let us say, $(L_\tau/\hbar\omega_\tau) > 5$). Their results show that different mechanisms lead to different, characteristic band shapes. Thus, as we have seen, linear vibronic coupling results in a symmetric triple band; spin–orbit coupling by itself gives rise to a number of asymmetric peaks, e.g. for $^3T_{1u}$ three peaks ($J' = 2, 1, 0$); linear vibronic coupling to an α_1- or an ε-mode broadens the bands; finally quadratic vibronic coupling causes asymmetry.

In general, all these effects act together. The calculations of References 111–114 refer specifically to the A- and C-bands of heavy ion phosphors[96,99,112] and for these it is clear that the linear $T \otimes \tau_2$ coupling is the most important single factor in determining the band shape. Quantum mechanical computations of absorption and emission bands based on the solutions [equation (3.72)] of the linear problem were shown in Reference 115. Their band shapes, reproduced here in Figure 3.30, show for weak and moderate linear coupling some agreement with the semiclassically predicted[96,111,114] bands and also some significant deviations from these.

The calculations of Reference 115 were for an optical transition which is allowed between the ground electronic singlet and one component, say $| \zeta \rangle$, of the upper electronic triplet. Parameters which were varied were the coupling coefficient L, the temperature T and the width α of a Gaussian curve by which the lines were broadened so as to simulate experimental curves. In absorption (the first and second row of drawings in Figure 3.30) the three peaked structure is apparent at low resolution $\alpha \geqslant 0.5$. The central peak in each of these drawings lies very near to $\hbar\omega = E_{el}$, as predicted from equation (3.75). The structure becomes more symmetric as the temperature or the coupling strength increases. There is a tendency in the calculated curves for the peak separations to follow a $\sqrt{(kT)}$ dependence at high temperatures, similarly to equation (3.75).

The comparison of Figure 3.30 with the experimental C-bands (Figure 3.29) is of only limited validity since in the latter the coupling strength is at least $10\hbar\omega_r$. In agreement with experimental bands in absorption the low energy wing in Figure 3.30 becomes lowered relative the central peak with increase of temperature. On the other hand the broader wing in Figure 3.30 is the high energy one, contrasting observation. This discrepancy is not surprising, since other vibrational modes (including the contribution of the lattice) have not been taken into account. It is however important to recognize in the curves a number of features (asymmetry, the rise and fall of the wings with temperature) which are absent in the semi-classical treatment of the linear $T \otimes \tau_2$ coupling.

In the emission spectra (the third row in Figure 3.30) the following points are noteworthy. At low temperature and high resolution ($\alpha < 0.5$) a progression with maximum near the $[E_{JT}/\hbar\omega]$th line is observed, as expected; at low temperatures and low resolution the details are mainly on the long wavelength side, while for high temperature and low resolution the details and reduced slope fall on the other side.

The characteristic $T \otimes \tau_2$ absorption band is also observed in colour centres in halides[116-8] and in the so-called U_2 bands, due to interstitial hydrogen in alkali bromides and iodides.[119] The similarly shaped band in the 19–23,000 cm⁻¹ region of the $^4T_{1g}(^4F) \to {}^4T_{1g}(^4P)$ absorption in $CoSO_4 . 7H_2O$ [47] is possibly also due to $T \otimes \tau_2$ coupling in the excited state. There are however cases when structure in the low resolution spectrum[120] was mistakenly ascribed to the J-T-E, while its true causes were vibrational side bands due to J-T inactive modes.[93]

4

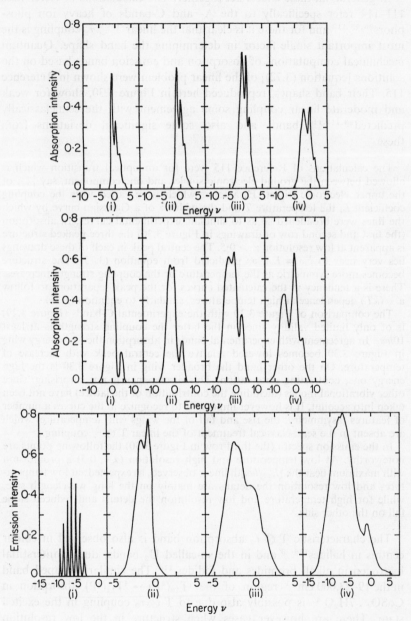

Figure 3.30

There are also some sophisticated models of vibronic coupling to an orbital triplet and involving an *odd* vibrational mode. In a U-centre (which is a localized system of H^- in an alkali halide) a linear interaction with a lattice mode and a quadratic interaction with an odd localized mode give rise to three separate distortions, each associated with one component of an electronic triplet.[121] In another case,[122] hyperfine interaction of an impurity Ag^{2+}-ion was observed with two of the eight surrounding fluorine ions in CaF_2 or SrF_2. This was explained[122] by the existence of a J-T-coupling of the $^2T_{2g}$ state to an odd mode, which displaces the silver in an $\langle 110 \rangle$-direction. Covalent admixture between the Ag- and fluorine-orbitals was postulated. It was then suggested that the fluorine-orbitals are also subject to a J-T-E which is not restricted by any parity rule, since the fluorine ions are in a T_d surrounding. It still remains questionable how such argument can dispose of the essentially centro-symmetrical position of the Ag-ion.

Zero phonon lines, in particular that of the excited state $^4T_{2g}$ of V^{2+} in MgO [123] for which $T \otimes (\varepsilon + \tau_2)$ coupling was invoked in various forms,[123-4] will be discussed later (§4.6) in connection with strain.

3.3.8 *Coriolis coupling in* $T \otimes \tau_2$

The relation shown in equation (7.29) for the Coriolis coefficient ζ_v of a spherical top in a $\Gamma_{v,j}$ vibronic state arising from an electronic triplet can be rewritten as follows:

$$\zeta_v(\Gamma_v, j) = K_{jj}^{\Gamma_v \Gamma_v}(T_1) + M_j^{\Gamma_v}(T_1)$$

Here the first term is the reduction factor given in equation (3.73), due to the electronic angular momentum operator $O^{T_{12}} = l_z$. The second term is a similar quantity associated with the angular momentum m_z of the vibrational mode. Its computed values are shown in Figure 3.31 for the

---

Figure 3.30 Calculated transition probabilities (per unit energy) *vs.* light frequency ω for absorption by a $T \otimes \tau_2$ system (Reference 115). All energies are in units of the vibrational energy ($\hbar \omega_\tau = 1$). The parameters of the figures are, with the drawings listed from left to right:

First row: Absorption, $L = -\sqrt{6}$, α (the broadening parameter) $= 0 \cdot 5$
   (i) $kT = 0$, (ii) $kT = 0 \cdot 5$, (iii) $kT = 1$, (iv) $kT = 4$
Second row: Absorption, $L = -6$, $\alpha = 1$
   (i) $kT = 0$, (ii) $kT = 0 \cdot 5$, (iii) $kT = 1$, (iv) $kT = 4$.
Third row: Emission
   (i) $L = -6$, $\alpha = 0 \cdot 1$, $kT = 0$
   (ii) $L = -\sqrt{6}$, $\alpha = 0 \cdot 5$, $kT = 0 \cdot 5$
   (iii) $L = -\sqrt{6}$, $\alpha = 0 \cdot 5$, $kT = 4$
   (iv) $L = -6$, $\alpha = 0 \cdot 5$, $kT = 4$

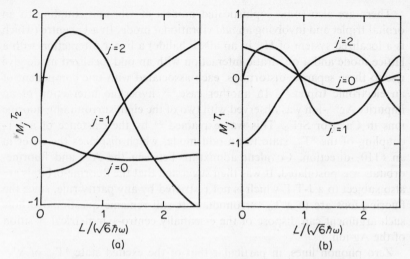

Figure 3.31 Reduction factors for vibrational operators $M_j^{\Gamma_v}(T_1)$ as function of the coupling strength for a T_2 electronic state coupled to a τ_2-mode. (a) $\Gamma_v = T_2$, (b) $\Gamma_v = T_1$. (From Reference 115.)

triplet states labelled by $j = 0, 1, 2 \ldots$ [It may be verified, that for a T_2 electronic state, for $L = 0$

$$M_0^{T_1}(T_1) = +\tfrac{1}{2}, \quad M_1^{T_2}(T_1) = -\tfrac{1}{2}]$$

The rather violent variation of the quantities, especially when compared to the smooth form of the analogous curve [Figure 3.9(b)] for $E \otimes \varepsilon$, is remarkable. It is perhaps connected to the congestion of triplets in Figure 3.26. As the coupling constant increases, the effective vibrational angular momentum of each eigenstate varies rather rapidly.

3.4 $G_{\frac{3}{2}} \otimes (\varepsilon + \tau_2)$

3.4.1 $G_{\frac{3}{2}} \otimes (\varepsilon + \tau_2)$

Fourfold degenerate states $G_{\frac{3}{2}}$ arise in cubic systems containing an odd number of electrons in which the spin–orbit coupling operates within either an orbital doublet or a triplet. Here we briefly discuss the eigenstates and eigenvalues following the work of Moffitt and Thorson,[7,125] of Child,[126] of Goodman and Weinstock[62] and of Morgan.[127] $G_{\frac{3}{2}}$ states are subject to interaction with ε- and τ_2-modes (in O symmetry). The vibronic part of the Hamiltonian is:

$$H_{JT} = \frac{1}{2\sqrt{2}}L(\varepsilon)(q_\theta\rho_1 + q_\epsilon\rho_2) + \frac{1}{2\sqrt{3}}L(\tau_2)(q_\xi\sigma_1 + q_\eta\sigma_2 + q_\zeta\sigma_3)$$

$$(3.76)$$

where the matrices are, as defined by Dirac (Reference 129, p. 257),

$$\rho_1 = \begin{pmatrix} 0 & 0 & 1 & 0 \\ 0 & 0 & 0 & 1 \\ 1 & 0 & 0 & 0 \\ 0 & 1 & 0 & 0 \end{pmatrix} \quad \rho_2 = \begin{pmatrix} 0 & 0 & -i & 0 \\ 0 & 0 & 0 & -i \\ i & 0 & 0 & 0 \\ 0 & i & 0 & 0 \end{pmatrix}$$

$$\rho_3 = \begin{pmatrix} 1 & 0 & 0 & 0 \\ 0 & 1 & 0 & 0 \\ 0 & 0 & -1 & 0 \\ 0 & 0 & 0 & -1 \end{pmatrix}$$

$$\sigma_1 = \begin{pmatrix} 0 & 1 & 0 & 0 \\ 1 & 0 & 0 & 0 \\ 0 & 0 & 0 & 1 \\ 0 & 0 & 1 & 0 \end{pmatrix} \quad \sigma_2 = \begin{pmatrix} 0 & -i & 0 & 0 \\ i & 0 & 0 & 0 \\ 0 & 0 & 0 & -i \\ 0 & 0 & i & 0 \end{pmatrix}$$

$$\sigma_3 = \begin{pmatrix} 1 & 0 & 0 & 0 \\ 0 & -1 & 0 & 0 \\ 0 & 0 & 1 & 0 \\ 0 & 0 & 0 & -1 \end{pmatrix}$$

(3.77)

The four basic vectors of these matrices transform in octahedral or tetrahedral symmetry as

$$\frac{1}{\sqrt{2}}(\phi_{\frac{1}{2}} + i\,\phi_{-\frac{3}{2}}), \; \frac{1}{\sqrt{2}}(-\phi_{-\frac{1}{2}} - i\,\phi_{\frac{3}{2}}), \; \frac{1}{\sqrt{2}}(-\phi_{\frac{1}{2}} + i\,\phi_{-\frac{3}{2}}), \; \frac{1}{\sqrt{2}}(\phi_{-\frac{1}{2}} - i\,\phi_{\frac{3}{2}})$$

(3.78)

respectively where the ϕ's are the components of the $G_{\frac{3}{2}}$ manifold, as in Koster *et al.*[130] The linear coupling coefficients are the reduced matrix elements

$$L(\varepsilon) = \langle G_{\frac{3}{2}} || H^{E} || G_{\frac{3}{2}} \rangle (= 2\sqrt{2}\hbar l_{\varepsilon}/\sqrt{M\omega_{\varepsilon}} \text{ in the notation of Reference 7})$$

$$L(\tau_2) = \langle G_{\frac{3}{2}} || H^{T_2} || G_{\frac{3}{2}} \rangle (= 2\sqrt{3}\hbar l_{\tau}/\sqrt{M\omega_{\tau}} \text{ in the notation of Reference 7})$$

The q's are the normal mode coordinates divided by the zero point vibrational amplitudes. To obtain the complete Hamiltonian, the kinetic energies and the elastic energies of the modes ε and τ_2 (involving, respectively, the masses M and frequencies ω_{ε} and ω_{τ}) have to be added to equation (3.76). The complete problem has not been solved yet, though approximations have been made.

The most notable of these is based on the circumstance that in the weak coupling limit, up to the second order in the coupling coefficient, the two types of vibronic coupling are separable. The energies to this order are[7,126]

$$(n_\varepsilon + 1)\hbar\omega_\varepsilon - \tfrac{1}{8}\frac{L^2(\varepsilon)}{\hbar\omega_\varepsilon}\,(1 \pm l)$$

and

$$(n_\tau + \tfrac{3}{2})\hbar\omega_\tau - \tfrac{1}{48}\frac{L^2(\tau_2)}{\hbar\omega_\tau}\,[3 + 4J(J + 1) - 4M(M + 1)]$$

Here l takes the values (as in the $E \otimes \varepsilon$ situation) $-n_\varepsilon$, $-n_\varepsilon + 2$, . . . n_ε. M is n_τ, $n_\tau - 2$ . . . 0 or 1, depending on whether n_τ is even or odd. $J = M \pm \tfrac{1}{2}$ (when $M \neq 0$) and $\tfrac{1}{2}$ for $M = 0$. J is an eigenvalue of the total angular momentum operator

$$\mathbf{\hat{J}} = \mathbf{\hat{J}}_{\text{vibr}} + \tfrac{1}{2}\boldsymbol{\sigma}.$$

$\mathbf{\hat{J}}$ satisfies the commutation rules of angular momenta $\mathbf{\hat{J}} \wedge \mathbf{\hat{J}} = i\,\mathbf{\hat{J}}$ and commutes with the $G_{\frac{3}{2}} \otimes \tau_2$ Hamiltonian.

The quadratic approximation appears to be sufficient (and necessary) to account[62] for the infrared spectra of technetium and rhenium–hexa-fluorides ($4d^1$ and $5d^1$ configurations, $G_{\frac{3}{2}}$ ground state). Weinstock and Goodman have also added a second order coupling term (analogous to the term in $K(E)$ for $E \otimes \varepsilon$) to the $G_{\frac{3}{2}} \otimes \varepsilon$ Hamiltonian.[62]

The Hamiltonian of equation (3.76) becomes tractable when either $L(\tau_2)$ or $L(\varepsilon)$ becomes so small as to be negligible. In the former case we are left with two independent problems of the type $E \otimes \varepsilon$, which was treated in detail in §3.2. This is readily recognized by considering the matrices ρ_1, ρ_2 after interchange of the second and third columns and rows.

The absorption spectra of F-centres in heavy alkali halides reflects the occurrence of $G_{\frac{3}{2}} \otimes \varepsilon$; except that in many cases the proximity of doublet $E_{\frac{1}{2}}$ states requires the inclusion in the theory of the spin–orbit interaction between these two states and $G_{\frac{3}{2}}$.[131] We shall discuss this subject in more detail in §3.4.3.

The second alternative, when $L(\varepsilon)$ vanishes, also partitions the Hamiltonian matrices to two non-overlapping submatrices as is evident from the definition of $\boldsymbol{\sigma}$ given above, equation (3.77). Each eigenvalue characterized by J (which, we recall, is a good quantum number in any approximation, as long as $L(\varepsilon) = 0$) is therefore $2(2J + 1)$-fold degenerate. The problem, namely $G_{\frac{3}{2}} \otimes \tau_2$, was numerically solved by Thorson and Moffit, whose paper, Reference 125, contains a wide set of eigenvalues as well as a

Fortran program for the computation of the eigenvalues and eigen-vectors. These authors also point out the mysterious occurrence of a multiple degeneracy of eigenvalue, at (nearly) periodic intervals of the coupling coefficient $L(\tau_2)$.

The static problem is easily solved[2] in the general case: there are a pair of potential surfaces given by

$$\pm\sqrt{(\tfrac{1}{8}L^2(\varepsilon)q^2 + \tfrac{1}{12}L^2(\tau_2)q_\tau^2)}$$

plus the elastic terms quadratic in q and q_τ. [q_τ denotes the radius vector in the (q_ξ, q_η, q_ζ) space, q is the amplitude of the ε-modes, equation (3.6)].

The distortions are either of the trigonal type corresponding to $q_\tau = |L(\tau_2)|/(2\sqrt{3}\hbar\omega_\tau)$ (the reduced symmetry is D$_3$ when higher order coupling terms are present; in their absence, the stable configuration is D$_3$ or C$_2$) or of the tetragonal type (symmetry group D$_4$ with higher order coupling, or D$_2$ without these) at $q = |L(\varepsilon)|/(2\sqrt{2}\hbar\omega_\varepsilon)$. The condition for the two alternatives is

$$\frac{L^2(\tau_2)}{24\hbar\omega_\tau} > \text{ or } < \frac{L^2(\varepsilon)}{16\hbar\omega_\varepsilon}$$

the two sides of the inequality being $E_{JT}(\tau)$ and $E_{JT}(\varepsilon)$, respectively. (Table 1.1.)

Quenching of non-diagonal matrix elements will occur just as for the cases studied previously in connection with the Ham-effect. (§§3.2.5, 3.3.3, 3.3.6.) We turn to this topic now.

3.4.2 *Reduction factors for* G$_{\frac{3}{2}}$ ⊗ (ε + τ$_2$)

In analogy with equation (3.36) we write out the matrix Hamiltonian of external forces operating in the manifold of a vibronic quartet (designated for brevity by v):

$$H(\text{vibr.}) = 1 \cdot G(A_1)\mathbf{I} + K^{(v)}(A_2)G(A_2)\rho_3^{(v)} + K^{(v)}(E)[G(E_\theta)\rho_1^{(v)} +$$
$$+ G(E_\varepsilon)\rho_2^{(v)}]$$
$$+ K^{(v)}(T_1^a)[G^{(v)a}(x)W_x^{(v)} + \ldots] + K^{(v)}(T_1^b)[G^{(v)b}(x)W_x^{(v)} +$$
$$+ \ldots]$$
$$+ K^{(v)}(T_2)[G^{(v)}(\xi)\rho_3^{(v)}\sigma_1^{(v)} + G^{(v)}(\eta)\rho_3^{(v)}\sigma_2^{(v)} + G^{(v)}(\zeta)\rho_3^{(v)}\sigma_3^{(v)}]$$
$$(3.79)$$

The representations of the angular momentum operator $\mathbf{W}$ and of a second linearly independent operator $\mathbf{W}'$ transforming in the same way are given by

$$W_x = \frac{\sqrt{3}}{2}\eta_1 + \tfrac{1}{2}\sigma_1 - \tfrac{1}{2}\rho_1\sigma_1$$

$$W_y = \frac{\sqrt{3}}{2}\eta_2 + \tfrac{1}{2}\sigma_2 - \tfrac{1}{2}\rho_1\sigma_2$$

$$W_z = -\frac{i}{2}\sigma_3 - i\,\rho_1\sigma_3$$

$$W'_x = -\frac{\sqrt{3}}{4}\eta_1 + \sigma_1 + \tfrac{1}{4}\rho_1\sigma_1$$

$$W'_y = -\frac{\sqrt{3}}{4}\eta_2 + \sigma_2 + \tfrac{1}{4}\rho_1\sigma_2$$

$$W'_z = -i\,\sigma_3 + \frac{i}{2}\rho_1\sigma_3$$

The matrices ρ_i, σ_i, have already been defined in equation (3.77). To these we have added, also in the representation of equation (3.78),

$$\eta_1 = \begin{pmatrix} 0 & 0 & 0 & -i \\ 0 & 0 & -i & 0 \\ 0 & i & 0 & 0 \\ i & 0 & 0 & 0 \end{pmatrix} \qquad \eta_2 = \begin{pmatrix} 0 & 0 & 0 & 1 \\ 0 & 0 & -1 & 0 \\ 0 & -1 & 0 & 0 \\ 1 & 0 & 0 & 0 \end{pmatrix}$$

In equation (3.79) we note the presence of the reduction factors $K^{(v)}(\Gamma)$ appropriate to the vibronic state v. In the purely electronic state ($L(\varepsilon) = L(\tau_2) = 0$), all K's are unity. In the presence of a magnetic field, the second line of equation (3.79) may be rearranged (with the omission of v) in the equivalent form[128]

$$\beta K[H_x W_x + \ldots] + \beta L[H_x W_x^3 + \ldots] \tag{3.80}$$

The reduction factors appearing in equations (3.79) and (3.80) were calculated by Morgan[127] (Figure 3.32 and 3.33) in various approximations of the vibronic Hamiltonian, equation (3.76): when $L(\varepsilon) = 0$, when $L(\tau_2) = 0$ and when $\omega_\tau = \omega_\varepsilon$, $\sqrt{2}L(\tau_2) = \sqrt{3}L(\varepsilon)$. In the latter case H_{JT} retains a spherical symmetry and the Schrödinger equation may be solved in the manner done for $T \otimes (\varepsilon + \tau_2)$ in Reference 81.

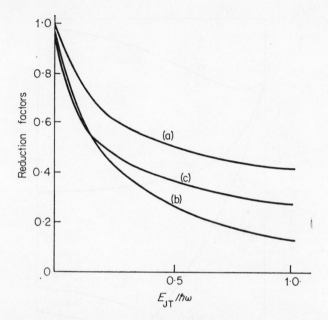

Figure 3.32 Reduction factors in equation (3.79) *vs.* coupling strengths in different approximations. (From Reference 127.)

(a) $K(T_2)$ for $L(\varepsilon) = 0$
(b) $K(E)$ for $L(\varepsilon) = 0$
(c) $K(E) = K(T_2)$ for $L(\tau_2)/(\hbar\omega_\tau) = \frac{3}{2}L(\varepsilon)/(\hbar\omega_\varepsilon)$, i.e.

$$E_{\mathrm{JT}}(\tau)/(\hbar\omega_\tau) = E_{\mathrm{JT}}(\varepsilon)/(\hbar\omega_\varepsilon)$$

The reduction factors $K(T_2)$ and $K(E)$ for $L(\tau_2) = 0$ are identical with p and q for $E \otimes \varepsilon$ [Figure 3.15].

3.4.3 *Optical effects in* $G_{\frac{3}{2}} \otimes \varepsilon$

For an interaction of $G_{\frac{3}{2}}$ with ε-modes alone $G_{\frac{3}{2}}$ behaves like E occurring twice (§3.2.7). However, in the F-bands in alkali halides the analysis becomes more involved than for $E \otimes \varepsilon$ alone, even though the vanishing of $L(\tau_2)$ is assumed. This occurs because the $G_{\frac{3}{2}u}$ term arises from the spin–orbit coupling splitting of $^2T_{1u}$ into a quartet $G_{\frac{3}{2}u}$ and a doublet $E_{\frac{1}{2}u}$ at longer wavelength; the two terms interact by vibronic coupling.

Only for Cs (which is the heaviest alkaline cation in the series for which observations have been made) is the spin–orbit coupling of sufficient strength to split off the $G_{\frac{3}{2}u}$ state from $E_{\frac{1}{2}u}$ to any significant extent ($\sim$250 cm^{-1}). Even so, the vibronic interaction between the two terms complicates the lineshape. This and the broadening mechanism due to the symmetric mode have been taken into account (within the limitations of

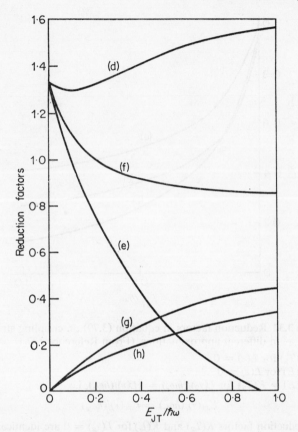

Figure 3.33 Zeeman energy reduction factors in $G_{\frac{3}{2}} \otimes (\varepsilon + \tau_2)$. (From Reference 127.)

(d) K in equation (3.80) for $L(\varepsilon) = 0$. At $L(\tau_2) = 0$, K is normalized to $4/3$, the Landé g-factor for a $p_{\frac{3}{2}}$ state.

(e) K in equation (3.80) for $L(\tau_2) = 0$

(f) K in equation (3.80) for $L(\tau_2)/(\hbar\omega_\tau) = \frac{3}{2}L(\varepsilon)/(\hbar\omega_\varepsilon)$.

(g) L in equation (3.80) for $L(\varepsilon) = 0$

(h) L in equation (3.80) for $L(\tau_2) = 0$

the configuration diagram method) by Moran,[131] whose curve is shown below (Figure 3.34). Using in his Monte-Carlo calculations the same physical parameters as Moran, Cho[114] has derived a band shape (also shown in Figure 3.34) which is significantly different from Moran's and from the experimental curve. Cho believes that the error lies with the perturbational treatment by Moran of the P-J-T-coupling between $G_{\frac{3}{2}}$ and $E_{\frac{1}{2}}$. A proper quantum-mechanical computation would be salubrious.

The double peak structure due to the J-T-E is perceptible in the F-band of cesium halides for the lighter halogens CsF[117] and CsCl,[116] while for the heavier compounds CsBr[116] and CsI[131] the two curves merge into a single broader band.[131]

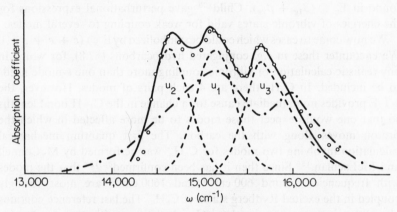

Figure 3.34 Absorption curve for F-centres in CsF.

Experimental points of Hughes and Rabin (Reference 117) are shown by circles and the theoretical curve of Reference 131 by full lines. This is the resultant of the three broken curves shown, of which u_1, u_2 arise predominantly from $G_{\frac{3}{2}u}$ and the asymmetric u_3 from $E_{\frac{1}{2}u}$. The calculated curve of Reference 114 is drawn by lines and dots.

At low temperatures ($T < 15$ °K) additional structure is observed on the short wavelength side of the absorption curve. (T.A. Fulton and D. B. Fitchen, *Phys. Rev.*, **179**, 846 (1969); **B1**, 4011 (1970)).

The bound hole states in covalent semi-conductors arise from $p_{\frac{3}{2}}$-type states near the top of the valence band (§2.4) and have $G_{\frac{3}{2}}$-character. The C-spectrum in GaP,[133] in which the hole is associated with a pair of electrons in a singlet state, also shows this character. However, the spectrum has not yet been interpreted definitively in terms of the J-T-E.

3.5 More complex cases

3.5.1 *Vibronic interaction by several modes*

In addition to the many-mode couplings $T \otimes (\varepsilon + \tau_2)$ [81] and $G_{\frac{3}{2}} \otimes (\varepsilon + \tau_2)$ [127] which were mentioned earlier, there are theoretical treatments of the coupling of an E-state in a square molecule with two (or more) non-degenerate modes (e.g. in $D_{4h} : E_g \otimes (\beta_{1g} + \beta_{2g})$). Ballhausen[134] solved the static problem for two vibrational modes and in the presence of spin-orbit coupling. He also showed some spectral patterns expected for the transitions $^2A \rightarrow {}^2E$ with one mode active and with spin–orbit coupling operative. The formulation is similar to the dimer problem (§3.6.1) and the qualitative features which we shall describe in that connection are also found in $E_g \otimes (\beta_{1g} + \beta_{2g})$. Child[135] gave perturbational expressions for the energies of vibronic states valid for weak coupling to several modes.

We now come to cases which could be symbolized by $E \otimes (\varepsilon + \varepsilon + \ldots)$. We encounter these in e.g. conjugated hydrocarbons (§7.8), for which in any realistic calculation of vibronic coupling more than one ε-mode needs to be included. In C_6H_6 there are 4 such pairs of modes. However, the J-T-E provides no immediate cause for a change in the C—H bond lengths so that one would expect those modes to be more affected in which the proton moves along with the carbon. The first quantum mechanical calculation involving two modes for $C_6H_6^-$ was performed by McConnell and McLachlan.[136] Since then it has been confirmed[63,137] that the modes with frequencies around 600 cm^{-1} and 1600 cm^{-1} are most strongly coupled in the excited Rydberg states of C_6H_6. The last reference contains also extensive quantum mechanical calculations with two modes for various cases of the J-T-E and P-J-T-E applicable to excited states of C_6H_6 and C_6H_5D. From these it appears that zero order vibronic states which are energetically close repel each other as soon as the vibronic coupling is turned on, even though the coupling between them is of a high order. In a different, classical approach[138] to the static problem in $C_6H_6^+$ and C_5H_5 the C—C bond distances $\Delta r_{s,s+1}$ were regarded as independent variables for the minimization of the energy. This is equivalent to the inclusion of all normal modes of the carbon skeleton.

3.5.2 *Coupling to the lattice modes*

The insufficiently justified use of a single (localized) normal mode of particular symmetry for an impurity centre in a solid provided a challenge to many workers. The attempts which have been made to include all lattice vibrational modes have not been conclusive so far. This applies to the many-body techniques,[139-40] which were employed to investigate the possibility of a splitting between the $(S_z =) -1 \rightarrow 0$ and $0 \rightarrow 1$ transitions in an $S = 1$ system. (See also Reference 141.) Ham[39] has however claimed on the basis of symmetry arguments, that for the cubic Hamiltonians used in the above references a splitting should not occur. As a general remark, the approximation methods employed in many-body theories (e.g. curtailment of diagram summation or decoupling of correlations) must be used for the J-T-E with great care so as not to spoil inadvertently the symmetry inherent in the situation. (It is known[142] that these approximation methods have the property of spoiling the detailed balance and they introduce irreversibility.) Thus the sum of the equal-time Green's functions

$$\ll \frac{\hbar}{i}\frac{\partial}{\partial q_\theta}, \sigma_\theta \gg + \ll \frac{\hbar}{i}\frac{\partial}{\partial q_\epsilon}; \sigma_\epsilon \gg \equiv G_\theta + G_\epsilon$$

involving the coordinates q_θ, q_ϵ and matrices σ_θ, σ_ϵ [equation (3.35)] of the $E \otimes \epsilon$ problem, is invariant under the transformation of the cubic group. The equation of motion for $G_\theta + G_\epsilon$ will involve the commutators of the momenta with the Hamiltonian, equation (3.4). The last term of this equation will give rise to terms proportional to

$$3 \ll (q_\theta^2 - q_\epsilon^2); \sigma_\theta \gg - 6 \ll q_\theta q_\epsilon, \sigma_\epsilon \gg.$$

This is also invariant. However, when a decoupling is applied in the random phase approximation we obtain

$$6 < q_\theta > [\ll q_\theta; \sigma_\theta \gg - \ll q_\epsilon; \sigma_\epsilon \gg] - 6 < q_\epsilon > [\ll q_\epsilon; \sigma_\theta \gg + \\ + \ll q_\theta; \sigma_\epsilon \gg]$$

where $\langle \rangle$ denotes the thermal average. In general, this expression will not be a cubic invariant.

It was already mentioned that the high-energy centrifugally stabilized states which were found by Slonczewski[13] would also persist when the electronic degeneracy is coupled to a continuum of lattice modes. The same problem $E \otimes \epsilon$ was given an alternative formulation by Bates et al.,[23] who suggested that the involvement of the lattice modes depended on the extent of the deformation, in the sense that for weak coupling the deformation is mainly confined to the neighbourhood of the centre, whereas for strong coupling the deformation spills over to the rest of the lattice. These authors have also formulated the problem in the Heisenberg picture and proposed that the non-linear operation of the (linear) J-T-E results in the addition of a beat-like low-frequency mode analogous to localized vibrations.

It was pointed out[39,66] that the rôle played by the reduction factors p, q or $K(\Gamma_v)$ [equation (3.58) and (3.60)] in the interpretation of the data (like ESR) on localized centres is the same whether we consider one localized vibration belonging to Γ_v of the point symmetry group or we include all lattice modes which have a component Γ_v at the centre. In the latter case there will be contribution to the reduction factors from all modes. The form of these contributions is not known, except in the cases noted below, but one would expect some resemblance to Debye–Waller factors. A poignant question is whether or not the inequalities derived for p and q in $E \otimes \epsilon$, especially the stringent inequalities $0.484 < q \leqslant 0.5$

found for non-linear coupling, would break down for several lattice modes contributing. This question is important since there is some experimental indication[143] that, in $MgO:Cu^{2+}$ at $4 \cdot 2$ °K, q is significantly below $0 \cdot 484$.

The equilibrium positions of an arbitrary number of modes and the stability conditions in a lattice were given in Reference 81 for $T \otimes (\varepsilon + \tau_2)$, thus providing a generalization of our Table 3.2.

In a further work,[144] in which the $T \otimes (\varepsilon + \tau_2)$ problem was considered (for $MgO:Fe^{2+}$) perturbationally, second order perturbation theory was extended to a Debye-spectrum of lattice phonons. It was found that the correct way to include these is that, if the localized model yields a function $F(\omega_\varepsilon)$ or $F(\omega_\tau)$ for a physical quantity, one must replace F by an average over all modes of the form

$$3 \int_0^{\omega_D} d\omega f^2(\omega) \left[1 - \frac{\sin (a\omega/c)}{a\omega/c} \right] \left(\frac{\omega}{\omega_D^2} \right) F(\omega)$$

where ω_D is the Debye frequency, $a/2$ is the Mg—O distance, c is the velocity of sound and $f(\omega)$ is a function which is not specified, except that $f \leqslant 1$. $T \otimes \varepsilon$, which was treated in References 39 and 66 is soluble to all orders. In the Debye model and with the approximations used by Ham, Schwarz and O'Brien,[144] one obtains

$$K(T_1) = K(T_2) = \exp \left\{ - \frac{1}{8\pi^2} \frac{L^2}{N\hbar\omega_D c^3} \int_0^{\omega_D} \frac{d\omega}{\omega} f^2(\omega) \left[1 - \frac{\sin a\omega/c}{a\omega/c} \right] \right\}$$

The $T \otimes \varepsilon$ problem was also considered by Stevens,[145] who used a different approach. He derived the expression $\propto \hat{z} \cdot \mathbf{R}/R^3$ for the magnitude of the static deformation in the lattice at a point $\mathbf{R}$, measured from the J-T-centre, for a $T \otimes \varepsilon$ static distortion, oriented along z. For more than one J-T-centre the deformations in the lattice are additive; however, the distortional energy contains a non-additive part which is then the interaction energy between the centres.[146] The use of a single vibrational mode was recently justified[93'] for the study of energy differences, vibronic or otherwise, which are much smaller than the phonon energies which are predominantly involved in vibronic coupling.

3.6 The pseudo Jahn-Teller Effect

3.6.1 $(A + B) \otimes \beta$

The simplest case with which we start our study is again a pair of states (the pair is designated for the sake of compactness and simplicity by $A + B$) separated in energy by $2W$ and coupled to a vibration [as in §3.1, equation (3.1)]. Physical examples for the use of the formalism will be described in the final Chapter: the resonant excitation of double molecules or dimers (§7.7), accidental degeneracies of electronic states (§7.6). The first theoretical study in the subject was by Öpik and Pryce[22] whose purpose was to investigate to what extent a spin–orbit coupling splitting of the otherwise degenerate orbitals will affect the nuclear motions. They showed that in the static problem at any rate, for a small splitting the pseudo J-T-E is essentially equivalent to the corresponding J-T-E, for an intermediate splitting the vibronic coupling introduces new and important results, while for large splittings the ground state is stable against distortions.[147] For an orbital doublet the proper measure of the splitting $2W$ is the P-J-T-E ratio [equation (1.1)]

$$S' = \frac{L^2}{(2W)(\hbar\omega)}$$

for which it was shown[22] that if $S' < 1$ the coefficient of the quadratic term in the distortion is positive and the system is stable, whereas for $S' > 1$ that coefficient is negative and instability occurs. We recall that L and $\hbar\omega$ were introduced in equation (3.1). What are the conditions of stability for a P-J-T-E in which the states are coupled by more than one mode? It turns out that it is not sufficient for stability that the coefficients of the quadratic terms in each mode be positive, but the quadratic form of the vibrations must be positive definite.[148] Recently, the observed symmetries of several molecules of the types X_2Y_2, X_2Y_4 and X_2Y_6 were explained[148'] from consideration of their stability with respect to $(A + B) \otimes \beta$.

We start our quantitative treatment of $(A + B) \otimes \beta$ with equation (3.1). We add to it the matrix

$$W\sigma_\epsilon \equiv W \begin{pmatrix} 0 & 1 \\ 1 & 0 \end{pmatrix}$$

as in equation (7.14). We note that the Hamiltonian is invariant under the combined operations, $q \to -q$ and interchange of the electronic states. The vectors having the form

$$\frac{1}{\sqrt{2}} \begin{pmatrix} \chi(q) \\ \pm\chi(-q) \end{pmatrix}, \tag{3.81}$$

with $\chi(q)$ quite general, are eigenvectors of these operations (combined), with eigenvalues ± 1. The solutions of the Hamiltonian can therefore be written in this form.

Frequently, for physical reasons, the Hamiltonian for $(A + B) \otimes \beta$

appears in the form which is obtainable from H' through a unitary transformation

$$H = UH'U^{-1} = -\tfrac{1}{2}\hbar\omega\left(\frac{\partial^2}{\partial q^2} - q^2\right) + \frac{1}{\sqrt{2}}Lq\begin{bmatrix} 0 & 1 \\ 1 & 0 \end{bmatrix} - W\begin{bmatrix} -1 & 0 \\ 0 & 1 \end{bmatrix}$$

(3.82)

where

$$U = \frac{1}{\sqrt{2}}\begin{pmatrix} 1 & 1 \\ 1 & -1 \end{pmatrix}$$

The matrices and vectors in this representation are printed in square rather than round brackets. The eigenvectors have the general form

$$\psi_s = \tfrac{1}{2}\begin{bmatrix} \chi(q) - \chi(-q) \\ \chi(q) + \chi(-q) \end{bmatrix} \quad \text{and} \quad \psi_a = \tfrac{1}{2}\begin{bmatrix} \chi(q) + \chi(-q) \\ \chi(q) - \chi(-q) \end{bmatrix}$$

(3.83)

The labels s and a refer to the symmetric and anti-symmetric character of the vibrational part of the electronic state having lower energy (for $W > 0$). The meanings of equation (3.81) and (3.83) are that in the round

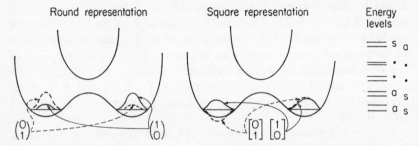

Figure 3.35 Sketch of the vibrational factors associated with the electronic components in two representations. On the left, the round representation [equation (3.81)] shows the localized character (right and left) associated with $\binom{1}{0}$ and $\binom{0}{1}$. In the middle, in the square representation [equation (3.83)] the two electronic components have vibrational factors of opposite parity. On the right the energy levels are shown for $W \ll L$.

representation the vibrations belonging to the electronic components $\binom{1}{0}$ and $\binom{0}{1}$ are mirror images (upright or inverted) of each other, while, in the square representations, $\begin{bmatrix} 1 \\ 0 \end{bmatrix}$ and $\begin{bmatrix} 0 \\ 1 \end{bmatrix}$ carry vibrational states of opposite symmetry. This is shown in the previous figure (Figure 3.35). Here the effective potentials, being the solutions of the static problem, are drawn together with the amplitudes of the vibrational functions belonging to each electronic state. Two lowest lying vibronic states are chosen for illustration (in full and broken lines, respectively). In the diagram of the energy levels on the extreme right the near-degenerate limit ($W \ll L$) is drawn. In dimers this limit is called 'weak exciton coupling'.[149] It is characterized by the splitting of a pair of oscillator levels, having quantum number n, into two levels

$$\text{Energy } (\pm) = (n + \tfrac{1}{2})\hbar\omega - \tfrac{1}{4}L^2/(\hbar\omega) \pm W \exp\left(-\tfrac{1}{2}L^2/(\hbar\omega)^2\right)L_n^0(L^2/(\hbar\omega)^2)$$

$$(3.84)$$

where L_n^0 is the Laguerre polynomial of order n. The dominant dependence in equation (3.84) on the parameters of the Hamiltonian is through the exponential. Similar magnitude splitting would be found in a deep double-well parabolic potential[150] whose minima are separated by a horizontal distance $\sqrt{2}L/(\hbar\omega)$. Returning to Figure 3.35 of the vibronic case, we see that the level s starts off below a, then, as the quantum number n increases, a gradually below s after which the order of levels oscillates forth and back (with a period of $\Delta n \sim \pi^2(\hbar\omega)^2/L^2$).

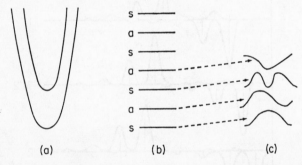

(a) (b) (c)

Figure 3.36 Schematic drawings, for $W \gg L$, of (a) the effective potential surface, (b) the energy level scheme and (c) vibrational factors for a few levels. (After Reference 152.)

The opposite limit ($W \gg L$, strong intermolecular coupling in dimers) has also its own characteristic energy level scheme, namely two vibrational bands with widely separated electronic origins ($2W$ apart) and having somewhat different vibrational spacings. (Figure 3.36.) For a few lowest-lying levels in either band, the vibrational functions $\chi(q)$ have nearly good parities, as can be seen in the drawing, so that in ψ_s and ψ_a of equation (3.83) either the upper or the lower vibrational function is virtually zero. Further calculated results for energies and wave-functions, which cover also the intermediate regions where $W \simeq L$, are found in the papers by Fulton and Gouterman.[151-2] Synthetic absorption and emission spectra between a non-degenerate electronic ground state and the vibronic levels were also calculated. These calculations show very much what is expected of them:

Consider transitions to either the $[^1_0]$ or the $[^0_1]$ state. The two transition probabilities are drawn in the following figure (Figure 3.37) above and

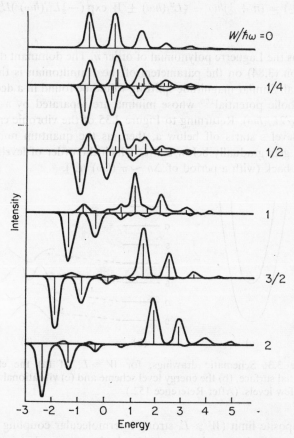

Figure 3.37 Calculated absorption curves for electronic transitions. (From Reference 151.) Above the baselines, transition $[^1_0]$ allowed.
Below the baselines, transition $[^0_1]$ allowed.
The line spectra and the broadened band spectra are shown. The parameter W is varied, as shown, whereas $L/\hbar\omega = 1$ throughout.

below the base-lines, respectively. In the near degenerate limit (which is illustrated in Figure 3.35) the transitions are accompanied by progressions familiar from displaced parabolic potentials. As the energy splitting $2W$ between the electronic states increases, the bands arising from transitions to the two electronic states split apart and their shapes also change. The band below the baselines, representing transitions to $[^0_1]$, changes smoothly from the 'displaced potential' case to a monotonically decreasing progression characteristic of transitions from a deeper to a shallower parabola. The upper parabola in Figure 3.35, pertaining to $[^1_0]$, is narrower than that of the non-degenerate ground state (assuming the absence of any quadratic vibronic coupling term), so that the structure of this band will show, for $W \gg L$, a progression rising to a maximum and then decreasing. Figure 3.37 shows this behaviour in detail.

Two useful, exact results are that the first moments of the $[^1_0]$ and $[^0_1]$ spectra are $+W$ and $-W$ respectively, and that the second moment of either is $W^2 + \frac{1}{2}L^2$.[151] Among more recent developments in the theory we note the application[153] of a truncated cumulant expansion method due to Kubo,[154] and a modified use of the adiabatic wave-functions for a two-well system.[155-6] The essence of the last method is that in the admixed wave-functions the amplitudes of the electronic states $[^1_0]$ and $[^0_1]$ are first derived from the static problem, then the Schrödinger equation for the purely vibrational motion is set up and solved with the lower effective potential in Figure 3.35 acting as a potential. It is expected that this method should provide a good approximation for the two limiting cases of W much larger or much smaller than L and that it should break down somewhere in the intermediate region. Somewhat remarkably, the breakdown occurs mainly for $W < L$.[155]

Returning to Figure 3.37 we note that for a fixed L the strength of the first transition (this is called the zero-phonon line) increases as W increases. In dimer terminology this means that due to inter-monomer coupling the first line will be stronger in the dimer than in the monomer. This was confirmed on the 'stable' dimer of anthracene by Chandross, Ferguson and McRae[157] (whose results together with the interpretation are shown in Figure 3.38) and on some cation dimers by Badger and Brocklehust,[158] who also noted a broadened vibrational structure due, presumably, to vibronic coupling.

The anomalously large vibrational intervals, about 800 cm^{-1} being about four times the normal spacing, found by Bron and Wagner[159] in the absorption spectra of Eu^{2+} in NaCl, KCl and RbCl were interpreted by them in terms of the P-J-T-E. It was suggested that the final state in the transition consists of the following system: an f^{6}d manifold (to which transitions from the f^7 ground state are allowed) and an f^7 state (transitions to which are forbidden by Laporte's rule) about one vibrational quantum lower. In the low symmetry C_{2v} of the site, which arises from a charge-compensating vacancy, the two states get coupled through a β_1 (or β_2)-mode. In a strong P-J-T-E the upper potential is squeezed together

(Figure 3.35) so that the vibrational quanta in that potential are increased (say, doubled). The selection rule on this mode is $\Delta n = 0, 2, 4, \ldots$ in agreement with the data. (There appears also another progression in the α_1 mode which suffers a displacement during transition. For this mode $\Delta n = 0, 1, 2, \ldots$.)

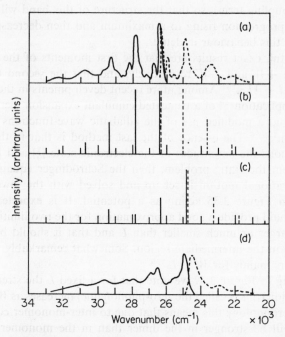

Figure 3.38 Comparison of the monomer and dimer spectra of anthracene. (From Reference 157.) (a) and (d) show the spectra of monomers and dimers respectively, in absorption (full lines) and emission (broken lines). The corresponding theoretical results are shown in (b) and in (c) (using $W = 1.35\hbar\omega, L = \hbar\omega$). Note the enhancement of the zero-phonon line in (c) and (d), due to inter-monomer coupling.

Unfortunately, the computed spectra of Fulton and Gouterman[151] show a quite different intensity pattern, owing to the fact that in the strong vibronic coupling case here envisaged, transitions to the lower potential also become allowed. (Figure 3.37 shows the nature of the predicted absorption curves. For comparison with Bron and Wagner's spectra both bands, below and above the base lines, should be considered.) In view of the discrepancy, the datum of Reference 159 must be regarded as still awaiting interpretation.

3.6.2 *Double degeneracy at a general point*

The matrix Hamiltonian

$$\begin{pmatrix} H^- & W \\ W & H^+ \end{pmatrix} \qquad (3.85)$$

$$H^\pm = \frac{\hbar\omega}{2}\left[-\frac{d^2}{dq^2} + (q \pm b)^2 \pm 2a \right]$$

expresses a simplified model for a pair of electronic states, which become degenerate at some general point $(q = -b/a)$ in a one-dimensional configuration space, and which interact with each other. This interaction is represented by the off-diagonal matrix element W. The simplifications made in arriving from the physical systems of Table 7.4 to the Hamiltonian above will be described in §7.6. In the particular case $a = 0$, equation (3.85) goes over to equation (3.82) for $(A + B) \otimes \beta$ and to equation (7.14) of the dimer case (upon writing $\hbar\omega b = -L/\sqrt{2}$, $q_- \equiv q$ and ignoring q_+).

The Hamiltonian equation (3.85) can be diagonalized numerically[160] by expanding the nth vibronic solution in the form

$$|n\rangle = \sum_m [\alpha_{nm}\chi_m^R(q)\binom{1}{0} + \beta_{nm}\chi_m^L(q)\binom{0}{1}] \qquad (3.86)$$

where χ_m^R and χ_m^L are harmonic oscillator wave-functions localized in the right and left wells which appear in Figure 7.5 (p. 230). $\binom{1}{0}$, $\binom{0}{1}$ are representatives of the electronic states, and α and β are numerical coefficients found by diagonalization of equation (3.85). Regarding α and β and again referring to Figure 7.5, we can expect that much below the intersection point β will be vanishing, near that point α and β will of comparable magnitude and much above it either all α's or all β's will be small in any particular vibronic state $|n\rangle$. All these expectations, which are borne out by calculations,[160] depend to some extent on the parameters a, b, W of the system and in the following paragraphs this dependence will be investigated.

Instead of quoting the eigenenergies associated with each state of equation (3.86), we present related but physically more relevant informations, namely the absorption and emission spectra. The following figure (Figure 3.39) shows line and band spectra for a set of parameters which are thought appropriate to a system, whose configuration diagram is sketched in Figure 7.5(b).

The absorption process $^4A_2 \rightarrow ^4T_2$ is much stronger than the intersystem transition $^4A_2 \rightarrow ^2E$. On the other hand, in emission the calculated $^4T_2 \rightarrow ^4A_2$ fluorescence and the $^2E \rightarrow ^4A_2$ phosphorescence are comparable in magnitude in our model system, a fact which is reflected in the decreased total emission strength of $^4T_2$, relative to absorption, clearly

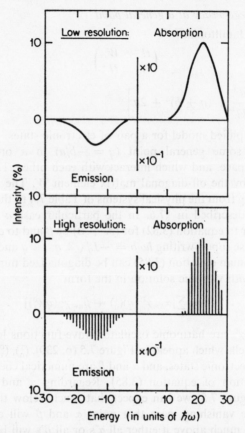

Figure 3.39 Spectra for a Cr^{3+} system at the temperature $T = 0.1\hbar\omega$. Absorption is drawn upwards and emission downwards. The origin of the abscissa is at the R-line. The lines (high resolution) show transition probabilities to and from individual states. The band (low resolution) arises from the superposition of the $^4T_2$ lines broadened by a Gaussian of width $1\hbar\omega$. Both the high and low resolution spectra are normalized to unity. (From Reference 160.)

shown in Figure 3.39. This loss of intensity is due to the radiationless intersystem crossing. $^4T_2 \rightarrow {}^2E$. (In the real cases of emerald and ruby phosphorescence dominates for $T = 0.1\hbar\omega$ and even at higher temperatures.)

The simplicity of this model enables us to study how the strengths of phosphorescence and fluorescence lines change with temperature. We suppose that the emission process starts at the instance that the excitation has reached the bottom of the $^4T_2$-well. It is reasonable to expect that, if the transfer of excitation from $^4T_2$ to ^2E (due to tunnelling) is more effective from the excited vibronic states of $^4T_2$ than from the ground vibronic state, then raising the temperature will increasingly populate the ^2E well and will enhance its emission. Of course, the enhancement of thermal activation will halt at such higher temperature that the thermal activation in the inverse direction $^2E \rightarrow {}^4T_2$ will also gain in effectiveness. At such temperatures ($T \simeq 0.5$) the fluorescent and phosphorescent intensities will level out as function of T. This behaviour is shown in the next figure (Figure 3.40). If, however, the tunnelling from the excited vibronic

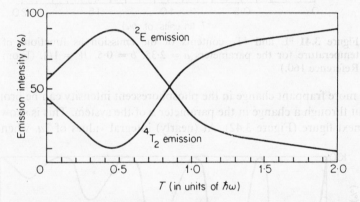

Figure 3.40 ^2E and $^4T_2$ contents of the emission as functions of temperature for the parameters $a = 2.06$, $b = 3$, $W = 1/3$. The ratio of the purely electronic $^4T_2 \rightarrow {}^4A_2$ and $^2E \rightarrow {}^4A_2$ transitions was taken as 100. (a and b are defined in Table 7.4, W in equation (3.85).) (From Reference 160.)

states of $^4T_2$ is not significantly more effective than from the ground vibronic state, then raising the temperature will not initially reduce the fluorescence strength. This type of behaviour appears in the next figure (Figure 3.41). The root of the differences between the two figures (3.40 and 3.41) lies in the different parameters which characterize the systems. We thus observe how the qualitative behaviour of the system may change with the values of the parameters.

A system for which the phosphorescence intensity at first increases with temperature and then going through a maximum decreases is Cr(urea)$_6$ (NO$_3$)$_3$.[161] From the emission and absorption spectra of this crystal Dingle identified the 14,196 cm^{-1} line as the R-line and found that its peak increases between $1.5\,°K$ and $3.7\,°K$ and subsequently decreases. The parameters of chromium hexaurea would be in our terminology $a = 0.04$ and $b = 2$.

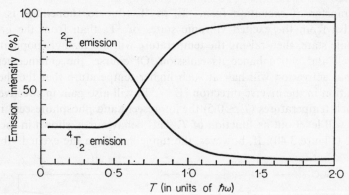

Figure 3.41 ^2E and $^4T_2$ contents of the emission as functions of temperature for the parameters $a = 2.35$, $b = 0.5$, $W = 1/3$. (From Reference 160.)

A more frappant change in the phosphorescent intensity can be brought about through a change in the parameter a of the system. This is shown in the next figure (Figure 3.42). At (nearly) integral values of $2a$, when the

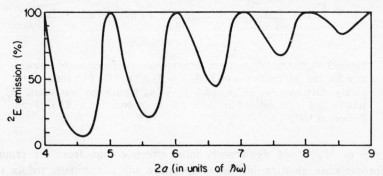

Figure 3.42 Total ^2E emission content at the temperature $T = 0.1\hbar\omega$ as a function of $2a$, the energy difference between the minima of the wells. This figure shows the resonance in phosphorescence as the zero-order vibrational levels coincide (for integral values of $2a$). (From Reference 160.)

sets of zero-order levels of the two parabolae become degenerate, there is a sudden increase of the ^2E component [i.e., a growth in the coefficients α of equation (3.86)] in the ground vibronic state of $^4T_2$ and the phosphorescence shoots up. It is just conceivable that the oscillatory behaviour (with temperature) of the E_2 emission observed by Cox et al.[162] in zinc and cadmium phosphors is also due to a sensitive change of the a parameter of this system. Since a change in the temperature changes the energy difference through anharmonic effects, these oscillations may be due to the levels of the two potential curves coming in and out of resonance as the energy difference increases. The required temperature-change of the energy difference would seem rather large, though.

The P-J-T-E between the singlet states $^1A_{1g}$ and $^1B_{1g}$ in cyclobutadiene (C_4H_4) is noteworthy in that it apparently causes the depression of a singlet state below the triplet $^3A_{2g}$, thus reversing the order of levels which exists in the undeformed situation (Figure 3.43 and §7.8.5.). (Other cases of P-J-T-E level reversals are mentioned in Appendix IX.)

We observe from the form of equation (7.25) that for $Lq_{1g} < 0$ both electrons go into the a-states of Figure 7.12 of Chapter 7. Other states not coupled by the matrix in equation (7.25), viz. $^1B_{2g}$ and $^3A_{2g}$, are also affected by the deformation because of the presence of the α_{1g}-mode in the deformation and these changes are also depicted, following References 163–4. The preceding figure agrees both with the expected triplet ground state in the regular square configuration and with the singlet nature of the stable ground state,[165] which has some support from observation.[166] More elaborate calculations[164] than those on which Figure 3.43

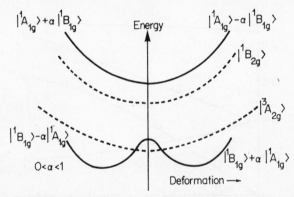

Figure 3.43 Energy levels for the four lowest states of cyclobutadiene taking into account configuration interaction within the lowest unfilled orbital doublet. (Based on References 163 and 164.) The characterization is in terms of the irreducible representations of D_{2h}. The coefficient $\alpha(>0)$ increases on moving from the centre outwards.

is based place the singlet lowest even in the square arrangement. In these calculations the stabilization energy of the deformed configuration is about 3000 cm^{-1}. Nevertheless, the singlet is only metastable since the energy of two acetylene (C_2H_2) molecules into which C_4H_4 dissociates is a further $15,000 \text{ cm}^{-1}$ lower. It has been known for some time[166] that cyclobutadiene has at best only a finite reaction-time.

A very different type of P-J-T-E was noted by Birgenau.[167] He showed that the reduction in g-factors of Ce^{3+} in four different hosts could be explained by the vibronically caused admixture of $|J = 5/2, J_z = \pm 1/2\rangle$ in the ground states, $|5/2, \pm 5/2\rangle$. The two Kramers doublets are separated by about 5 wave-numbers. The reduction in the g-factor amounted to 10% in some cases.

3.6.3 Pseudo Jahn–Teller Effect on a doublet

Hobey[168] made some calculations on several hydrocarbons for the case when one electronic component (ϕ_ϵ, to be specific) is lowered by an energy $2W$ with respect to the other component. Some relevant curves which arise from the addition of $-W\sigma_\theta$ to the linear Hamiltonian (the first line in equation (3.4)) are shown in Figure 3.44. In curve (a), where the energy separation between the two lowest vibronic state is plotted, the non-linear variation is due to the repulsion between the $|n_\theta = 0, n_\epsilon = 0 > |E_\theta\rangle$ and

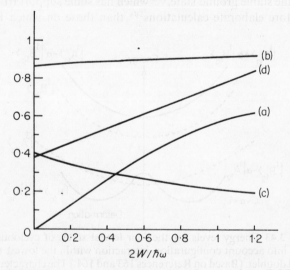

Figure 3.44 Properties of the lowest vibronic state in a P-J-T-E as function of the electronic energy separation $2W/\hbar\omega$.

(After Reference 168.) The calculations were made using a coupling strength $L/\hbar\omega = 1550/1421 = 1\cdot09$.

(a) Splitting, $E_\theta - E_\epsilon$, of the lowest pair of vibronic levels in units of $\hbar\omega$.

(b) The amplitude of $|n_\theta = 0, n_\epsilon = 0\rangle|E_\epsilon\rangle$ in the $|\Psi_\epsilon\rangle$ ground vibronic state.

(c) The amplitude of $|01\rangle|E_\theta\rangle$ in $|\Psi_\epsilon\rangle$.

(d) The amplitude of $|01\rangle|E_\epsilon\rangle$ in the first excited vibronic state $|\Psi_\theta\rangle$.

the $|01\rangle|E_\epsilon\rangle$ levels as the former approaches the latter from below. Similar effects are found in other pseudo-J-T systems.[151]

It was suggested by Bolton *et al.*[169] and by H. C. Longuet-Higgins (unpublished) that the contribution by the symmetric state ϕ_θ, which has a large positive spin density ρ_4 on the 4-proton, can account for the positive ρ_4 generally found in mono-substituted anions. (Figure 7.11 and Table 7.5 may be viewed for a more detailed description.) This contribution is partly a statistical effect, arising from a temperature dependent population of the excited state and is partly due to a vibronic mixing of $|01\rangle|E_\theta\rangle$ in the vibronic ground state. For the statistical effect the energy separation of the first excited vibronic state from the ground state is required. [Figure 3.44(a).] The vibronic correction to the negative spin density operates first through a reduction of the $|00\rangle|E_\epsilon\rangle$ component and secondly through the increase of the $|01\rangle|E_\theta\rangle$ component by vibronic coupling. The amplitudes of both are plotted in the preceding figure as function of the pseudo J-T-energy splitting. The amplitude of the $|01\rangle|E_\epsilon\rangle$ component in the first excited state is also plotted, for the sake of completeness, although it has no bearing on the spin density. Other components, in the range of $2W$ plotted, contribute to the spin density by an amount at least one order of magnitude weaker. Comparison between Hobey's vibronic corrections and the experimental data (which can be seen in Table 7.5) shows a discrepancy for toluene⁻ and ethylbenzene⁻ and fair agreement for the other two compounds. In recent, more refined calculations[170] all significant modes (not just a single pair of ε-modes) were included in the vibronic calculations and a more sophisticated method (the Antisymmetrized Molecular Orbital–Configuration Interaction method)[171] was used to obtain the electronic states and the coupling coefficients. At $T = 200$ °K, $\rho_4 = 0.024$ was found for the toluene anion, in good agreement with experiment.

For disubstituted benzene anions the hyperfine splitting constants of the protons on the substituents provide a means of comparison with the vibronic theory. It is found[172] that whether one calculates vibronic contribution for these anions from scratch or one extrapolates[173] from the mono-substituted benzene ion, the vibronic contribution turns out to be too low (by a factor of about 3). This may possibly be remedied by getting the values of W lowered through the inclusion of higher order vibronic couplings which would quite likely lower $|\Psi_\theta\rangle$ with respect to $|\Psi_\epsilon\rangle$.[174] Alternatively, the vibronic coupling may be intensified by employing, more realistically, lower vibrational C—C bond stretching frequencies for the anion than the average of 1421 cm⁻¹ [168] which is appropriate for the neutral molecule.

A dynamical reduction of the D_{6h}-symmetry will be described in §7.8.4 in connection with mono-deuterated benzene. As a consequence, a change in the para-proton spin density ρ_4 [Figure 7.11(c), (d)] is envisaged. Anticipating the discussion we note that in §7.8.4 the J-T-E is not brought into the picture except that we mention the degeneracy of the electronic state. In fact, a dynamic coupling will tend to reduce the difference in the spin densities between the states. There is an exception to this, namely the regime of extremely strong vibronic coupling when the difference in the zero-point motion energies exceeds the frequency of tunnelling between the three equivalent wells. If the difference is also greater than the temperature then a reduction of the para-spin densities can occur in the frozen-in state.

An explanation in terms of this last possibility was given[175] for the observed reduction, about 10%, in the ratio of the para-proton hyperfine splitting to that of the other protons.[176] Notably, no such reduction was observed in the mono-deuterated cyclooctatetraene⁻. In this molecular ion (of originally D_{4h} symmetry) the vibrational mode β_{2g}, which presumably participates in the J-T-E, produces two frozen-in states, either of which leaves the para-proton with the original spin-density. It must be stated, however, that the foregoing interpretation invites the objection that the static frozen-in limit is unlikely to apply to the benzene ion (Appendix IX).

There are some alternative interpretations how an isotopic mass substitution affects the para-proton splitting. Thus, the resonance integral β, equation (7.21) is changed by deuteration[177] and this leads, through the way β appears in the electronic energy, to a lowering of the vibronic $|\Psi_\theta\rangle$ state with respect to the vibronic $|\Psi_\epsilon\rangle$. β depends on the isotopic substitution, since the resonance integral is in fact an average over the out-of-plane vibrational motions and the amplitudes of these vibrations are reduced when deuterium replaces hydrogen. The energy difference between the vibronic states was estimated by Lawler and Fraenkel[178] from the temperature dependence of the splittings, assuming a Boltzmann distribution of the occupation of the two states. In mono-deuterated benzene the energy difference ranges between (-23) and (-10) cm⁻¹, where the uncertainty arises from the difference of opinion among the various theories, how much spin density belongs to the para-proton (the 4-proton) in the antisymmetric state. Other deuterated benzene anions have stabilization energies shown in Table 3.7.

The vibrational frequencies of the ϵ-modes also split (and shift) by the isotopic replacement. This, as well as changes in the Jahn–Teller coupling coefficients, lead to a stabilization of the $|\Psi_\epsilon\rangle$ *vibronic* state, by about 10 cm⁻¹, even when the *electronic* states are regarded as degenerate.[137] In another work (D. Purins and M. Karplus, unpublished) both the initial electronic level splitting and the effects of vibronic coupling, in which the

Table 3.7 Splitting, in cm^{-1}, of the lowest pair of vibronic energy levels of mono- and bi-deuterated benzene anions. The energy of the $|\Psi_\epsilon\rangle$ state minus that of the $|\Psi_\theta\rangle$-state is shown.

Method Anion:	Benzene $1 - d^-$	Benzene $1,4 - d_2^-$	Benzene $2,6 - d_2^-$
Empirical	$(-23) - (-10)$	$(-40) - (-20)$	$12 - 25$
Empirical, based on Hückel MO spin densities[a]	-20 ± 2	-38 ± 1	$+22 \pm 4$
Vibronic, for degenerate $\phi_\theta, \phi_\epsilon$[b]		-9	
Vibronic for non-degenerate $\phi_\theta, \phi_\epsilon$[c]	$-18\cdot3$	$-46\cdot0$	$19\cdot7$

[a]Lawler and Fraenkel[178]
[b]B. Sharf[137]
[c]Purins and Karplus, quoted in [a]

θ, ϵ-modes were supposed inequivalent, were calculated. The resulting vibronic energy difference for the bi-deuterated benzene anion was $-46\cdot0$ cm^{-1}.

3.6.4 *Pseudo Jahn–Teller Effect on a triplet*, $(E + A) \otimes \varepsilon$

For the relevant matrix Hamiltonian the reader is referred to equation (7.24). To this the kinetic and elastic energies have to be added [as in equation (3.4)]. Fragments of this problem were considered in the literature in three different contexts but anything like a complete solution was not yet given.

In Al_2O_3 the sizeable trigonal crystal field splits the excited triplet $^3T_2$ state of V^{3+} into a doublet and singlet, whose reduced separation (measured by the no-phonon absorption lines) was explained[79,88] through the Ham-effect operating in a doubly degenerate mode. Originally, an interpretation in terms of a $T \otimes \varepsilon$-coupling in octahedral symmetry was undertaken, then, more properly, the problem was set up as an $(E + A) \otimes \varepsilon$ P-J-T-E in trigonal symmetry.[89] The reductions of the trigonal field and of the spin–orbit coupling were calculated by second order perturbation theory. Only the linear coupling terms of equation (2.24) were included; however, the spin–orbit coupling matrix elements were also added.

The matrix of equation (7.4) arises also in connection with trimers. A somewhat specialized form of the problem, considered by Gouterman and

Perrin,[179] can be formulated, as in equation (7.17) and (7.18), in terms of
the matrix Hamiltonian

$$\begin{bmatrix} Xq_A & -W & -W \\ -W & Xq_B & -W \\ -W & -W & Xq_C \end{bmatrix}$$

with $q_A + q_B + q_C = 0$. These authors gave contours of wave-functions
and synthesized absorption and emission spectra. The limiting cases W/X
small or large compared to unity, also discussed in Reference 179, can
be easily understood by reference to $E \otimes \varepsilon$ coupling. For W large, by
second or third order perturbation theory, the coupling of the doublet to
the singlet will give rise to terms equivalent to the higher order coupling
terms in equation (3.4). For W small, the situation becomes equivalent to
the right hand extreme of Figure 3.8 namely to the region of three equivalent wells, along q_A, q_B and q_C respectively, with small tunnelling $W = -\Gamma$
(note the minus sign and the inversion of the levels) between them.

The P-J-T-matrix of equation (3.24) was studied numerically by Van der
Waals and his associates[180] in connection with the lowest phosphorescent
triplet state of benzene and its derivatives[181] (described in §7.8.2). They
based their theory on the (interterm) P-J-T coupling and regarded the
intra-term $E \otimes \varepsilon$ coupling as a perturbation (at most).

The ESR spectra of the excited $^3B_{1u}$-term (populated by ultraviolet
radiation) show a loss of the three- or sixfold axis in benzene or several of
its derivatives imbedded in solids. The interpretation of these spectra in
terms of a P-J-T-E to the nearby $^3E_{1u}$ (depicted in Figure 7.10) was
pursued in several papers.[182-3]

The phosphorescent spectrum of $^3B_{1u}$ [181] also shows signs of reduced
symmetry. Sym.-trimethylbenzene in the host B-trimethylborazole[183] has
a strongly temperature dependent phosphorescent spectrum. This is indica-
tive[180] of a much smaller level separation between the lowest excited state
and its first vibrational level than the vibrational frequency of about
1600 cm^{-1}. The data were analysed in terms of a separation of 280 cm^{-1}
and were explained by the depression of an excited vibronic $^3B_{1u}$-$^3E_{1u}$
level due to the P-J-T-E between these two terms. The sudden decrease of
the phosphorescent decay rate of benzene in rigid glass[182] as the temperature
is lowered below 80 °K shows, by reference to Figures 3.40 and 3.41, the
presence of vibronic effects. (Relaxation from optically excited states will
be discussed in §6.6.) Traces of strong vibronic admixture due to second
and third order vibronic coupling were also seen in the $^3B_{1u}$-lines of
benzene in cyclohexane by Leach and Lopez–Delgado.[184]

In triptycene (a trimer) a 'frozen-in' ESR spectrum was observed at
20 °K, which changed at 77 °K to a spectrum due to excitation shared out

between the monomers.[182] Rather than accepting the proposal[185] that this transition must necessarily be due to a low frequency (of the order of 77 °K) vibration, we refer to the interpretations, in §6.4.5, of similar occurrences in the ESR of d^n ions.

3.6.5 *Pseudo Jahn–Teller Effect on quartets*

There are no quantum mechanical calculations on this problem or on problems of similar or greater complexity; however, the static problem was treated by Öpik and Pryce[22] and Baltzer.[186] Their problem was the existence of distortions, due to vibronic interaction between (say) an A_{1g} and a T_{1g} term, which arise from spin–orbit coupling on $^3T_{1g}$. For a

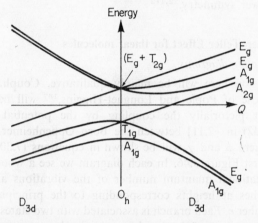

Figure 3.45 Energy levels *vs.* a trigonal distortion coordinate Q, arising from the linear Jahn–Teller Hamiltonian for a $^3T_{1g}$ term subject to spin–orbit coupling.[22] The abscissa and ordinate are in arbitrary units, which scale linearly with the spin–orbit coupling coefficient.

discussion we need the irreducible representations of $^3T_{1g}$ in three symmetries: O_h, D_{4h} and D_{3d}. The latter two arise from the first by distortions along an ε_g and a τ_{2g}-mode.

$$^3T_{1g} \rightarrow$$

O_h: $A_{1g} + E_g + T_{1g} + T_{2g}$

D_{4h}: $A_{1g} + (A_{1g} + B_{1g}) + (A_{2g} + E_g) + (B_{2g} + E_g)$

D_{3d}: $A_{1g} + E_g + (A_{2g} + E_g) + (A_{1g} + E_g)$

The energy levels drawn in Figure 3.45 can be understood in terms of these representations. A trigonal distortion is shown. A tetragonal D_{4h} distortion gives somewhat different results since there are now two two-dimensional representations of the type E_g which interact. In the figure shown the accidental degeneracy of E_g and T_{2g} in O_h is notable. It is lifted through spin–spin coupling[186] in O_h.

The addition of the quadratic terms in the vibration-coordinates results in non-symmetric stationary points. An exhaustive discussion of the various situations that exist is bound to be very tiresome. Instead of this we recall the conclusions of Öpik and Pryce that for moderate spin–orbit coupling strengths the nature and positions of the minima in the ground state will be similar to those in the purely orbital problem $T \otimes \tau_2$ or for tetragonal distortion, to those in $T \otimes \varepsilon$. On the other hand, for strong spin–orbit coupling the singlet will be stable in the symmetric configuration; $Q_\xi = Q_\eta = Q_\zeta = 0$, etc. The excited states may have metastable minima of lower symmetry.[22,112]

3.7 The Renner–Teller Effect for linear molecules

3.7.1 *The coupling diagrams*

Our discussion here will be mainly qualitative. Coupling diagrams, similar to those of Pople and Longuet-Higgins,[187] will be introduced. These exhibit pictorially the coupling by the potential $V(q, \phi; r, \theta)$ [equation (7.32) in §7.11] between the Born–Oppenheimer states, $\phi_\Lambda(r)$ $\chi_{v,|l|}(q, \phi)$, where ϕ and χ will be shown in equations (7.30) and (7.31).

Consider first Figure 3.46. In each diagram we see a wood of l-trees, l being the rotational quantum number of the vibrations and each tree having branches at heights corresponding to the principal vibrational quantum number v. Each branch is associated with two states $+1$ and -1, these being the values of Λ, the electronic angular momentum quantum number. Each gap between the trees belongs to a particular $K = l + \Lambda$, which is in the following four diagrams the only good quantum number. $K = 0, 1, 2, \ldots$ give rise to $\Sigma, \Pi, \Lambda, \ldots$ vibronic states. Anticipating the discussion following equation (7.32), only even order terms in the expansion of the potential $V(q, \phi; r, \theta)$ enter, so that $n =$ even and odd are not coupled. For the sake of clarity we present these two cases separately in Figure 3.46.

Interactions between states are of two types, those denoted in the figure by zig-zagged lines between a pair of zero-order degenerate states and those denoted in the figure by broken lines between non-degenerate states. As is well known, coupling between degenerate states splits these in the first order; this splitting will occur (and for small to moderate coupling will dominate the level schemes) for all states excepting those which are not attached by zig-zagged lines (the lowest $|K| = 1$ states). The relation of the coupling diagram to the energy levels may be seen by comparing the Figures 3.46 and 3.47.

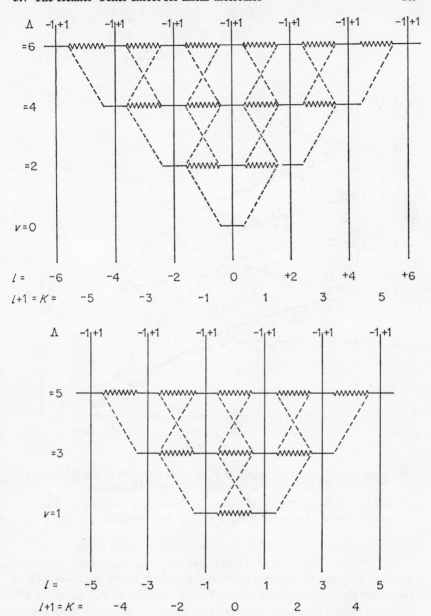

Figure 3.46 Coupling diagrams for Π electronic states ($|\Lambda| = 1$). ($\sim\!\!\sim$: interaction between degenerate states, — — interaction between non-degenerate states).

Upper figure: v and l even.   Lower figure: v and l odd.

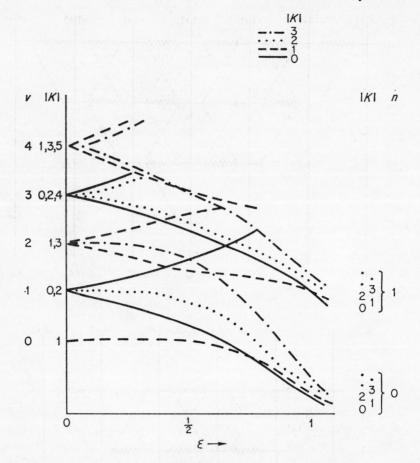

Figure 3.47 The energy levels of a triatomic molecule in the Renner–Teller effect *vs.* the coupling strength ε. (Schematic.)

On the left are the levels of a linear molecule in a Π-electronic state, on the right are the lower-lying levels of a bent molecule.

The coupling diagrams of Figure 3.46 apply to degenerate electronic states of the type Π. We consider only coupling due to the lowest-order even term in the potential, equation (7.32). We refer to this equation and note that this term behaves for small q as $b(r)q^2 \cos 2(\theta - \phi)$. With the aid of the coupling diagrams the illustrated energy levels (Figure 3.47) can be easily understood. For $K \neq 0$ (i.e. for the vibronic states labelled by Π, Δ, Φ, . . .) all even v (or odd v) levels are intermixed and therefore avoid each other by the non-crossing rule; for $K = 0$ all states of the same parity in the angular coordinate $\theta - \phi$ move in the same direction, either up, in which case they belong to the upper surfaces V^+ which can be seen in Figure 7.16, or down, when they are associated with V^-. In Figure 3.47 the abscissa is the dimensionless coupling strength $\langle b(r) \rangle/$ $(4\pi^2 M \nu_2^2) = \varepsilon$. ν_2 is the vibrational frequency of the distorting mode, ε is the parameter in Renner's basic work[188] used also by Sponer and Teller.[189] For low values of ε the case (a) of Figure 7.16 applies. As soon as $\varepsilon \to 1$, the lower potential flattens out and the molecule becomes unstable. In order to reach cases (b) or (c), higher order anharmonic terms have to be included in the potential. Such terms were absent in earlier work on the subject. In the figure shown they are included by linking the levels at $\varepsilon \sim 0.5$ to the level scheme of an anharmonic potential of cylindrical symmetry. The details of this level scheme depend of course on the choice of the postulated anharmonic potential. For example, the stabilization in the bent configuration may be achieved by retaining higher order terms (in q^n) in the potential *unperturbed* by vibronic coupling. We have already met such a term in the case of E $\otimes$ ε. [The last term in equation (3.4).] Two expressions, which include higher than second order coupling, were studied: $\alpha(1 + \beta q^2)^{-1}$ in Reference 190 and $\alpha\, e^{-\beta q^2}$ (Reference 191), where α and β are constants. The level ordering for the lower vibrational and rotational states, such as are shown in Figure 3.47 above, does not much depend on the choice of the potential, only the spacing and the depths of the levels do. Alternatively, we could have worked with higher order *vibronic* coupling terms (analogous to the penultimate term in equation (3.4)). With these terms present, the $K = 0$ levels still cross.

In case (b) of Figure 7.16 with deep minima we can use a 'Dixon plot' as a diagnostic. In this[25] the intervals between two adjacent levels on the lower surface are plotted as function of the mean level height. A sudden break is notable at just those levels whose heights match the barrier (Figure 3.48).

While Dixon's results were obtained for a particular model of a double-well potential and not for the Renner effect where there are excitations to the upper potential, the results apply very likely to the Renner effect, especially for $K = 0$. For $K \neq 0$, the plot has only qualitative relevance to the Renner-effect. There are some simple physical reasons for the occurrence of minima near the barrier height, as noted by Herzberg (Reference 76). Much below this height, the motion is in a narrow potential well (actually, moat), while much above it where the vibronic coupling is no longer important, the interval is about twice the fundamental frequency ν_2. We have seen earlier, §3.2.2, that the numerical data of E $\otimes$ ε also show some interesting drops in the level intervals at and around the barrier height.[8]

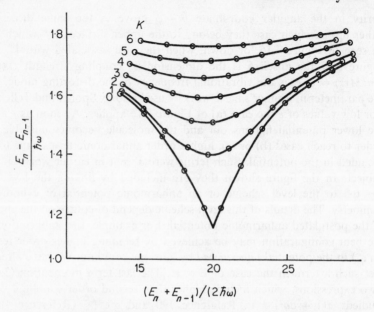

Figure 3.48 Energy level differences *vs.* mean energies for various values of the angular momentum quantum number K. After Dixon.[25] (By permission of The Faraday Society.)

The inclusion in the coupling diagrams of further couplings, such as spin–orbit coupling[191-2] or of correction terms proportional to l^2,[193] is quite easy and rewarding for anyone interested in qualitative effects. As an example we depict in Figure 3.49 the changes in the coupling-diagram when a spin–orbit coupling term in the form $\xi\Lambda\Sigma$ splits the levels of a $^2\Pi$-electronic state. Here ξ is the spin–orbit coupling coefficient and $\Sigma = \pm\frac{1}{2}$. Only $P = \Lambda + \Sigma + l$ gives an exact description of the state. However, to a first approximation, i.e. neglecting the couplings between non-degenerate states (denoted by broken lines) we readily see that the level at $v = 0$ is split by spin–orbit and vibronic couplings combined into two levels, $v = 1$ splits into four levels and so on.

The addition of spin–orbit coupling to the coupling diagrams for $^3\Pi$ changes these in analogous ways.[192]

Vibronic effects on Δ-electronic states were observed in an excited state of CNC. The results of Merer and Travis[194] can be understood by constructing the appropriate coupling diagrams. In this case it can be shown that the states which are not coupled by the zig-zagged lines, and therefore do not suffer first order splitting, arise from the zero order states with $K = v$ or $v + 2$.

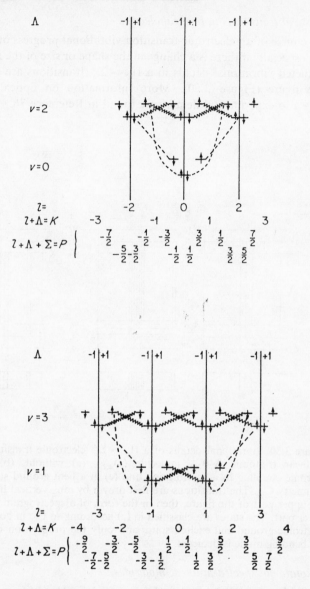

Figure 3.49 Coupling diagram, including spin–orbit coupling, in a $^2\Pi$-electronic state. For clarity (four) broken line couplings within $K = 0$ have been omitted from the lower figure.

3.7.2 *Optical transitions in linear molecules*

In the course of an electronic transition vibrational progressions will be induced, especially if there is a change in the shape or size of the molecule. The predicted vibrational details in a $\Pi_g \to \Sigma_u^+$ transitions are shown in the next figure (Figure 3.50). More information on optical spectra, including also experimental details, are found in Reference 76.

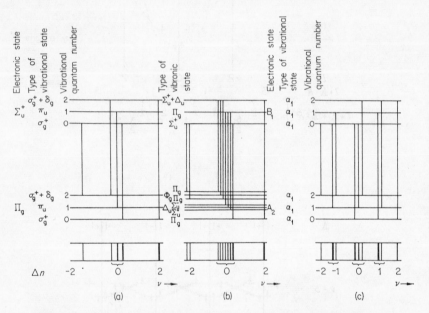

Figure 3.50 Vibrational details of a $\Pi_g \to \Sigma_u^+$ electronic transition in a linear triatomic A-B-A molecule ($D_{\infty h}$), (a) without, (b) with vibronic coupling to a π_u-vibration and (c) in a bent ground state of symmetry C_{2v}. The transitions are first drawn by long vertical lines in the upper part of the figure, then in the form of a spectrogram in the lower part. The states are classified in (c) according to C_{2v} in both the electronic ground and excited states and only the progression due to the bending mode is depicted.

3.7.3 *Rotational spectra in linear molecules*

This subject has been extensively discussed by Herzberg in his book (Reference 76, Chapter 2) especially with reference to the available experimental data, so we shall be brief here. The nature of selection rules for rotational states has been a subject of discussion[195-6] (see also our treatment of rotation–vibration coupling in §3.2.8). For linear molecules the species of the rotational states in the appropriate molecular symmetry group ($D_{\infty h}$ or $C_{\infty v}$) can only be given[196] for negligible rotational-vibronic interactions, that is, precisely for the conditions when the effect of vibronic coupling on the rotational levels is insensible. Therefore the selection rules for the rotational spectra of linear-bent or bent-bent transitions cannot be determined within the molecular symmetry group but special procedures must be adopted.[195] Experimentally, rotational details are present in the bent-linear transition of HCN, DCN,[197] of C_2H_2 and C_2D_2, [198–200] of CS_2,[201] in the linear-bent transitions of NO_2 [202] and in the transitions between the two states of case (b) (Figure 7.16 in §7.11) in HCO and DCO,[203] in NH_2 [204–6] and in BH_2 and CH_2.[207] Some experimental and theoretical data on the deformation of linear molecules are collected in Appendix IX.

3.7.4 *The Walsh-diagrams*

Since it is a matter of first importance to decide which of the three cases of Figure 7.16, (a), (b) or (c), in §7.11 is applicable in each situation, we shall now describe some methods whose purpose is to settle this question. An early, systematic approach was taken by Walsh in a series of papers[208] (though the first attempt of this kind was made by Mulliken,[209] who drew figures similar to the ones below on an empirical basis) (Figure 3.51).

Walsh drew diagrams consisting of energy levels of molecular orbitals as functions of some 'distortion' parameter. The building-up principle was then applied to each molecule. If in the course of this building up an orbital was reached which got significantly lowered in energy upon distortion, then a distorted state of the whole molecule was predicted.

Thus, hydrid molecules with 5–8 valence electrons and other molecules with 17–20 valence electrons are expected to be bent. This is indeed the case for BH_2 (five valence electrons), CH_2 (6 electrons; CH_2 is bent in an excited singlet state, the triplet ground state is spin-stabilized and linear), NH_2 (7), PH_2 (7), OH_2 (8). Among the bent non-hydride molecules, we name NO_2 (17), NOCl (18), NOBr (18), NO_2^- (18), O_3 (18), SO_2 (18), ClO_2 (19), F_2O (20), Cl_2O (20), Cl_2S (20) and Br_2Te (20). Further details on individual molecules will be given presently. The Walsh diagrams were originally constructed by correlating the orbitals in the two extreme configurations, namely, for the linear molecule and one bent into a right angle, finding the relative energies in these two configurations and then smoothly joining the extremes. The diagrams were very useful in predicting a number of configurations and they gained enthusiastic support, especially among experimentalists. However, the theoretical standing of the diagrams is regarded as suspect[210–3] and indeed the justifications originally given are not in line with modern ideas. Nonetheless, the diagrams *can*

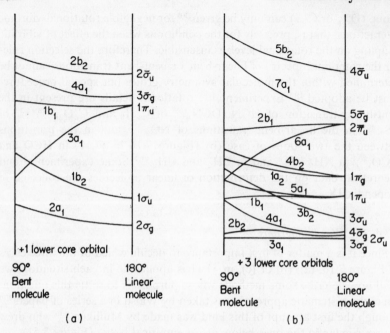

Figure 3.51 Walsh diagrams for (a) triatomic hydrid molecules XH_2, (b) triatomic molecules XY_2. The energy levels of the molecular orbitals are shown as functions of the HXH or YXY angle with the appropriate classifications of the state. Diagram (b) is presented with the modification due to Herzberg (Reference 76). Antibonding orbitals are denoted by a bar.

be, if not rigorously justified, at least explained in the simple terms of the LCAO theory of Molecular Orbitals. (Reference 76, Figures 123–4.)

Diagrams similar to Figure 3.51 can also be drawn for other larger molecules, e.g. for XH_3 or AB_3. From these an interesting prediction of the Walsh diagrams may be noted, that degenerate states of planar XH_3 or AB_3 are not stabilized by out of plane bending. Accordingly, the pyramidal structure of e.g. NH_3 never arises from a Renner–Teller distortion but from the structural preference of non-degenerate states. Of course, some in-plane deformation will remove the degeneracy of these molecules by a J-T-E of the $E \otimes \varepsilon$ type.

A semi-empirical calculation based on the angular variation of directed valence bonds in XH_2 was carried out by Longuet-Higgins and Jordan,[211] employing sp-hybridized wave-functions on X (X = N, C or B) and spd-hybridized orbitals on X = P by Jordan.[213] Bent molecular configurations were 'predicted' for N, B and C (in its $^1A_1$ excited state), in agreement with the experimental data, but a linear configuration was found calculationally for the $^2\Pi_u$ ground state of BH_2. This was since also found to be bent, $\widehat{HBH} = 131°$ (Reference 207), as required by the correlation diagrams of Walsh. From the theoretical point of view, the situation is made more complicated by the circumstance that in the valence-bond scheme these molecules differ from each other only by having occupied orbitals whose effect on the angular disposition is secondary. These orbitals are the non-bonding orbitals, of which BH_2 has one and zero in the plane of the bent molecule and perpendicular to it respectively; NH_2 has two and one; CH_2 has two and zero. It turns out therefore that the linearity of BH_2, which is from our point of view the most interesting prediction of the theory, arises from the empirical part ploughed into the theory. This has the consequence that the nature of any failings of the Walsh diagrams is not theoretically established.

An informative way of presenting theoretical results relating to the bending of XY_2 molecules dates back to 1936. Hirschfelder et al.[214] drew the contours of the potential energy as function of the coordinates of a Y atom in the plane in which X and the other Y were held rigidly fixed.

3.8 References

1. C. J. Ballhausen, *Theor. Chim. Acta*, **3**, 368 (1965).
2. A. D. Liehr, *J. Phys. Chem.*, **67**, 389 (1963).
3. R. N. Porter, R. M. Stevens and M. Karplus, *J. Chem. Phys.*, **49**, 5163 (1968).
4. R. N. Porter and M. Karplus, *J. Chem. Phys.*, **40**, 1105 (1964).
5. M. S. Child and H. C. Longuet-Higgins, *Phil. Trans. Roy. Soc. A*, **254**, 259 (1961).
6. W. Moffitt and A. D. Liehr, *Phys. Rev.*, **106**, 1195 (1957).
7. W. Moffitt and W. Thorson, *Phys. Rev.*, **108**, 1251 (1957); also in *Calcul des functions d'onde moleculaires* (Ed. R. A. Daudel, Paris: CNRS, 1958) p. 141.
8. H. C. Longuet-Higgins, U. Öpik, M. H. L. Pryce and R. A. Sack, *Proc. Roy. Soc. A*, **244**, 1 (1958).
9. M. S. Child, *Mol. Phys.*, **5**, 391 (1962).
10. H. Uehara, *J. Chem. Phys.*, **45**, 4536 (1966).
11. I. B. Bersuker, *Zh. Eksperim. i Teor. Fiz.*, **43**, 1315 (1962); **44**, 1239 (1963) [English transl.: *Soviet Phys. JETP*, **16**, 933 (1963); **17**, 836 (1963)].
12. M. C. M. O'Brien, *Proc. Roy. Soc. A*, **281**, 323 (1964).
13. J. C. Slonczewski, *Phys. Rev.*, **131**, 1596 (1963).
14. K. A. Müller, 'Jahn–Teller effects in magnetic resonance' in *Magnetic Resonance and Relaxation* (Ed. R. Blinc, Amsterdam: North Holland, 1967) p. 192.
15. M. H. L. Pryce, K. P. Sinha and Y. Tanabe, *Mol. Phys.*, **9**, 33 (1965).
16. F. S. Ham, *Phys. Rev.*, **166**, 307 (1968).
17. L. L. Lohr, Jr., *Inorg. Chem.*, **6**, 1890 (1967).

18. R. Englman, *Phys. Letters*, **2**, 227 (1961).
19. B. Halperin and R. Englman, *Solid State Commun.*, **7**, 1579 (1969).
20. J. S. Griffith, *The Irreducible Tensor Method for Molecular Symmetry Groups* (London: Prentice Hall, 1962).
21. L. Pauling and E. B. Wilson, *Introduction to Quantum Mechanics* (New York: McGraw-Hill, 1935).
22. U. Öpik and M. H. L. Pryce, *Proc. Roy. Soc. A*, **238**, 425 (1957).
23. C. A. Bates, J. M. Dixon, J. R. Fletcher and K. W. H. Stevens, *J. Phys. C. (Proc. Phys. Soc.)*, **1**, 859 (1968).
24. J. C. Slonczewski and V. L. Moruzzi, *Physics*, **3**, 237 (1967).
25. R. N. Dixon, *Trans. Faraday Soc.*, **60**, 1363 (1964).
26. U. Höchli, O. S. Leifson and K. A. Müller, *Helv. Phys. Acta*, **36**, 484 (1964).
27. U. Höchli and K. A. Müller, *Helv. Phys. Acta*, **37**, 209 (1964); *Phys. Rev. Letters*, **12**, 730 (1964); **13**, 565 (1964).
28. U. Höchli, K. A. Müller and P. Wysling, *Phys. Letters*, **15**, 1 (1965).
29. P. Wysling, U. Höchli and K. A. Müller, *Helv. Phys. Acta*, **37**, 629 (1964).
30. P. Wysling, K. A. Müller and U. Höchli, *Helv. Phys. Acta*, **38**, 358 (1965).
31. D. P. Breen, D. C. Krupka and F. I. B. Williams, *Phys. Rev.*, **179**, 241 (1969).
32. G. Herzberg and H. C. Longuet-Higgins, *Disc. Faraday Soc.*, **35**, 77 (1963).
33. A. H. Wilson, *The Theory of Metals* (Cambridge: The University Press, (1954)).
34. F. I. B. Williams, D. C. Krupka and D. P. Breen, *Phys. Rev.*, **179**, 255 (1969).
35. U. T. Höchli, *Phys. Rev.*, **162**, 262 (1967).
36. U. T. Höchli and T. L. Estle, *Phys. Rev. Letters*, **18**, 128 (1967).
37. L. L. Lohr, Jr., *Inorg. Chem.*, **6**, 1890 (1967).
38. R. Englman and B. Halperin, *Phys. Letters*, **31A**, 473 (1970).
39. F. S. Ham, *Jahn–Teller Effects in Electron Paramagnetic Resonance Spectra* (New York: Plenum Press, 1971).
40. R. E. Coffman, *Phys. Letters*, **19**, 475 (1965); **21**, 381 (1966); *J. Chem. Phys.*, **48**, 609 (1968).
41. Yu. E. Perlin, *Fiz. Tverd. Tela*, **10**, 1941 (1968) [English transl.: *Soviet Phys.-Solid State*, **10**, 1531 (1969)].
42. M. Wagner, *Phys. Letters*, **29A**, 472 (1969).
43. F. A. Cotton and M. D. Meyers, *J. Am. Chem. Soc.*, **82**, 5023 (1960).
44. G. D. Jones, *Phys. Rev.*, **150**, 539 (1960).
45. D. S. McClure, *J. Chem. Phys.*, **36**, 2757 (1962).
46. H. Hartman, H. L. Schläfer and K. H. Hansen, *Z. Anorg. Allg. Chem.*, **284**, 153 (1956).
47. O. G. Holmes and D. S. McClure, *J. Chem. Phys.*, **26**, 1686 (1957).
48. C. A. Beevers and H. Lipson, *Proc. Roy. Soc. A*, **146**, 570 (1934).
49. D. Reinen, *Z. Naturforsch.*, **23a**, 521 (1968).
50. H. D. Koswig and I. Kunze, *Phys. Stat. Sol.*, **3**, 81 (1963); **9**, 451(1965).
51. H. D. Koswig, U. Retter and W. Ulrici, *Phys. Stat. Sol.*, **24**, 605 (1967).
52. P. P. Sorokin, M. J. Stevenson, J. R. Lankard and G. D. Pettit, *Phys. Rev.*, **127**, 503 (1962).
53. R. E. Dietz, H. Kamimura, M. D. Sturge and A. Yariv, *Phys. Rev.*, **132**, 1559 (1963).
54. M. D. Sturge, *Solid State Phys.*, **20**, 91 (1967).

55. M. S. Child and H. L. Strauss, *J. Chem. Phys.*, **42**, 2282 (1965).
56. J. C. Slonczewski, *Solid State Commun.*, **7**, 519 (1969).
57. J. Gaunt, *Trans. Faraday Soc.*, **50**, 209 (1954).
58. H. H. Claassen, J. G. Malm and H. Selig, *J. Chem. Phys.*, **36**, 2890 (1962).
59. H. H. Claassen, H. Selig and J. G. Malm, *J. Chem. Phys.*, **36**, 2888 (1962).
60. B. Weinstock and H. H. Claassen, *J. Chem. Phys.*, **31**, 262 (1959).
61. M. S. Child, *Mol. Phys.*, **3**, 605 (1960).
62. B. Weinstock and G. L. Goodman, *Adv. Chem. Phys.*, **9**, 169 (1965).
63. B. Sharf and J. Jortner, *Chem. Phys. Letters*, **2**, 68 (1968).
64. A. A. Kaplyanski and A. K. Przhevuskii, *Opt. i Spektroskopiya*, **19**, 597 (1965) [English transl.: *Opt. Spectry.*, **19**, 331 (1965)].
65. L. L. Chase, *Phys. Rev. Letters*, **23**, 275 (1969); *Phys. Rev.*, **B2**, 2308 (1970).
66. F. S. Ham, *Phys. Rev.*, **138**, A 1727 (1965).
67. R. Englman and D. Horn in *Paramagnetic Resonance*, Vol. 1, Ed. W. Low (New York: Academic Press, 1963) p. 329.
68. A. Abragam and M. H. L. Pryce, *Proc. Phys. Soc.*, **A63**, 409 (1950).
69. W. Low, *Paramagnetic Resonance in Solids* (New York: Academic Press, 1960).
70. J. S. Alper and R. Silbey, *J. Chem. Phys.*, **51**, 3129 (1969).
71. J. Yamashita and T. Kurosawa, *J. Phys. Chem. Solids*, **5**, 34 (1958).
72. H. Fröhlich, S. Machlup and T. K. Mitra, *Phys. Kondens. Materie*, **1**, 359 (1963).
73. S. Davidov, *The Theory of Molecular Excitons* (New York: McGraw-Hill 1962).
74. H. M. McConnell and A. D. McLachlan, *J. Chem. Phys.*, **34**, 1 (1961).
75. C. J. Ballhausen, *Theor. Chim. Acta*, **3**, 368 (1965).
76. G. Herzberg, *Molecular Spectra and Molecular Structure*, Vol. 3 (Princeton: Van Nostrand, 1966).
77. A. E. Douglas and J. M. Hollas, *Can. J. Phys.*, **39**, 479 (1961).
78. J. H. Van Vleck, *J. Chem. Phys.*, **7**, 72 (1939).
79. W. C. Scott and M. D. Sturge, *Phys. Rev.*, **146**, 262 (1966).
80. H. Kamimura and S. Sugano, *J. Phys. Soc. Japan*, **14**, 1612 (1959).
81. M. C. M. O'Brien, *Phys. Rev.*, **187**, 407 (1969).
82. J. Owen and K. W. H. Stevens, *Nature*, **171**, 836 (1953).
83. J. Owen, *Proc. Roy. Soc. A.*, **227**, 183 (1955).
84. J. H. E. Griffiths and J. Owen, *Proc. Roy. Soc. A*, **226**, 96 (1954).
85. J. Owen and J. H. M. Thornley, *Repts. Progr. Phys.*, **29**, 675 (1966).
86. I. Broser, U. Scherz and M. Wöhlecke, *Bull. Am. Phys. Soc.*, **14**, 863 (1969); *J. Luminescence*, **1**, 39 (1970).
87. L. Schiff, *Quantum Mechanics* (New York: McGraw-Hill, 1955).
88. P. J. Stephens and M. Lowe-Pariseau, *Phys. Rev.*, **171**, 322 (1968).
89. P. J. Stephens, *J. Chem. Phys.*, **51**, 1995 (1969).
90. R. M. McFarlane, J. Y. Wong and M. D. Sturge, *Bull. Am. Phys. Soc.*, **12**, 709 (1967); *Phys. Rev.*, **166**, 250 (1968).
91. R. Englman and B. Barnett, *J. Luminescence*, **3**, 37 (1970).
92. A. F. J. Cox and W. E. Hagston, *J. Phys. C: Solid State Phys.*, **3**, 1954 (1970).
93. G. A. Slack, F. S. Ham and R. M. Chrenko, *Phys. Rev.*, **152**, 376 (1966).
93′. F. S. Ham and G. A. Slack, *Phys. Rev.*, **B4**, 777 (1971).
94. M. Y. Chen and D. S. McClure, *Bull. Am. Phys. Soc.*, **14**, 79 (1969).
95. M. D. Sturge, *Phys. Rev.*, **B1**, 1005 (1970).

96. Y. Toyozawa and M. Inoue, *J. Phys. Soc. Japan*, **20**, 1289 (1965); **21**, 1663 (1966).
97. C. C. Klick and W. D. Compton, *J. Phys. Chem. Solids*, **7**, 170 (1958).
98. R. Edgerton, *Phys. Rev.*, **138**, A85 (1965).
99. A. Fukuda, S. Makishima, T. Mabuchi and R. Onaka, *J. Phys. Chem. Solids*, **28**, 1763 (1967).
100. M. Tomura and H. Nishimura, *J. Phys. Soc. Japan*, **20**, 1536 (1965); **21**, 2081 (1966).
101. M. Spiller, V. Gerhardt and W. Gebhardt, *Bull. Am. Phys. Soc.*, **14**, 872 (1969); *J. Luminescence*, **1**, 651 (1970).
102. A. Fukuda, *Phys. Rev. Letters*, **26**, 314 (1971).
103. N. N. Kristoffel, *Opt. i Spekt.*, **18**, 798 (1965) [English transl. *Opt. Spectry.*, **18**, 448 (1965)].
104. B. G. Vekhter and B. S. Tsukerblat, *Fiz. Tverd. Tela*, **10**, 1574 (1968) [English transl.: *Soviet Physics—Solid State*, **10**, 1250 (1968)].
105. I. B. Bersuker and B. G. Vekhter, *Fiz. Tverd. Tela*, **5**, 2432 [English transl.: *Soviet Phys.—Solid State* **5**, 1772 (1964)]; *Phys. Stat. Sol.*, **16**, 63 (1966).
106. P. Wysling and K. A. Müller, *Phys. Rev.*, **173**, 327 (1968).
107. M. D. Sturge, *Phys. Rev.*, **140**, A880 (1965).
108. M. Caner and R. Englman, *J. Chem. Phys.*, **44**, 4054 (1966).
109. B. Weinstock, H. H. Claassen and C. L. Chernick, *J. Chem. Phys.*, **38**, 1470 (1963).
110. S. L. Altman and A. P. Cracknell, *Rev. Mod. Phys.*, **37**, 19 (1965).
111. A. Honma, *J. Phys. Soc. Japan*, **24**, 1082 (1968).
112. A. Fukuda, *J. Phys. Soc. Japan*, **27**, 96 (1969); *Phys. Rev.*, **B1**, 4161 (1970).
113. A. Fukuda, K. Inohara and R. Onaka, *J. Phys. Soc. Japan*, **19**, 1274 (1964).
114. K. Cho, *J. Phys. Soc. Japan*, **25**, 1372 (1968).
115. R. Englman, M. Caner and S. Toaff, *J. Phys. Soc. Japan*, **29**, 306 (1970).
116. H. Rabin and J. H. Schulman, *Phys. Rev. Letters*, **4**, 280 (1960); *Phys. Rev.*, **125**, 1584 (1962).
117. F. Hughes and H. Rabin, *J. Phys. Chem. Solids*, **24**, 586 (1963).
118. F. C. Brown, B. C. Cavenett and W. Hayes, *Proc. Roy. Soc. A*, **300**, 78 (1967).
119. F. Fischer, *Z. Phys.*, **204**, 351 (1967).
120. A. S. Marfunin, A. N. Platonov and V. E. Fedorov, *Fiz. Tverdogo Tela*, **9**, 3616 (1967) [English transl.: *Soviet Physics—Solid State*, **9**, 84 (1968)].
121. M. Wagner in *Localized Excitations in Solids*, Ed. R. F. Wallis (New York: Plenum Press, 1968) p. 551.
122. R. C. Fedder, *Bull. Am. Phys. Soc.*, **14**, 62 (1969), *Phys. Rev.*, **B2**, 32, 40 (1970).
123. M. D. Sturge, *Phys. Rev.*, **140**, A 880 (1965).
124. F. S. Ham in *Optical Properties of Ions in Crystals*, Ed. Crosswhite and Moss (New York: Interscience, 1967) p. 357.
125. W. Thorson and W. Moffitt, *Phys. Rev.*, **168**, 362 (1968).
126. M. S. Child, *J. Mol. Spectrosc.*, **10**, 357 (1963).
127. T. N. Morgan, *J. Luminescence*, **1**, 1 (1970); *Phys. Rev. Letters*, **24**, 887 (1970).
128. J. M. Luttinger, *Phys. Rev.*, **102**, 1030 (1956).
129. P. A. M. Dirac, *The Principles of Quantum Mechanics* (Oxford: Clarendon Press, 1958).

130. G. F. Koster, J. O. Dimmock, R. G. Wheeler and H. Statz, *Properties of the 32 Point Groups* (Cambridge, Mass.: M.I.T. Press, 1963).
131. P. R. Moran, *Phys. Rev.*, **137**, A1016 (1965).
132. D. W. Lynch, *Phys. Rev.*, **127**, 1537 (1962) and unpublished.
133. D. G. Thomas, M. Gershenzon and J. J. Hopfield, *Phys. Rev.*, **131**, 2397 (1963).
134. C. J. Ballhausen, *Theor. Chim. Acta*, **3**, 368 (1965).
135. M. S. Child, *Mol. Phys.*, **3**, 601 (1960).
136. H. M. McConnell and A. D. McLachlan, *J. Chem. Phys.*, **34**, 1 (1961).
137. B. Sharf, *Thesis, Ph.D.*, Weizmann Institute of Science, Rehovoth, 1969.
138. L. C. Snyder, *J. Chem. Phys.*, **33**, 619 (1960).
139. K. W. H. Stevens and F. Persico, *Nuovo Cimento*, **B41**, 37 (1966).
140. H. Böttger, *Phys. Stat. Sol.*, **23**, 325 (1967); **26**, 681 (1968).
141. H. A. M. van Eekelen and K. W. H. Stevens, *Proc. Phys. Soc.*, **90**, 199 (1967).
142. H. Callen, R. H. Swendsen and R. Tahir-Kheli, *Phys. Letters*, **25**, 505 (1967).
143. K. Ždánský, *Phys. Rev.*, **177**, 490 (1969).
144. F. S. Ham, W. M. Schwarz and M. C. M. O'Brien, *Phys. Rev.*, **185**, 548 (1969).
145. K. W. H. Stevens, *J. Phys. C: Solid State Phys.*, **2**, 1934 (1969).
146. P. Novak and K. W. H. Stevens, *J. Phys. C: Solid State Phys.*, **3**, 1703 (1970).
147. H. A. Jahn, *Proc. Roy. Soc. A*, **164**, 117 (1938).
148. A. A. Kiselev, *Optics and Spectry.*, **23**, 195 (1967).
148'. R. G. Pearson, *J. Chem. Phys.*, **52**, 2167 (1970).
149. W. T. Simpson and D. L. Peterson, *J. Chem. Phys.*, **26**, 588 (1957).
150. E. Merzbacher, *Quantum Mechanics* (New York; J. Wiley 1961) Sec. 6.
151. R. L. Fulton and M. Gouterman, *J. Chem. Phys.*, **35**, 1059 (1961); **41**, 2280 (1964).
152. M. Gouterman, *J. Chem. Phys.*, **42**, 351 (1965).
153. J. H. Young, *J. Chem. Phys.*, **49**, 2566 (1968).
154. R. Kubo, *J. Phys. Soc. Japan*, **17**, 1100 (1962).
155. R. Lefebvre and M. Garcia-Sucre, *Intern. J. Quant. Chem.*, **1**, 339 (1967).
156. M. Garcia-Sucre, F. Gény and R. Lefebvre, *J. Chem. Phys.*, **49**, 458 (1968).
157. E. A. Chandross, J. Ferguson and E. G. McRae, *J. Chem. Phys.*, **45**, 3546 (1966).
158. B. Badger and B. Brocklehurst, *Trans. Faraday Soc.*, **65**, 2576, 2582, 2588 (1969).
159. W. E. Bron and M. Wagner, *Phys. Rev.*, **145**, 689 (1966).
160. B. Barnett and R. Englman, *J. Luminescence*, **3**, 55 (1970).
161. R. Dingle, *J. Chem. Phys.*, **50**, 1952 (1969).
162. A. F. J. Cox, W. E. Hagston and C. J. Radford, *J. Phys. C. (Proc. Phys. Soc.)*, **1**, 1746 (1968).
163. M. Gouterman and G. Wagnière, *J. Chem. Phys.*, **36**, 1188 (1962).
164. R. J. Buenker and S. D. Peyerimhoff, *J. Chem. Phys.*, **48**, 354 (1968).
165. M. J. S. Dewar and G. J. Gleicher, *J. Am. Chem. Soc.*, **87**, 3255 (1965).
166. L. Watts, J. D. Fitzpatrick and R. Pettit, *J. Am. Chem. Soc.*, **87**, 3253 (1965); **88**, 623 (1966).
167. R. J. Birgenau, *Phys. Rev. Letters*, **19**, 160 (1967).
168. W. D. Hobey, *J. Chem. Phys.*, **43**, 2187 (1965).

169. J. R. Bolton, A. Carrington, A. Forman and L. E. Orgel, *Mol. Phys.*, **5**, 43 (1962).
170. D. Purins and M. Karplus, *J. Chem. Phys.*, **50**, 214 (1969).
171. L. Salem, *Molecular Orbital Theory of Conjugated Systems* (New York: Benjamin, 1966) Chapter 8.
172. E. de Boer and J. P. Colpa, *J. Phys. Chem.*, **71**, 21 (1967).
173. W. D. Hobey, *J. Chem. Phys.*, **45**, 2718 (1966).
174. C. A. Coulson and A. Golebiewski, *Mol. Phys.*, **5**, 71 (1962).
175. A. Carrington, H. C. Longuet-Higgins, R. E. Moss and P. F. Todd, *Mol. Phys.*, **9**, 187 (1965).
176. R. G. Lawler, J. R. Bolton, G. K. Fraenkel and T. H. Brown, *J. Am. Chem. Soc.*, **86**, 520 (1964).
177. M. Karplus, R. G. Lawler and G. K. Fraenkel, *J. Am. Chem. Soc.*, **87**, 5260 (1965).
178. R. G. Lawler and G. K. Fraenkel, *J. Chem. Phys.*, **49**, 1126 (1968).
179. M. H. Perrin and M. Gouterman, *J. Chem. Phys.*, **46**, 1019 (1967).
180. J. H. Van der Waals, A. M. D. Berghuis and M. S. de Groot, *Mol. Phys.*, **13** 301 (1967); **21**, 497 (1971).
181. G. C. Nieman and D. S. Tinti, *J. Chem. Phys.*, **46**, 1432 (1967).
182. M. S. de Groot and J. H. Van der Waals, *Mol. Phys.*, **6**, 545 (1963).
183. M. S. de Groot, I. A. M. Hesselmann and J. H. Van der Waals, *Mol. Phys.*, **10**, 91 (1965); **10**, 241 (1966); **16**, 45 (1969); **16**, 61 (1969).
184. S. Leach and R. Lopez-Delgado, *J. Chim. Phys.*, **61**, 1636 (1964).
185. Y. Maréchal, *J. Chem. Phys.*, **44**, 1908 (1966).
186. P. K. Baltzer, *J. Phys. Soc. Japan*, **17**, Suppl. B.1, 192 (1962).
187. J. A. Pople and H. C. Longuet-Higgins, *Mol. Phys.*, **1**, 372 (1958).
188. R. Renner, *Z. Physik*, **92**, 172 (1934).
189. H. Sponer and E. Teller, *Rev. Mod. Phys.*, **13**, 75 (1941).
190. W. Thorson and I. Nakagawa, *J. Chem. Phys.*, **33**, 994 (1966).
191. J. A. Pople, *Mol. Phys.*, **3**, 16 (1960).
192. J. T. Hougen, *J. Chem. Phys.*, **36**, 1874 (1962).
193. J. T. Hougen and J. P. Jesson, *J. Chem. Phys.*, **38**, 1524 (1963).
194. A. J. Merer and D. N. Travis, *Canad. J. Phys.*, **44**, 353 (1966).
195. J. T. Hougen, *J. Chem. Phys.*, **37**, 1433 (1962); **39**, 358 (1963).
196. B. J. Dalton, *J. Chem. Phys.*, **44**, 4406 (1966).
197. G. Herzberg and K. K. Innes, *Can. J. Phys.*, **35**, 842 (1957).
198. J. K. G. Watson (unpublished, quoted in Ref. 76).
199. C. K. Ingold and G. W. King, *J. Chem. Soc.*, **1953**, 2702 (1953).
200. K. K. Innes, *J. Chem. Phys.*, **22**, 863 (1954).
201. B. Kleman, *Can. J. Phys.*, **41**, 2034 (1963).
202. R. W. Ritchie and A. D. Walsh, *Proc. Roy. Soc. A*, **267**, 395 (1962).
203. G. Herzberg and D. A. Ramsay, *Proc. Roy. Soc. A*, **233**, 34 (1955).
204. D. A. Ramsay, *J. Chem. Phys.*, **25**, 188 (1956).
205. K. Dressler and D. A. Ramsay, *Phil. Trans. Roy. Soc. A*, **251**, 553 (1959).
206. D. R. Eaton, J. W. C. Johns and D. A. Ramsay (unpublished, quoted in Ref. 76).
207. G. Herzberg and J. W. C. Johns, *Proc. Roy. Soc. A*, **295**, 107 (1966); **298**, 142 (1967).
208. A. D. Walsh, *J. Chem. Soc. (London)*, **1953**, 2260, 2266, 2288, 2296, 2301, 2306 (1953); *Ann. Rep. Chem. Soc.*, **63**, 44 (1966).

209. R. S. Mulliken, *Rev. Mod. Phys.*, **14**, 204 (1947).
210. H. H. Schmidtke and H. Preuss, *Z. Naturforschg.*, **16a**, 790 (1961).
211. P. C. H. Jordan and H. C. Longuet-Higgins, *Mol. Phys.*, **5**, 121 (1962).
212. C. A. Coulson and A. H. Neilson, *Disc. Faraday Soc.*, **35**, 71 (1963).
213. P. C. Jordan, *J. Chem. Phys.*, **41**, 1442 (1964).
214. J. O. Hirschfelder, H. Eyring and B. Topley, *J. Chem. Phys.*, **4**, 170 (1936).

4 Distorted J–T systems

Outline

In this Chapter we take a step towards reality by considering the effect of a distortion on the ideal vibronic systems of the previous Chapter. The distortion, if small, can be regarded either as a perturbation on or as a diagnostic of the vibronic system. Examples of distortions are: externally induced macroscopic strains, low symmetry crystal fields and local random disturbances.

4.1 The description of strain in vibronic systems

The application of axial stress has been one of the most powerful means in the study of J–T centres. The reason for this is that a moderate stress generally suffices to render the minima of the centres markedly inequivalent, without affecting most other parameters of the centre. Thus for a uniaxial compression of about half a kilobar and elastic moduli of the order of 10^{-12} cm²/dyne ($\sim 10^{-6}$ bar^{-1}) as in silicon the minima of the Jahn–Teller centre formed by the group V atom and a vacancy (§7.4) become split by about 30 cm^{-1}. [1-2] With this magnitude of splitting, in many cases the ground state becomes localized. It may be remarked, however, that frequently even much smaller differences suffice to localize the system in one well. Thus for the J–T-system of $Cu^{2+} \cdot (H_2O)_6$ calculated by Öpik and Pryce[3] (according to whom the parameters of §3.2.3 have the values $\alpha = 8$ cm^{-1}, $\beta = 600$ cm^{-1}) a splitting of 10^{-3} cm^{-1} achieves in the ground state a 99·99% localization in one well.

In spite of the importance of strains for the J-T-E, the theory is in its initial stages. We shall outline a commonly adopted approach. Figure 4.1 demonstrates the philosophy behind this approach.

It is supposed that the external force is transmitted by the lattice (represented by the outer circle in the figure) to the J-T centre through the localized modes (middle circle). A simple representation of the interaction between the lattice and localized modes is by a displaced harmonic oscillator

$$\tfrac{1}{2}\hbar\omega_\varepsilon(q_\theta - q_\theta^0)^2 + \tfrac{1}{2}\hbar\omega_\varepsilon(q_\epsilon - q_\epsilon^0)^2 + \tfrac{1}{2}\hbar\omega_\tau[(q_\xi - q_\xi^0)^2 + \cdots]$$

The term expressing the coupling is found, after expanding the first two parentheses, as

$$-\hbar\omega_\varepsilon(q_\theta q_\theta^0 + q_\epsilon q_\epsilon^0) \tag{4.1}$$
$$\equiv -(\gamma_\theta q_\theta + \gamma_\epsilon q_\epsilon)$$

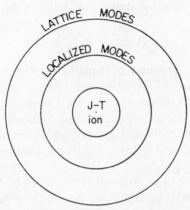

Figure 4.1 The mutual interactions of the lattice, the localized modes and the electron cloud according to the model described in the text. The localized modes are acted upon by the lattice modes from the front and (independently) by the electrons from the rear.

As defined here $|\gamma_\theta|$, or $|\gamma_\epsilon|$ will be equal to the strain energy for a deformation whose magnitude equals the zero point motion amplitude of the θ or ϵ-mode. (This strain energy is a much more 'physical' quantity than the usually employed strain energy for unit strain.)

There are similar expressions for the τ-modes, which shall however be disregarded for the present. If we assume that the force constants of the J-T centre are the same as those of the host material then the displacements q_θ^0, q_ϵ^0 can be expressed in terms of the strains e_θ, e_ϵ that would be present in the absence of the J-T centre.[4]

$$q_\theta^0 = \sqrt{\frac{M\omega D^2}{\hbar}}\, e_\theta, \tag{4.2}$$

$$q_\epsilon^0 = \sqrt{\frac{M\omega D^2}{\hbar}}\, e_\epsilon,$$

where D is a geometric factor. [For octahedral, tetrahedral, cubic complexes D is $3^{-\frac{1}{2}}2$, $2^{-\frac{1}{2}}2/3$, $4/3$ times the ligand-metal distance, respectively.]

The other interaction affecting the modes of the centre is due to the electrons (the inner circle). This is supposed to act independently of the localized modes-lattice coupling and is expressed, as usual, by the J-T-Hamiltonian [e.g., for $E \otimes \varepsilon$ in cubic symmetry, by equation (3.4)].

It was suggested by Ham[4] that the coupling shown by equation (4.1) can be expressed in an equivalent way through an interaction between the outer and inner circles in Figure 4.1 in terms of the splitting of the doublet state caused by the external strain. This splitting can be written as

$$\frac{L}{2}\sqrt{\frac{M\omega D^2}{\hbar}}\,[e_\theta\sigma_\theta + e_\epsilon\sigma_\epsilon] \tag{4.3}$$

[L is the coupling constant of equation (3.4).] The equivalence was shown by Ham to hold for purely linear vibronic coupling and for strong linear coupling in the presence of non-linear terms. [In one sense equation (4.3) is superior, since an expression like equation (4.1) does not allow the separation of the radial and angular motions in the q_θ, q_ϵ-plane, which was achieved in equation (3.13) by the transformation matrix S applied to equation (3.5).]

An amended form of equation (4.1) was given by Williams et al.[5] who related the coefficients γ_θ and γ_ϵ to the elastic properties of the centre rather than to those of the host. But even if the two happen to be the same there are difficulties with the formulation proposed above. These reflect the inadequacy of the localized mode picture (a subject discussed in §3.5.2), which is the basis of our treatments of J-T-centres in solids.

It must be realized that a vibronic state in a J-T-centre will not have the elastic properties of the host, even if the force constants of the centre are the same. In other words, a J-T-centre is necessarily an elastic inhomogeneity. Moreover, the deformations due to the stress and to the J-T-E are not additive. To see this let us return to Figure 1.1. This figure showed calculated curves for the mean local strain around the J-T-centre as function of the strain energy which would exist in the absence of the J-T-centre. The curves were drawn for two cases, which we shall presently discuss. Let us note now that the curves do not have a constant slope but are strongly non-linear (in γ_θ), although for the curves pertaining to strong coupling there exist two linear regions: an initial soft region and a later normal one. The non-linearity will also be apparent in the analytical expressions to be given later.

The two parts of the figure are representative of the two classes of simple J-T-systems: the curves in Figure 1.1(a) are typical of cases in which the electronic states in different potential wells are not orthogonal, e.g. in $E \otimes \varepsilon$ (for which the curves were calculated), $T \otimes \tau$, etc. Figure 1.1(b) typifies $(A + B) \otimes \beta$ (the case actually computed), $T \otimes \varepsilon$, etc., in which electronic states in different wells are originally orthogonal; there are, however, perturbations (spin–orbit coupling, trigonal fields) whose strength is represented in the figure by $W (=1)$, which admix the orthogonal states so that the proper eigenstates will be combinations of the original states possessing certain symmetry properties.

For either class the steep, initial portion of the curves appropriate to strong coupling bears witness to what was termed in the introduction (Chapter 1) one of the three basic properties of strongly coupled J-T-systems: the sudden, nearly spontaneous deformation of the system in response to a small perturbation. Physically, this deformation arises from the rapid escape of the system from the higher wells to a lower well. (In the $E \otimes \varepsilon$ and $T \otimes \varepsilon$ cases this deformation is representable as an orientation in the q_θ, q_ϵ-plane.) The slope of the curve is then inversely proportional to the overlap between the vibrational states in different wells, thus it is highly non-linear in the J-T distortion. For values of the strain energy (the abscissa) of the order of the tunnelling energy there is a break in the slope of the curve: at this point the angular orientation is complete and from now on the deformation is a radial process, i.e. the normal process whereby a strain is set up by an uniaxial stress in a non-degenerate situation.

We continue with the study of the effects of stress on $E \otimes \varepsilon$ systems. For purely linear coupling in the presence of a tetragonal strain the following zero-order states should be used [see §3.2.1].

$$\begin{pmatrix} \cos \\ \sin \end{pmatrix} j\phi, \qquad j = \tfrac{1}{2}, \tfrac{3}{2}, \dots$$

The expectation values of the strain and of the tetragonal strain–energy, equation (4.1) with $\gamma_\epsilon = 0$, are:

$$\left.\begin{aligned} \langle q_\theta \rangle = (\pm)\tfrac{1}{2}q_0, \ \langle E \rangle = (\pm)\tfrac{1}{2}\gamma_\theta q_0 \ \text{ in } \begin{pmatrix} \cos \\ \sin \end{pmatrix}\tfrac{1}{2}\phi \\ = 0 \qquad\qquad = 0 \qquad\qquad \text{ in all other states} \end{aligned}\right\} \quad (4.4)$$

(q_0 is given in equation (3.15)).

There are also non-diagonal terms of the energy connecting $\begin{pmatrix} \cos \\ \sin \end{pmatrix} j\phi$ and $\begin{pmatrix} \cos \\ \sin \end{pmatrix} j \pm 1)\phi$. A complete solution of the vibronic problem which includes these has not yet been given for linear coupling.

We return now to strong linear and non-linear coupling. In the following figures (Figures 4.2 (a), (b) and (c)) we show what happens to the cubic levels of Figure 3.8 when a tetragonal stress is applied. The curves

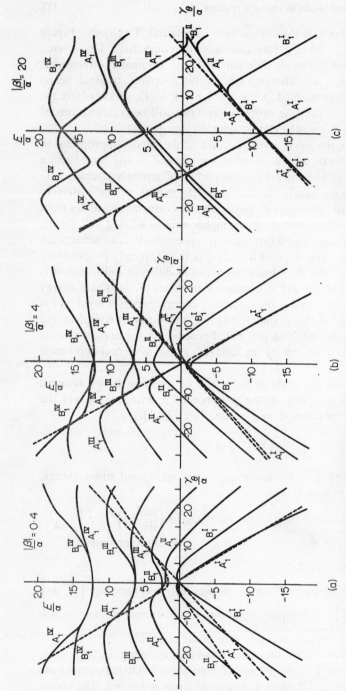

Figure 4.2 Vibronic energy levels *vs.* tetragonal strain energy γ_θ for $E \otimes \varepsilon$. Energies in units of α, equation (3.22), are plotted as function of γ_θ/α, for three values of the non-linear coupling strength $|\beta|/\alpha$:

0.4	(a)
4.0	(b)
20.0	(c)

On the right one potential well (that at $\phi = 0$) is depressed and two are raised, on the left two are lowered and one is raised. The states are classified in D_4 symmetry. When more than one state belongs to the same species, a roman numeral superscript is added for distinction. The results of the three state approximation are shown by broken lines. (R. Englman and B. Helperin, *Phys. Rev.*, **B2**, 75 (1970).)

are shown for the values $|\beta|/\alpha = 0.4$, 4 and 20 of the non-linear coupling strength.

Let us interpret the curves of Figure 4.2, noting that the downward or upward slope of each curve at any value of the abscissae gives the relative weight for the system being, for that state and that value of γ_θ, in the well $\phi = 0$ or $\phi = \pm 2\pi/3$. Consider for example in Figure 4.2(a) the excited state whose level is at $E/\alpha \sim 2.5$. For zero or low tetragonal field strength this level has negligible slope: The state is approximately evenly distributed in the three wells. As the tetragonal field strength grows, the asymmetry of the potential wells becomes perceptible and the state becomes localized in one $(\phi = 0)$ or in two $(\phi = 2\pi/3)$ wells. The state in question acquires an upward tend because it belongs to the latter class. Subsequently, the level sinks down into the well at $\phi = 0$ and the curve bends over to a steady downward slope. In this last region the downward sloping levels are the uniformly spaced harmonic oscillator levels of the $\phi = 0$ well.

An experimental replica of Figure 4.2(b), in which $|\beta|/\alpha = 4$, is seen in the absorption lines of SrF_2: Eu^{2+} when a [001]-stress P is applied[6] (Figure 4.3). Here the zero stress splitting 3Γ equals $6.5\ cm^{-1}$. Then, assuming $|\beta|/\alpha = 4$, α (the rotational quantum in the q_θ, q_ϵ-plane) works out from Figure 3.9 to be about $8\ cm^{-1}$. Writing $\gamma_\theta = AP$, from a comparison of the theoretical and experimental curves one obtains for the quantity $A \sim 1\ cm^{-1}/(kg/mm^2)$, which is reasonable.[6] The value derived for the coupling constant L is of the order of $10^3\ cm^{-1}$.

The stress spectrum of $KMgF_3$:V^{2+} [7] is discussed on p. 144.

Before going on with the consideration of any real cases let us return to the three state model (§3.2.4), in which only the lowest three states $|\Psi_\theta\rangle$, $|\Psi_\epsilon|$, $|\Psi_{A_1}\rangle$ are included. (The energy levels of this model were shown by broken lines in Figure 4.2.) The matrix which arises from the presence of the strain components $q_\theta^0 = \gamma_\theta/\hbar\omega$ and $q_\epsilon^0 = \gamma_\epsilon/\hbar\omega$ is

$$
\begin{array}{c}
|\Psi_{A_1}\rangle \\
|\Psi_\theta\rangle \\
|\Psi_\epsilon\rangle
\end{array}
\begin{bmatrix}
3\Gamma & \gamma_\theta r & \gamma_\epsilon r \\
\gamma_\theta r & -\gamma_\theta q & \gamma_\epsilon q \\
\gamma_\epsilon r & \gamma_\epsilon q & \gamma_\theta q
\end{bmatrix}
\tag{4.5}
$$

Here q and r are the reduction factors defined in equations (3.30) and (3.31), 3Γ is the E–A_1 separation in O-symmetry.

We consider strains of symmetries D_4 and D_2 separately.

D_4:

When $\gamma_\epsilon = 0$ the matrix of equation (4.5) can be diagonalized analytically to yield the states ψ_1, ψ_2 and ψ_3 whose energies (measured from the

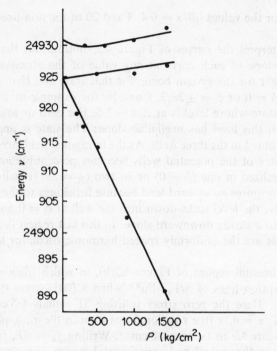

Figure 4.3 The three lowest levels of the $4f^6 5d$ configuration of Eu^{2+} in SrF_2 under a uniaxial compression P in the [001]-direction. (From Reference 6.) The points are experimental points, the lines represent the (3-state) phenomenological theory of Reference 6.

level of the unstressed doublet level) and expectation values are

$$E_1 = \gamma_\theta q$$

$$\begin{pmatrix} E_2 \\ E_3 \end{pmatrix} = \tfrac{1}{2}(3\Gamma - \gamma_\theta \cdot q) \pm \tfrac{1}{2}[\gamma_\theta^2(q^2 + 4r^2) + 2(3\Gamma) \cdot \gamma_\theta \cdot q + (3\Gamma)^2]^{\tfrac{1}{2}}$$

$$\langle \psi_1 | \cos\phi | \psi_1 \rangle = -q$$

$$\langle {}^{\psi_2}_{\psi_3} | \cos\phi | {}^{\psi_2}_{\psi_3} \rangle = \tfrac{1}{2}q \mp \frac{(3\Gamma) \cdot q + \gamma_\theta(q^2 + 4r^2)}{2[\gamma_\theta^2(q^2 + 4r^2) + 2(3\Gamma)\gamma_\theta \cdot q + (3\Gamma)^2]^{\tfrac{1}{2}}}$$

The following limiting expressions for the three energies are of interest.
For strong non-linear coupling ($q \sim \tfrac{1}{2}$, $|r| \sim 1/\sqrt{2}$) and weak tetragonal field, $3\Gamma \gg \gamma_\theta$:

$$E_i = (3\Gamma, +\tfrac{1}{2}\gamma_\theta, -\tfrac{1}{2}\gamma_\theta)$$

For very strong coupling ($q \sim \tfrac{1}{2}$, $|r| \sim 1/\sqrt{2}$) and strong tetragonal field, $3\Gamma \ll \gamma_\theta$

$$E_i = (2\Gamma + \tfrac{1}{2}\gamma_\theta, \tfrac{1}{2}\gamma_\theta, \Gamma - \gamma_\theta) \qquad (4.6)$$

From the last expression we recapture (if we neglect the small tunnelling energy 3Γ) the levels of three systems localized in the wells $\phi = 0$, $+2\pi/3$, $-2\pi/3$, in other words, along the directions z, x, y, respectively. When a tetragonal z-oriented stress is applied, two systems rise as $\frac{1}{2}\gamma_\theta$ and one sinks as $-\gamma_\theta$.

Computed reduction factors q in D_4-symmetry are shown as function of the tetragonal strain in the next figure (Figure 4.4). Since in D_4 the three matrix-elements in equation (3.37) are no longer equivalent we have to define three q's as follows:

$$q_1 = -\langle \Psi'_\theta | \sigma_\theta | \Psi'_\theta \rangle$$

$$q_2 = \langle \Psi'_\epsilon | \sigma_\theta | \Psi'_\epsilon \rangle$$

$$q_{12} = -\langle \Psi'_\epsilon | \sigma_\epsilon | \Psi'_\theta \rangle$$

These are shown in Figure 4.4.

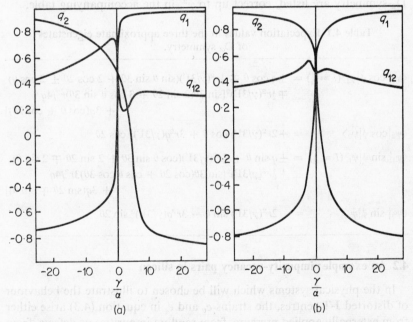

Figure 4.4 The reduction factors q_1, q_2, q_{12} defined in the text, as function of the tetragonal strain energy. The curves were computed (by B. Halperin, 1969) using the eigenstates of the Schrödinger equation, [equation (3.21)], to which the tetragonal strain energy [equation (4.1) with $e_\epsilon = 0$] was added.

(a) $\beta/\alpha = 0.4$, (b) $\beta/\alpha = 4$.

Feher and coworkers[8] observed an anisotropic ESR absorption due to holes bound to group III acceptors (B, Al, Ga or In) in silicon when this was compressed axially. As will be pointed out in §7.4, the holes belong to the $G_{\frac{3}{2}}$-representation (of T_d) which, if subject to $G_{\frac{3}{2}} \otimes \varepsilon$ only, behaves as an E-multiplet. The line in Reference 8 was found to shift to lower magnetic field as the stress was increased. This was interpreted[9] as an increase in the anisotropic part of g or in q in accordance with Figure 4.4. Numerically, g changed by only about a few per cent, as the stress increased three-fold to 900 kg/cm^2.

D_2:

In the presence of an orthorhombic distortion

$$\left.\begin{aligned}\gamma_\theta &= \gamma \cos \theta \\ \gamma_\epsilon &= \gamma \sin \theta\end{aligned}\right\} \tag{4.7}$$

the expectation values in the three approximate eigen-states ψ, ψ_2 and ψ_3 of D_2-symmetry are listed, correct up to γ^2, in the accompanying table.

Table 4.1 Expectation values in the three approximate eigenstates of D_2 symmetry.

$$\begin{aligned}
\langle\psi_i|\cos\phi|\psi_i\rangle(i=\tfrac{1}{2}) &= \pm q \cos\theta \mp \tfrac{1}{2}r^2(\gamma/3\Gamma)(\sin\theta\sin 3\theta + 2\cos 2\theta \pm 2\cos\theta) \\
&\quad \mp\tfrac{1}{2}r^2(\gamma/3\Gamma)^2[\sin 3\theta(2\sin 2\theta + 3\cos\theta\sin 3\theta)r^2/4q \\
&\quad\quad\quad\quad + 3q(\cos\theta \pm \cos 2\theta)] \\
\langle\psi_3|\cos\phi|\psi_3\rangle &= +2r^2(\gamma/3\Gamma)\cos\theta + 3r^2q(\gamma/3\Gamma)^2\cos 2\theta \\
\langle\psi_i|\sin\phi|\psi_i\rangle(i=\tfrac{1}{2}) &= \pm q\sin\theta \pm \tfrac{1}{2}r^2(\gamma/3\Gamma)(\cos\theta\sin 3\theta + 2\sin 2\theta \mp 2\sin\theta) \\
&\quad +\tfrac{1}{2}r^2(\gamma/3\Gamma)^2[\sin 3\theta(\cos 2\theta + \cos\theta\cos 3\theta)3r^2/4q \\
&\quad\quad\quad\quad + 3q(\sin 2\theta \mp \sin\theta)] \\
\langle\psi_3|\sin\phi|\psi_3\rangle &= 2r^2(\gamma/3\Gamma)\sin\theta - 3r^2q(\gamma/3\Gamma)^2\sin 2\theta
\end{aligned}$$

4.2 An example: impurity-vacancy pairs in silicon

In the physical systems which will be chosen to illustrate the behaviour of distorted J-T centres, the strains e_θ and e_ϵ in equation (4.3) arise either from externally applied pressure, from random impurities or defects, from macroscopic spontaneous deformation of the host crystal or from a low symmetry (e.g. trigonal) crystal field.

In Chapter 7 we shall describe from the point of view of the J-T-E the centre, formed upon irradiation of a diamond-type lattice, which consists of a vacancy trapped by a group V atom (e.g., P, As, Sb) (§7.4). The application of a stress in the [110]-direction of the Si lattice makes [−112] a natural direction of quantization in the plane of the triangle formed from the three silicon atoms, (b) (c) (d) in Figure 7.3. A small stress would shift (in compression: lower) the E_θ-vibronic level and raise by different amounts the singlet and E_ϵ levels. In the experiments of Watkins and coworkers[1-2] the crystal was compressed by a stress of the order of 500 kg/cm² and it was found that for each ESR line that gained in intensity there were two lines which, as far as could be ascertained, lost intensity by equal amounts. One concludes that the case of equation (4.6) is applicable and that the tunnelling splitting 3Γ is smaller than the strain energy γ_θ. (The latter was about 15 cm⁻¹ for As and Sb for the quoted magnitude of stress.)

By measuring the temperature dependence of the coalescence of the anisotropic ESR lines (this subject will be discussed in Chapter 6) Watkins and coworkers were already in the possession of information that their systems were subject to a strong non-linear vibronic coupling. They found[1-2] that the distortion of the silicon triangle, caused by random inhomogeneties near the impurity-vacancy centre, reoriented at a rate which depended on the temperature T exponentially. One derives from this reorientation rate a barrier of about 500 cm⁻¹. This quantity corresponds to 2β of §3.2.3, if the case of E ⊗ ε is strictly applicable to these systems. (This is questionable, since P-J-T coupling is likely to connect the ground state to crystal field terms of other symmetries or to other configurations of the same symmetry). The vibrational energy $\hbar\omega$ was estimated in Reference 2 as 270 cm⁻¹ and α $(=(\hbar\omega)^2/4E_{JT}) = 5$ cm⁻¹, so that the use of the strong non-linear coupling model is consistent with the data. For further numerical details, Appendix IX may be read.

4.3 The *g*-factor for strong non-linear E ⊗ ε coupling and in the presence of strain

The treatment of the spin Hamiltonian in §3.2.5 is frequently subject to the following modifications. First, when the non-linear coupling is strong, then the first vibronic singlet is brought down to a small distance 3Γ above the ground vibronic doublet. If 3Γ is comparable to (or smaller than) the anisotropic part of the Zeeman energy, i.e. comparable to the coefficient $G(E_\theta)$ or $G(E_\epsilon)$ introduced in equation (3.34), then the presence of the singlet cannot be neglected, since the anisotropic part will couple the singlet to the doublet. For a magnetic field 4 kilogauss strong, the anisotropic term is of the order of 0·02 cm⁻¹ and this is expected to exceed 3Γ for $\alpha \sim 10$ cm⁻¹, as long as $\beta/\alpha > 35$ (Table 3.1). In general there exists, however, a more forceful argument for the inclusion of the nearby singlet state, namely, the coupling of this state to the doublet states by a low-symmetry strain field acting on the paramagnetic centre. For example, a predominantly tetragonal field associated with a defect along or near a cubic axis of the centre, or an extended defect oriented along a {100} direction will couple the A_2 state with E_ϵ or A_1 (if this is the nearby singlet) with E_θ. Quantitatively (as is seen from the low-lying curves of Figure 4.2), the tetragonal strain energy of the defect will have to be comparable to or stronger than 3Γ for the effect of the singlet on the doublet to be felt. For strong non-linear coupling, such that $\beta/\alpha > 15$ and $3\Gamma \lesssim 1$ cm^{-1} this condition is generally met, since the characteristic energies associated with random static strains are a few wave-numbers in the commonly employed crystals. But even if 3Γ is larger than the strain energy, it is still necessary to include the strain in the study of the angular dependence of ESR, since the anisotropic Zeeman energy is likely to be smaller than the strain energy. [This point of view was advocated by Chase (1967, unpublished) and by Ham.[4]] Summarizing this part we say, that in the prediction of the ESR spectra in an $E \otimes \varepsilon$ situation random strain practically always needs to be included. The truly variable (and often questionable) parameter is the ratio of the tunnelling energy 3Γ to the low symmetry field strength. Below, we shall show some graphs (Figure 4.6) for the angular dependence of some ESR lines, in which this ratio is varied in some rather restricted manner. To give an account of the angular variation in a comprehensive manner could be very involved indeed.

To include the strain field and the singlet state it is best to start with H(elect.) given in equation (3.34), and to extend the meaning of two G coefficients as follows:

$$G(E_\theta) = \tfrac{1}{2}\beta\sum_{ij}S_i(g_{ij}^{\epsilon\epsilon} - g_{ij}^{\theta\theta})H_j + \gamma_\theta q_0$$

$$G(E_\epsilon) = \beta\sum_{ij}S_i g_{ij}^{\upsilon\epsilon}H_j + \gamma_\epsilon q_0 \tag{4.8}$$

where q_0 is the displacement in the reduced normal coordinate due to the linear J-T-E [equation (3.15)] and γ_θ, γ_ϵ are strain energies associated with the strains e_θ, e_ϵ [equations (4.1) and (4.2)]. The definitions of $G(E_\theta)$ and $G(E_\epsilon)$ will be later further extended when the nuclear spins are included (§6.5).

We use now the equivalence between equations (4.1) and (4.3) and recall the definitions of g_1 and g_2 from equation (3.43). We derive:

$$H(\text{elect.}) = g_1\beta(\mathbf{S}\cdot\mathbf{H}) - (2/3)g_2\beta S_z H\left\{\left(n^2 - \frac{l^2 + m^2}{2}\right)\cos\phi\right.$$

$$\left. + \frac{\sqrt{3}}{2}(l^2 - m^2)\sin\phi\right\} - q_0(\gamma_\theta\cos\phi + \gamma_\epsilon\sin\phi) \tag{4.9}$$

In this equation the magnetic field was written in terms of its direction cosines with respect to the z-axis, i.e. $\mathbf{H} = H(l, m, n)$. The spin in the anisotropy term was quantized along the direction, say z', of the magnetic field and the components of the spin operator perpendicular to this direction were neglected. This assumption will be revoked later in a particular case (when $H\|(1, 1, 1)$ as in §6.5). We note that the curly brackets can also be written in a more symmetric form.

$$\left\{ l^2 \cos\left(\phi - \frac{2\pi}{3}\right) + m^2 \cos\left(\phi + \frac{2\pi}{3}\right) + n^2 \cos\phi \right\} \qquad (4.10)$$

At this stage we recall the definitions of the reduction factors q[equation (3.30)] and r' [equation (3.31)] and rewrite H(elect.) as a matrix in the function space of the lowest three vibronic states (§3.2.4)

$$
\begin{array}{l}
|\Psi_{A_2}\rangle: \\
\\
\\
|\Psi_\varepsilon\rangle: \\
\\
\\
|\Psi_\theta\rangle:
\end{array}
\left[
\begin{array}{l}
3\Gamma \qquad g_2\beta S_{z'}H\left(n^2 - \dfrac{l^2 + m^2}{2}\right)r' + \gamma_\theta q_0 r' \\
\\
\qquad\qquad\qquad -g_2\beta S_{z'}H\dfrac{\sqrt{3}}{2}(l^2 - m^2)r' - \gamma_\varepsilon q_0 r' \\
\\
g_2\beta S_{z'}H\left(n^2 - \dfrac{l^2 + m^2}{2}\right)q + \gamma_\theta q_0 q \\
\\
\qquad\qquad\qquad g_2\beta S_{z'}H\dfrac{\sqrt{3}}{2}(l^2 - m^2)q + \gamma_\varepsilon q_0 q \\
\\
-g_2\beta S_{z'}H\left(n^2 - \dfrac{l^2 + m^2}{2}\right)q - \gamma_\theta q_0 q
\end{array}
\right]
$$

$$(4.11)$$

The matrix given above is for the case of d^9 in six co-ordinated surroundings, when the neighbouring singlet is A_2. If, instead, this singlet is A_1 then the matrix elements of this state (in the first row or column) with E_θ and E_ε have to be interchanged compared to those of A_2 shown above, replacing also r' by r [equation (3.31)] and exchanging the negative signs in the (A_2, E_θ) position above for pluses [cf. equation (4.5)]. We now recall from our earlier discussion that one has very often

$$g_2\beta H \ll |\gamma|q_0, \qquad g_2\beta H \ll 3\Gamma$$

Regarding the relative values of the strain energies γ_θ and γ_ε, when the strains are due to distant point defects we can expect a random distribution of the angle θ associated with the strain orientations [equation (4.7).] The line shape due to such inhomogeneous strain was derived by Ham[4] for a vibronic doublet only, i.e. when $|\gamma|q_0 \ll 3\Gamma$, and for linear coupling.

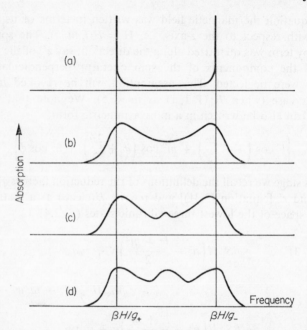

Figure 4.5 The theoretical line shape in ESR of a doublet subject to linear $E \otimes \varepsilon$ and random strain. Curve (a) is the superposition of spectra having g-values[4] $g = g_1 + (g_\pm - g_1) \cos(\alpha - \theta)$, where $g_\pm$ are given in equation (3.42)

$$\alpha = \tan^{-1} \frac{\sqrt{3}(l^2 - m^2)}{2n^2 - l^2 - m^2}$$

is the magnetic field orientation angle and the strain orientation (θ) is randomly distributed. Curve (b) is a broadened version of curve (a). Curve (c) shows the beginning of the coalescence and motional narrowing of the vibronic-Zeeman levels due to relaxation.[10] In (d) the coalescence is more advanced.

This line-shape is shown in Figure 4.5 together with a broadened version of it[10] which will be closer to reality owing to time dependent processes. The central hump on the lower curves is also due to these processes, i.e. to relaxation effects which will be discussed later in Chapter 6.

The extremities of the band are given by the two g-values in equation (3.43). It was suggested by Ham[4] that the line shapes of the anisotropic ESR-spectra of Cu^{2+} in the octahedral environment of MgO^{11} and of the cubically coordinated Sc^{2+} in CaF_2 and SrF_2 [12-3] arise from a weak or moderate linear J-T-E whose doublet ground states Ψ_θ, Ψ_ϵ are split by a random strain distribution.

Somewhat similar line shapes to that shown in Figure 4.5 were also found in the acoustic paramagnetic resonance (a.p.r.) spectrum of Cr^{2+} in MgO.[14] In a.p.r. experiments the transitions between the lines split by the magnetic field are induced by an acoustic wave (instead of an r.f. magnetic field), whose attenuation is then measured. The line shapes were interpreted by Fletcher and Stevens[15] as due to a distribution of tetragonal strains e_θ, with the maxima of the attenuation appearing at the 'critical points' of the H vs. e_θ curve. These were found by numerical diagonalization within the spin quintets of the vibronic singlet and doublet. F. S. Ham (1971, private communication) working within the vibronic doublet only and using a perturbational approach (in H) derived critical points in the distribution of e_θ and e_ε, agreeing with peaks of the a.p.r. data.[14]

The line shape due to the three-states of equation (4.11) with a generally oriented strain field has been computed by Chase[16] for Eu^{2+} in fluorite.

For extended defects, e.g. edge dislocations, which are preferentially oriented along (say) {100} the tetragonal component of the strain is larger on the average than the orthorhombic component, although the direction of the tetragonal strain is equally distributed among the three cubic axes. Considering this case, we can factor the matrix of equation (4.11).

$$|\Psi_{A_2}\}: \left[3\Gamma \quad g_2\beta S_{z'}H\left(n^2 - \frac{l^2+m^2}{2}\right)r' + \gamma_\theta q_0 r' \atop g_2\beta S_{z'}H\left(n^2 - \frac{l^2+m^2}{2}\right)q + \gamma_\theta q_0 q \right] + g_1\beta(\mathbf{S}.\mathbf{H})$$

$$|\Psi_\varepsilon\}: \tag{4.12}$$

$$|\Psi_\theta\}: \quad g_1\beta(\mathbf{S}.\mathbf{H}) - g_2\beta S_{z'}H\left(n^2 - \frac{l^2+m^2}{2}\right)q - \gamma_\theta q_0 q \tag{4.13}$$

The anisotropic parts $g - g_1$ of the calculated g-factors arising from such a situation are shown in the following three drawings (Figure 4.6). The magnetic field lies in the (110)-plane, its direction is given by the abscissa, the angle ϕ, which is 0° for $H\|[001]$ and 90° for $H\|[1-10]$. The figures are for a d^9 configuration with $q = \frac{1}{2}$ and $r' = 0.707$, and using the parameters

$$\gamma_\theta q_0 = -1 \text{ cm}^{-1}$$

$$\beta H = 0.2 \text{ cm}^{-1} \quad \text{or} \quad H = 4 \text{ kilogauss}$$

$$3\Gamma = 0.1, 1, 15 \text{ cm}^{-1}$$

The three choices of the tunnelling splitting are appropriate to hydrated $La_2Mg_3(NO_3)_{12}:Cu^{2+}(3\Gamma = 0.1 \text{ cm}^{-1})$,[4,17] $Al_2O_3:Ni^{3+}(3\Gamma = 1 \text{ cm}^{-1})$[4,18] and $CaF_2:Eu^{2+}(3\Gamma \sim 15 \text{ cm}^{-1})$.[6,16]

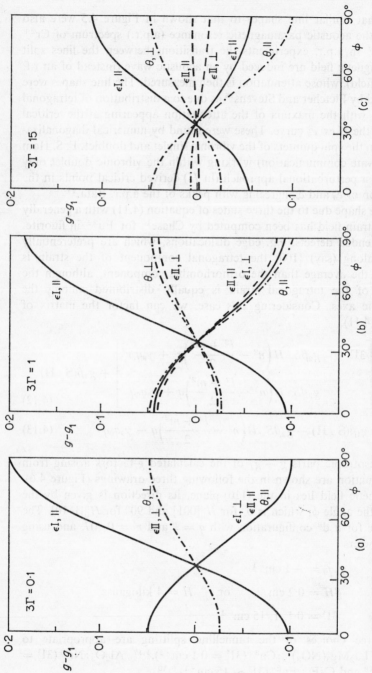

Figure 4.6 Calculated g-factors for Cu²⁺ vs. the orientation of the magnetic field. The three state model is used and a tetragonal field given by $\gamma\theta\theta_0 = -1\ \mathrm{cm}^{-1}$ is assumed to be present. The angle ϕ describes the deflection of the magnetic field in the (110)-plane from [001]. Other parameters are $\beta H = 0.2\ \mathrm{cm}^{-1}$ and

$3\Gamma = 0.1\ \mathrm{cm}^{-1}$ $3\Gamma = 1\ \mathrm{cm}^{-1}$ $3\Gamma = 15\ \mathrm{cm}^{-1}$
(a)   (b)   (c)

There are three sets of g-values (designated εI, εII and θ) in accordance to the equations (4.12–3) from which they arise.

For each choice of parameters there are three sets of g-values (labelled in the figures ϵI, θ, ϵII), each set consisting of an in-plane g-value (corresponding to the tetragonal strain field lying along [001]) and two out-of-plane g-value (strain along [100] and [010]). The strengths I of the sets of lines can be calculated. They satisfy the relations

$$I(\epsilon\mathrm{I}):I(\theta):I(\epsilon\mathrm{II}) = 1 : \exp - \frac{1}{2kT}\{-3\Gamma + 3\gamma_\theta q_0 +$$

$$+ \sqrt{[(3\Gamma + \tfrac{1}{2}\gamma_\theta q_0)^2 + 2(\gamma_\theta q_0)^2]}\} :$$

$$\exp - \frac{1}{kT}\sqrt{[(3\Gamma + \tfrac{1}{2}\gamma_\theta q_0)^2 + 2(\gamma_\theta q_0)^2]}$$

It will be noted that the $3\Gamma = 0\cdot1$ curves are indistinguishable from those in a typical tetragonal spectrum [due to a pair of electronic states $(\phi_\theta, \phi_\epsilon)$ not coupled vibronically and subject to strong, {001}-oriented tetragonal fields]. In the drawing with $3\Gamma = 15$ (Figure 4.6 (c)) the middle, flat curves are associated with states which possess strong A$_2$ components.

The four lines observed for Al$_2$O$_3$:Ni^{3+} show[18'] similar angular behaviour to ϵI$\|$, ϵI$\perp$, $\theta\|$ and $\theta\perp$ in Figure 4.6(b) in the central portion of that figure, but apparently deviate from it for other angles.

4.4 Ni^{3+} in tetragonal SrTiO$_3$

The ESR studies on Ni^{3+} in SrTiO$_3$ [19-21] are of particular interest not only because in these experiments the low symmetry field in the J-T-centre was supplied by the host crystal itself, but also because the local stress and strain (the abscissa and ordinate of Figure 1.1) were established independently. The perovskitic crystal SrTiO$_3$ undergoes a cubic to tetragonal phase transformation near 106 °K, in the course of which the TiO$_6$ units rotate slightly around the new tetragonal axis and the oxygen octahedra become elongated by an amount of $c/a - 1 \approx 1 \times 10^{-4}$, at $T = 77$ °K. The temperature dependent strain $e_\theta(T)$ in the TiO$_6$ octahedra was determined by a combination of sources,[20-3] which occasionally checked one another, and is shown in Figure 4.7.

If now the J-T ion Ni^{3+} (t^{6}e configuration, ^{2}E state) is introduced substitutionally for Ti^{4+}, the ESR spectrum will show a behaviour in accordance with the spin-Hamiltonian of equation (4.9). Note however that for a t^{6}e ion we have $L > 0$, $\beta > 0$ so that (Table 3.2) the electronic ground state in a stable distortion ($\phi = 0$) is $|E_\theta\rangle$. Also the isotropic g-factor g_1 has a slightly changed definition from that given in equation (3.43).[21]

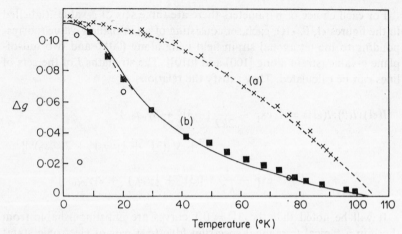

Figure 4.7 'Stress and strain' at the Ni^{3+} site in $SrTiO_3$.
(a) Experimental points (×) and a fitted curve – – for the tetragonal
local field (stress) as derived from the rhombic splitting of ESR of a
Fe^{3+}-vacancy centre. (From Reference 21.)
(b) Experimental points (■ from Reference 23, ○ from Reference 24)
for Δg (difference between perpendicular and parallel g-values) of
$SrTiO_3$: Ni^{3+}. The curve – · – is a two parameter fit for linear J-T-
coupling, the curve ⎯⎯ for strong non-linear coupling. The data of
(a) and (b) are matched at low temperatures.

From equation (4.9) one finds for the difference of the g-values perpendicular and parallel to the tetragonal axis of $SrTiO_3$

$$\Delta g = g_2 \langle q_\theta \rangle / q_0 = |4\lambda/\Delta| \langle \cos \phi \rangle \qquad (4.14)$$

where the last form results from the substitution for g_2 from equation (3.43)
(for Cu^{2+} or Ni^{3+}). Also, $\langle \sin \phi \rangle = 0$, by symmetry.

Müller and coworkers[20-1] measured Δg for temperatures up to the
transition point. Their results enable an immediate evaluation of the
magnitude of the local strain $\langle \cos \phi \rangle$ which arises jointly from the J-T-E
and the tetragonal stress in $SrTiO_3$.

The thermal average $\langle \cos \phi \rangle$ can also be evaluated in the limiting cases
of §4.1:

For linear coupling only and utilizing equation (4.4)

$$\Delta g = \left| \frac{2\lambda}{\Delta} \right| \frac{e^{\gamma_\theta q_0/kT} - 1}{e^{\gamma_\theta q_0/kT} + 1}, \qquad (4.15)$$

while for strong non-linear coupling $(3\Gamma \ll |\gamma_\theta|)$, from equation (4.6),

$$\Delta g = \left| \frac{4\lambda}{\Delta} \right| \frac{e^{3\gamma_\theta q_0/(2kT)} - 1}{e^{3\gamma_\theta q_0/(2kT)} + 2} \qquad (4.16)$$

Rewriting the last two results by means of $\gamma_\theta = \hbar\omega\sqrt{\dfrac{M\omega D^2}{\hbar}}e_\theta(T)$,

so that Δg depends on the temperature also through the strain $e_\theta(T)$ that would exist in the absence of the J-T-centre, Slonczewski et al.[21] fitted the temperature dependence of the experimental points (Figure 4.7) by regarding λ/Δ and $D(M\omega^3/\hbar)^{\frac{1}{4}}$ as adjustable parameters. $|\lambda/\Delta|$ was found to be 0·027 if one used equation (4.15) and 0·056 if equation (4.16) was used. Since $|\lambda/\Delta|$ is unlikely to be less than 0·04, the strong non-linear coupling model is preferable. In this model the rather high value of 5×10^3 cm^{-1} is derived for the linear coupling strength L.

4.5 A trigonal field

The six oxygen ions which surround substituted d^n metal ions in Al_2O_3, are distorted from the regular octahedral shape and give rise to a trigonal field at the centre.[25-6] In the infrared spectrum of the d^1 ion Ti^{3+}, splittings δ_1 and δ_2 of magnitudes 37·8 and 107·5 cm^{-1} were observed[27] between the three Kramers doublets into which the ground state term $^2T_2$ is decomposed by the spin–orbit coupling and the trigonal field. The splittings δ_1 and δ_2 were interpreted by Macfarlane et al.[28] in a calculation which included a second order perturbation theory of the Ham effect in $T \otimes \varepsilon$ coupling. (§3.3.3.)

A useful check on the theory is to measure the change δg of the g-factor in the ground state Kramers doublet as a linear function of an electric field $\mathcal{E}$ applied perpendicular to the trigonal axis ($\hat{z}$). Information is gained about the tensor T_{ijk} where i, j and k denote the components of $\mathcal{E}$, of H and of the electronic spin S, respectively. Experimentally[29] δg was found to be 10^{-4} per (kilovolt/cm), with the most significant tensor components being T_{113} and T_{111}. In the theory,[29] δg arises from the cross-term, proportional to $\mathcal{E}H$, in a second order perturbation theory which admixes the ground state Kramers doublets with the two doublets δ_1 and δ_2 higher up. The results thus obtained were comparable to the experimental values, although by a factor 3–4 smaller.

It is reasonable to start a calculation of this nature, as was apparently also done in Reference 29, with zero-order eigenstates which are linear combinations of the three states in equation (3.57) with $v_\theta = v_\epsilon = 0$. Nevertheless, we must be wary of calculating second order contributions to δg with these basic states alone. Instead, wave-functions containing admixtures by the spin orbit coupling (ζ) and the trigonal field (v) from $v_\theta \neq 0 \neq v_\epsilon$ should be included. This type of correction can be made perturbationally in $\zeta/(\hbar\omega)$ and $v/(\hbar\omega)$, as in Reference 29. Alternatively, corrections to all orders in these ratios may be required.[30] This may be prohibitively difficult for $T \otimes \varepsilon$, but we recall from the discussion in

§4.1 the existence of the analogue-system $E \otimes \beta$ for which numerical computation is feasible.

We compare now two calculations for $E \otimes \beta$: The first (i) is based on the set $n_\beta = 0$ (which is analogous to the set $v_\theta = v_\epsilon = 0$) and is perturbational; the second (ii) involves the exact diagonalization of the matrix-Hamiltonian, equation (3.82). In this Hamiltonian W mixes the two states of E and is analogous to either ζ or v. We shall seek then the cross term in second order perturbation theory arising from the perturbation

$$(aE + bH)q \qquad (4.17)$$

where a and b are considered (to keep this exercise as simple as possible) constants. In (i) this perturbation will admix in the ground state the state which lies

$$\delta_1 = 2W \exp\left[-\tfrac{1}{2}L^2/(\hbar\omega)^2\right]$$

above it. To make the comparison to the $T \otimes \varepsilon$ case significant parameters derived from $Al_2O_3:Ti^{3+}$ must be used, e.g. $W/\hbar\omega = 1\cdot2$, $E_{JT}/\hbar\omega = L^2/(2\hbar\omega)^2 = 1$. Then the cross terms evaluated by the two methods are

$$\text{(i)} \quad 18 \times ab \times EH$$
$$\text{(ii)} \quad 9 \times ab \times EH$$

The discrepancy between the two calculations is significant and increases with increasing E_{JT} for a fixed W.

4.6 Zero-phonon lines

For the study of spectral effects in solids which are much smaller than the widths of optical bands, in fact smaller than any expected characteristic structure (due generally to phonons or, possibly, to magnons) on the bands, one turns to zero-phonon lines. These lines, analogous to the Mössbauer-line, appear in solids: the rigider the solid and the weaker the electron-vibrational coupling the more prominent the lines are.

The investigation of zero-phonon lines for the study of the J-T-E was initiated by M. D. Sturge and his review[31] also provides a very useful introduction to much of the earlier work. What manifestations of the J-T-E satisfy the above mentioned criterion of smallness? (This criterion ensures that the spectral details of interest are kept separate from the irrelevant structure on the band). The effects so far investigated consist of stress induced splittings, trigonal field or spin–orbit splitting quenched by the Ham-effect, Zeeman-effect, tunnelling splitting and second order J-T-E. In this section we briefly sketch some of the work on the first two effects.

The trigonal field splitting of the zero phonon line was found to be reduced in $MgO:Ni^{2+}$ [32] through the Ham-effect (operating in $T \otimes \varepsilon$) by a factor of $0\cdot1$. A similar reduction is said to occur in $KMgF_3$ (R. E. Dietz, A. Misetich and F. R. Merritt (unpublished), quoted by Reference 31, p. 204). In the $^3T_2$ state of V^{3+} in Al_2O_3 an even stronger reduction of the trigonal field was seen.[33-4] The factor $0\cdot022$ was quoted by Scott and

Sturge.[34] Later work on polarized spectra of the zero-phonon lines split by the Zeeman-effect[35] led to one-half of this estimate for the quenching factor of the trigonal field and to twice this value for the quenching of spin–orbit coupling. (This was discussed in §3.3.4). The zero phonon lines in Al_2O_3 are differently polarized in the $^4A_2 \rightarrow {}^4T_2$ transition on Cr^{3+} (σ-polarization) and in the $^3T_1 \rightarrow {}^3T_2$ transition on V^{3+} (both σ- and π-polarizations). The difference was first explained[33] as depending on whether the doublet or the singlet is the vibronically stabilized state which is depressed by the trigonal field. In a subsequent interpretation the quantum interference (constructive for Cr^{3+} and destructive for the excited triplet in V^{3+}) between the second order J-T-E and the static trigonal field of Al_2O_3 was proposed tentatively as the cause for the different polarization behaviours.[36] At present, the question may still be regarded as open (M. D. Sturge, private communication, 1971).

The application of unaxial stress on the zero phonon lines of the excited state $^4T_{2g}$ of V^{2+} in MgO[37] gives curious results. In the unstressed crystal two zero-phonon lines (about $40\ cm^{-1}$ apart) are observed. Applying stress one sees that these belong to the same absorbing system (since they borrow intensity from each other) but that they are not the split components of the cubic orbital triplet T_{2g} (the spin–orbit coupling is evidently quenched by the Ham-effect) in a locally arising D_{4h} symmetry, since then a {100} type stress would yield at least four lines instead of the observed three, or the split components in D_{3d}, since then stressing along {111} would show four lines instead of two as seen.[38] Sturge[37] suggested therefore a phenomenological model which reproduced his results, but whose physical origin was and has remained mysterious. According to this model, one sees the T_{2g}-states subject to a strong $T \otimes \varepsilon$ Ham-effect, which cuts out both spin–orbit coupling and the externally applied {111} stress (so that the two lines seen in the absence of this do not get further split) but which still tolerates an effective, and randomly oriented, trigonal strain of the order of $40\ cm^{-1}$. Ham,[39] on the other hand, suggested that the system observed was the mixture of a zero-phonon T_{2g} set and of a one (τ_2)-phonon-T_{2g} vibronic set, altogether 12 states, of which 9 states (three triplets) participate in the absorption process. Detailed calculations based on a quasi-molecular model, assuming a linear $T \otimes \tau_2$ coupling and no interaction with $T \otimes \varepsilon$ coupling,[40] did not support this idea; it is possible, however, that there is a resonant vibrational mode, arising from the substitution of the heavier vanadium ion in place of Mg and having sufficient ε_g-character locally, which can make up the required 9 states (the phonon-less ground state T_{2g}, and one-phonon T_{1g} and T_{2g} vibronic states) within about $40\ cm^{-1}$ of the ground state. Apparently, only in the excited state is this resonant mode sufficiently long-lived for detection.

The position and intensities of the Zeeman spectra of the $^4T_2$ excited state in $KMgF_3:V^{2+}$ were found to agree well with a theory of the Ham-effect based on $T \otimes \varepsilon$ $(E_{JT}/\hbar\omega \sim 0.9)$ with spin–orbit coupling added as a perturbation on the vibronic states.[7] In the double group $^4T_2$ decomposes into two doublets and two quartets $(G_{3/2}^{(1)}, G_{3/2}^{(2)})$, which spread out to an overall separation of $40\ cm^{-1}$ and a $G^{(1)} - G^{(2)}$ separation of $10\ cm^{-1}$. The separations are small compared to the vibronic interaction between the double group terms, so it is not possible to approach this case from the point of view $G_{\frac{3}{2}} \otimes \varepsilon$ $(=E \otimes \varepsilon)$ and indeed a $\{100\}$-stress leads to splittings rather different from those of Figure 4.2.

4.7 Dichroism

Stress induced dichroism was first observed on R-centres in KCl by Silsbee.[41] (The R-centre is described in §7.5 and depicted in Figure 7.4.) In transitions from the ground state ^2E to a number of doublets and singlets a preferred direction of polarization was observed after a uniaxial compressive stress was applied. This had a component along the local y-direction (the crystallographical $[-112]$-direction) and no component along the local x-direction (the $[110]$-direction). In the following figure (Figure 4.8) a portion of the absorption spectrum is shown, together with the difference of intensities, $I_x - I_y$, in the two polarizations and the assignments given by Silsbee.[41]

Comparing the strong y-polarization for the transition to A_2 and the x-polarization for the transition to A_1 with the selection rules in Appendix VII we see that the state (ϕ_ϵ) which is anti-symmetric with respect to reflection in yz-plane is stabilized in the ground state ^2E by the stress.

We will now ask to what extent would the results be modified if the J-T-E were absent? The first answer, the one essentially given by Silsbee, is that in the absence of the J-T-E the polarization would be much stronger, in fact practically 100%, whereas in the observed case it was (for the R_2-band) about 30%. This situation comes about, because without vibronic coupling the applied stress would split apart the purely electronic states and the stabilized state would be entirely $|E_\epsilon\rangle$ (at any rate, at low temperatures); while with vibronic coupling there would inevitably be some admixture of the state $|E_\theta\rangle$ in the vibronic ground state $|\Psi_\epsilon\rangle$. The amount of admixture was shown in curve (c) of Figure 3.44, as function of $2W/(\hbar\omega)$. In the present context $2W$ stands for the splitting of the electronic states by the applied stress.

On the basis of our already quoted values of 0.3 for the polarization Silsbee concluded that $E_{JT}/\hbar\omega > \frac{1}{2}$. (On the other hand, the absence of any thermal excitation to the $j' = \pm\frac{3}{2}$ states of §3.2 or to the vibronic singlets derived from them, indicates that $E_{JT}/\hbar\omega < 4$). Another potential diagnostic for the stabilization of a vibronic state is the application of a stress of opposite sign (namely, an extension along the y-direction). Either in the absence of the J-T-E, or with only linear coupling, the states $|\Psi'_\theta\rangle$ and $|\Psi'_\epsilon\rangle$ simply exchange roles when the stress is reversed, and equal and opposite polarizations are expected. On the other hand, with non-linear

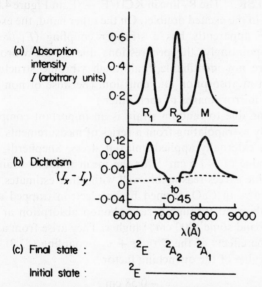

Figure 4.8 (a) Total absorption strengths of R-centres in KCl at 2 °K in arbitrary units. R_1, R_2, M are conventional labels of the bands.
(b) Polarization (i.e. difference of intensities in x and y polarizations) with a compressive stress of $2\cdot 5$ kg/mm² in the y-direction.
(c) Assignment of the bands.
(From Reference 41.)

coupling, the drawings of Figure 4.2 show that the two vibronic states behave non-equivalently.

It is appropriate to point out that the dichroism in the transition to a ²E state is due to the inequivalence of the x and y-polarizations, which is indicated by our writing in the selection rules of Appendix VII $M_x + M_y$ rather than $M_{x,y}$. This dichroism is expectedly smaller than those to the singlets (Figure 4.8).

Magnetic circular dichroism (MCD) was observed in R-centres[42-3] and in a negatively charged R-centre (called R'-centre) in which a transition occurs from a ground state singlet (probably ³A₂) to a doublet (³E).[44-5] (In MCD one measures the difference in the imaginary parts of the complex refractive indices for right and left circularly polarized light. Reference

46 describes the subject in detail). The observed zeroth and first moments of the zero-phonon line yield, by an extension of the moment analysis method,[47] the reduction factor p (§3.2.3, equation (3.29)) and consequently the strength of the linear J-T-coupling. The values $E_{\rm JT}/\hbar\omega = 1\cdot15$ (in KF),[42] $1\cdot6$ (in KCl),[42] $1\cdot8$ (in KCl)[43] derived from the dichroism of the R_2-line ($^2E \to {}^2A_2$) agree well with the estimate $E_{\rm JT}/\hbar\omega = 1\cdot5$ (in KCl) arrived at by ESR.[48] The R_1 line in KCl ($^2E \to {}^2E$, in Figure 4.8) indicates[43] a weak J-T-E in the excited doublet. On the other hand, the excited doublet of R′ in LiF apparently has a stronger coupling ($E_{\rm JT}/\hbar\omega = 1.5$–3).[45] Rather disappointingly, the progressions due to the vibronic levels in Figure 3.4 are not identifiable,[44] although vibronic structures are well resolved both in absorption and emission. (Because of non-linear effects these are not mirror images of each other.)

The splitting due to random strain is an important component in the theory.[42–3] By extrapolating from a series of measurements of the zeroth moment with externally applied uniaxial stress, Shepherd[42] was able to deduce the value of $\sim 1\cdot5$ cm^{-1} for the mean random strain splitting in KCl. This value agrees reasonably well with other estimates.[41,49]

The F^+ centre in CaO, formed by an electron trapped at an oxygen vacancy, is observed through a zero-phonon absorption at 28,102 cm^{-1} and a broad band some 1000 cm^{-1} higher. They arise from a $^2A_{1g} \to {}^2T_{1u}$ transition. The effects of the $T \otimes (\varepsilon + \tau_{2g})$ coupling (§3.3) are apparent in the small value of the quenching factor

$$K(T_1) = \frac{-0\cdot58 \text{ cm}^{-1}}{-31 \text{ cm}^{-1}} \sim 0\cdot02$$

for the spin–orbit coupling constant as observed by MCD and Faraday rotation.[50] The couplings to the ε_g and τ_{2g} modes are regarded[50] as having very nearly equal strengths (in which case $K(E) \sim K(T_2) \sim 0\cdot4$ for strong coupling),[51] in agreement with Hughes[52] interpretation of the quenching factors for strain which are large ($\sim \frac{1}{2}$, for either {001} or {111} oriented strain), and his moment analysis of the band. Recently a theoretical band shape[53] based on $T \otimes (\varepsilon + \tau_2)$ coupling and using the parameters of the moment analysis[52] was shown to fit closely the observed band. In another calculation,[54] the MCD band shape was obtained for the previously discussed R′-centres, using the semi-classical approximation.

4.8 References

1. G. D. Watkins and J. W. Corbett, *Phys. Rev.*, **121**, 1001 (1961); **134**, A1359 (1964).
2. E. L. Elkin and G. D. Watkins, *Phys. Rev.*, **174**, 881 (1968).
3. U. Öpik and M. H. L. Pryce, *Proc. Roy. Soc. A.*, **238**, 425 (1957).

4. F. S. Ham, *Phys. Rev.*, **166**, 307 (1968); also in *Electron Paramagnetic Resonance*, Ed. S. Geschwind (New York: Plenum Press, 1971).
5. F. I. B. Williams, D. C. Krupka and D. P. Breen, *Phys. Rev.*, **179**, 255 (1969).
6. A. A. Kaplyanski and A. K. Przhevuskii, *Opt. i Spektroskopiya*, **19**, 597 (1965) [English transl.: *Opt. Spectry.*, **19**, 331 (1965)].
7. M. D. Sturge, *Phys. Rev.*, **B1**, 1005 (1970).
8. G. Feher, J. C. Hensel and E. A. Gere, *Phys. Rev. Letters*, **5**, 309 (1960).
9. T. N. Morgan, *Phys. Rev. Letters*, **24**, 887 (1970).
10. J. R. Herrington, T. L. Estle, L. A. Boatner and B. Dischler, *Phys. Rev. Letters*, **24**, 984 (1970).
11. R. E. Coffman, *Phys. Letters*, **19**, 475 (1965); **21**, 381 (1966); *J. Chem. Phys.*, **48**, 609 (1968).
12. U. T. Höchli and T. L. Estle, *Phys. Rev. Letters*, **18**, 128 (1967).
13. U. T. Höchli, *Phys. Rev.*, **162**, 262 (1967).
14. G. Marshall and V. W. Rampton, *J. Phys. C(Proc. Phys. Soc.)*, **1**, 594 (1968).
15. J. R. Fletcher and K. W. H. Stevens, *J. Phys. C. Solid State Phys.*, **2**, 444 (1969).
16. L. L. Chase, *Phys. Rev. Letters*, **23**, 275 (1969); *Phys. Rev.*, **B2**, 2308 (1970).
17. D. P. Breen, D. C. Krupka and F. I. B. Williams, *Phys. Rev.*, **179**, 241 (1969).
18. M. D. Sturge, J. T. Krause, E. M. Gyorgy, R. C. Le Craw and F. R. Merritt, *Phys. Rev.*, **155**, 218 (1967).
18.′ S. Geschwind and J. P. Remeika, *J. Appl. Phys.*, **33**, 370 (1962).
19. K. A. Müller, W. Berlinger, J. C. Slonczewski and R. S. Rubins, *Bull. Am. Phys. Soc.*, **13**, 433 (1968).
20. K. A. Müller, W. Berlinger and F. Waldner, *Phys. Rev. Letters*, **21**, 814 (1968).
21. J. C. Slonczewski, K. A. Müller and W. Berlinger, *Phys. Rev.*, **B1**, 3545 (1970).
22. B. Alefeld, *Z. Phys.*, **228**, 454 (1969).
23. H. Unoki and T. Sakudo, *J. Phys. Soc. Japan*, **23**, 546 (1967).
24. R. S. Rubins and W. Low in *Paramagnetic Resonance*, Ed. Low (New York: Academic Press, 1963), p. 59.
25. E. B. Royce and N. Bloembergen, *Phys. Rev.*, **131**, 1912 (1963).
26. C. A. Bates and J. M. Dixon, *J. Phys. C: Solid State Phys.*, **2**, 2209 (1969).
27. E. D. Nelson, J. Y. Wong and A. L. Schawlow, *Phys. Rev.*, **156**, 298 (1967), also in *Optical Properties of Ions in Crystals*, Ed. Crosswhite and Moos (New York: Wiley, 1967) p. 375.
28. R. M. Macfarlane, J. Y. Wong and M. D. Sturge, *Phys. Rev.*, **166**, 250 (1968).
29. C. A. Bates and J. P. Bentley, *J. Phys. C: Solid State Phys.*, **2**, 1947 (1969).
30. P. J. Stephens, *J. Chem. Phys.*, **51**, 1995 (1969).
31. M. D. Sturge, *Solid State Phys.*, **20**, 91 (1967).
32. R. Pappalardo, D. L. Wood and R. C. Linares, *J. Chem. Phys.*, **35**, 1460 (1961).
33. D. S. McClure, *J. Chem. Phys.*, **36**, 2757 (1962).
34. W. C. Scott and M. D. Sturge, *Phys. Rev.*, **146**, 162 (1966).
35. P. J. Stephens and M. Lowe-Pariseau, *Phys. Rev.*, **171**, 322 (1968).
36. M. D. Sturge, *Bull. Am. Phys. Soc.*, **11**, 886 (1966).
37. M. D. Sturge, *Phys. Rev.*, **140**, A880 (1965).
38. A. E. Hughes and W. A. Runciman, *Proc. Phys. Soc.*, **86**, 615 (1965).
39. F. S. Ham in *Optical Properties of Ions in Crystals*, Ed. Crosswhite and Moos (New York: Wiley, 1967) p. 357.

40. M. Caner and R. Englman, *J. Chem. Phys.*, **44**, 4054 (1966).
41. R. H. Silsbee, *Phys. Rev.*, **138**, A180 (1965).
42. I. W. Shepherd, *Phys. Rev.*, **165**, 985 (1968).
43. W. Burke, *Phys. Rev.*, **172**, 886 (1968).
44. H. B. Fetterman and D. B. Fitchen, *Solid State Commun.*, **6**, 501 (1968).
45. J. A. Davis and D. B. Fitchen, *Solid State Commun.*, **6**, 506 (1968).
46. A. D. Buckingham and P. J. Stephens, *Ann. Rev. Phys. Chem.*, **17**, 399 (1966).
47. C. H. Henry, S. E. Schnatterly and C. P. Slichter, *Phys. Rev.*, **137**, A583 (1965).
48. D. C. Krupka and R. H. Silsbee, *Phys. Rev.*, **152**, 816 (1966).
49. P. Duval, J. Gareyte and Y. Merle d'Aubigné, *Phys. Letters*, **22**, 67 (1966).
50. Y. Merle d'Aubigné and A. Roussel, *Phys. Rev.*, **B3**, 1421 (1971).
51. M. C. M. O'Brien, *Phys. Rev.*, **187**, 407 (1969).
52. A. E. Hughes, *J. Phys. C: Solid State Phys.*, **3**, 627 (1970).
53. M. C. M. O'Brien, *J. Phys. C: Solid State Phys.* (in the press) (1971).
54. R. Sati, M. Inove and S. Wang, *Can. J. Phys.*, **48**, 2769 (1970).

5 The Jahn–Teller Effect in extended systems

Outline

Corrections to the Born–Oppenheimer approximation in extended systems (as opposed to the 'localized centres' of the previous chapters) can also have effects which are non-perturbational and therefore qualify for the name of 'Jahn–Teller effects'. Some of these effects result in low-temperature phases of reduced symmetry. Not very much is gained by appending the label J-T-E to these phenomena in solids, however, the question may be and has been asked 'why does the J-T-E not operate in this or that case in solids?' and the answer is that it does operate under the circumstances described in this chapter.

5.1 Jahn–Teller effect in band structure

5.1.1 J-T-E in band structure

The removal of degeneracy in an electronic band of a solid by lattice distortion can have consequences ranging from the insignificant to the drastic. One question that arises is whether or not this should be called J-T-E. This question we shall not discuss (often the naming of the effect depends on whether a better alternative nomenclature is found); we shall, however, try to throw the formulation in such a shape as to make the relation to the molecular J-T-E evident.

There are two types of (non-accidental) electronic degeneracies in a band, which we shall discuss now. One is the degeneracy of the energy $E(\mathbf{k})$ with respect to the arms in the star of the wave-vector $\mathbf{k}$, which occurs for any general wave-vector in the Brillouin zone. There is also the degeneracy at special points or lines in the zone. A third type of (accidental) degeneracy due to the overlapping of bands, as in semimetals, will be studied later.

For those who prefer the parlance of group theory, the first type arises because of the relation (Reference 1, p. 39)

$$RE(\mathbf{k}) = E(\mathbf{k})$$

where R is an operator of the point group. The second type is due to the degeneracy of a small irreducible representation of the factor group

$$G(\mathbf{k})/T,$$

where $G(\mathbf{k})$ is the space group of $\mathbf{k}$, i.e. the little group of the second kind belonging to the $\mathbf{k}$-representation of T (T being the subgroup of translations).

The question of the removal of a degeneracy of the second type through a distortion of the crystal was considered from the point of view of symmetry by Birman[2] and by Kristofel.[3] Since these special, high-symmetry points or lines are vanishingly small in number compared to points of a general wave-vector, their J-T-effect is unimportant, except so far as to give some indication of the occurrences of pseudo J-T effects at nearby general points. Mention should, however, be made of Kristofel's distinction between the cases when the ratio Band Width/Stabilization Energy is large or small. We shall see that, as a general rule, this ratio must be small in order for the Jahn–Teller effect to be observable.

There are two kinds of lattice distortions which can effectively split the bands. One is a homogeneous, non-isotropic distortion of the lattice, i.e. a shear strain; the other is a superperiodicity in the lattice, which may or may not reduce the point symmetry of the crystal. Both kinds have long histories. The first is usually associated with the names of Hume–Rothery (mainly private communications and dialogues), Jones[4] and Goodenough;[5] the second appeared in an early work of Peierls[6] (also Reference 7, p. 108) and was re-invented for the bond alternation of long chain molecules by Kuhn[8] and by Salem and Longuet-Higgins.[9] The dormant theoretical predictions have recently received new relevance by experimental findings, which will be described in due course. Some incipient progress has also been made in putting details into the theory. It is clear, however, that (at the time of writing) we are just beginning to see a new application of the J-T-E being opened up.

5.1.2 Uniform shearing strain in metals and alloys

We recall the definition of the basis vectors K_i ($i = 1, 2, 3$) of the reciprocal lattice in terms of the primitive basis vectors a_j ($j = 1, 2, 3$) of the direct lattice (Reference 10, p. 49).

$$\mathbf{K}_i = \mathbf{a}_j \wedge \mathbf{a}_k / |\mathbf{a}_{i'} \cdot \mathbf{a}_{j'} \wedge \mathbf{a}_{k'}| \quad (i, j, k \text{ and } i', j', k' \text{ cyclic})$$

We also recall that non-vanishing matrix elements of the crystal potential V will exist between electronic states characterized by the wavevectors $\mathbf{k}$ and $\mathbf{k}'$ which are such that $\mathbf{k}' - \mathbf{k}$ is some integral multiple of a $\mathbf{K}_i$. For a cubic lattice we can write this condition formally as

$$\langle \mathbf{k}'|V|\mathbf{k}\rangle = V_0 \delta_{\mathbf{k}'-\mathbf{k},0} + V_1 \sum_i \delta_{\mathbf{k}'-\mathbf{k},\pm\mathbf{K}_i} + V_2 \sum_{ij} \delta_{\mathbf{k}'-\mathbf{k},\pm\mathbf{K}_i\pm\mathbf{K}}$$
$$+ V_3 \sum_{ijk} \delta_{\mathbf{k}'-\mathbf{k},\mathbf{K}_i\pm\mathbf{K}_j\pm\mathbf{K}_k} \quad (i,j,k \text{ cyclic}) \tag{5.1}$$

The V_0, V_1, . . . are conventional Fourier components of the crystal potential; they have been brought outside the summation signs, because the lattice has cubic symmetry. Naturally, there will be additional terms of the form $V_x \sum_{ijk} \delta_{\mathbf{k}'-\mathbf{k},l\mathbf{K}_i+m\mathbf{K}_j\pm n\mathbf{K}_k}$, where l, m, n are positive or negative integers and V_x depends on l, m, n. We shall neglect these terms since the importance of the coefficients V_x presumably decreases with increasing l, m, n.

Let us suppose now that the lattice originally possessed a simple cubic structure, so that

$$|\mathbf{a}_1| = |\mathbf{a}_2| = |\mathbf{a}_3| = a$$

$$|\mathbf{K}_1| = |\mathbf{K}_2| = |\mathbf{K}_3| = K^0 = (a)^{-1}$$

If we now impart to the lattice uniform strains, denoted by e_θ and e_ϵ, these will change the magnitudes of the basis vectors of the direct lattice, making them in general unequal, and will therefore (by virtue of the relations written out above) also alter the reciprocal lattice vectors in the manner

$$\mathbf{K}_i = \hat{\mathbf{I}} K^0 (1 + \alpha_{i\theta} e_\theta + \alpha_{i\epsilon} e_\epsilon)$$

where $\hat{\mathbf{I}}$ is a unit vector along the i-axis and the α's are coefficients depending on the geometry of the lattice. In effect, it is clear, that for an $\mathbf{a}_i$ elongated by the strain, $\mathbf{K}_i$ will be shortened and conversely.

The potential due to the deformed lattice will have non-zero matrix elements between a general $\mathbf{k}$ in the Brillouin zone and some other $\mathbf{k}'$ in the extended Brillouin zone scheme such that the two wave-vectors differ by some multiple of the altered $\mathbf{K}_i$; e.g., if the general point $\mathbf{k}$ is in the undeformed lattice near a corner of the Brillouin-zone, then for small distortions there will be just seven other states to which $\mathbf{k}$ will get coupled and which will have similar energies. The matrix connecting the eight states will have eight (in general) different eigenvalues each being dependent through $\mathbf{K}_i = \mathbf{K}_i(e_\theta, e_\epsilon)$ on the distortion. The physically important quantity is the free energy or, at low temperatures, the sum of the electronic energies up to the Fermi-energy $E_F(e_\theta, e_\epsilon)$. Even in the absence of analytical solution, which will be given for some models below, it is clear what is physically happening (Figure 5.1).

Consider the interaction energy between the ions and a nearly free electron in a state **k** (whose position is near a boundary plane of the Brillouin-zone) in the (a) undistorted, (b) extended and (c) compressed form of the lattice. Real combinations of the wave-functions will catch the ions (supposed to exert an attraction on the electron) at different amplitudes (Figure 5.1(a)). In an extended state of the lattice, (b), the energy of the same wave will be lowered; in a compressed state of the lattice the energy will be little changed, if at all, since the wave still meets the ions at all amplitudes as in (c). Let us now suppose, for definiteness

Figure 5.1 The change of the wave function-ion overlap due to lattice distortion.

(a) Undistorted Lattice.
(b) Extended Lattice: Energy is lowered.
(c) Compressed Lattice: Energy is unchanged.

that in the cubic situation the Fermi energy is just inside, but close to, the boundary planes of the first Brillouin-zone. Interpreting the above figure for a three dimensional lattice about to distort in an e_θ strain-mode, we anticipate a lowering of energy in the direction of one cubic axis and little change along two others, altogether a negative anisotropy energy. Clearly the Fermi-surface will no longer have cubic symmetry. Whether or not the distortion will actually occur, depends on a number of factors so far neglected, e.g. the dependence of the Fourier coefficients V_i (including V_0) in equation (5.1) on the distortion and the resistance of the deeper lying electrons to it. The sum of these restoring forces can probably be well represented by an elastic energy term of the form $\frac{1}{2}A(e_\theta^2 + e_\epsilon^2)$ whose main characteristic is that it is positive and quadratic in the strain.

We must now consider the nature of the term in the total energy which promotes the distortion. This term will be called the distortive (or anisotropy) term. Arguing both by physical intuition and on the basis of the models for which analytical solutions exist,[11-2] one can say that for distortions large compared to the bandwidth the energy is linear, for distortions smaller than the bandwidth the energy is quadratic in the distortion. This follows,[13] since the energy gain near the Fermi-surface is proportional to the distortion and the fraction of states benefiting from the distortion is initially of the order of the ratio of the distortional energy divided by the band width. Later, for large distortion, the number of states involved reaches saturation and then one reaches the linear regime. Now with the anisotropy energy linear in the distortion a distorted state is necessarily stable. Nevertheless, it would be wrong to examine the linear regime for the stability of the distorted state. (One must also beware of getting the impression from Adler and Brooks[12] that this is legitimate). What in fact ensures a distorted configuration in the present single band model, is that the distortive energy in the quadratic regime exceeds the elastic energy. In the contrary case, the linear, large distortion regime is not reached before the positive elastic energy, quadratic in the distortion, outweighs the gain in the distortional energy. We repeat that this appears to be true, for any reasonably simple physical model with a *single* energy band. (Here we note a difference from localized centres.) With multiple bands (such as used by Labbé and Friedel[11] and discussed by us at some length in §5.1.3) a discontinuity in slope of the anisotropy energy may occur just as the distortion empties a band. The various situations are depicted in the following figure (Figure 5.2).

In the distorted situation, as the temperature is raised, excited states will be occupied whose energy increases with the distortion; such a state is the 'anti-bonding' state which can be obtained for case (b) of Figure 5.1 by adding a phase of 90° to the wave-function. This will lead to a decrease in the (absolute value of the) distortive energy and to a transition to the undistorted phase at some elevated temperature. The phase transition will be of the first or second order in case II of Figure 5.2 and of first order in cases resembling III.

Supposing the distortion to be axial, random factors will determine along which one of the three cubic axes will the distortion take place. No tunnelling, or dynamic J-T-E, can exist between equivalent directions of a homogeneous uniform strain, which can be regarded as the limiting case of a long-wavelength phonon having an infinite mass. It seems reasonable, however, to stretch the formalism to include also distortions whose wavelength is long but finite. The effect of such distortions on an electron in the crystal is similar to that of a long wavelength acoustic phonon. Locally, on the scale of electronic wavelengths, the distortion will be uniform; the Fermi surface will also be distorted locally, provided that the electronic relaxation is faster than the long-wavelength phonon. For a full discussion of the events which occur as the acoustic wave passes, one should take into account all the elements (mirrors, surfs, etc.) which enter Pippard's

treatment of ultrasonic attenuation.[14] For us it is important to realize
that there will be tunnelling between different many-electronic states
corresponding to instantaneous distortions at different orientations and
phenomena similar to the dynamic J-T-E may occur. The time required

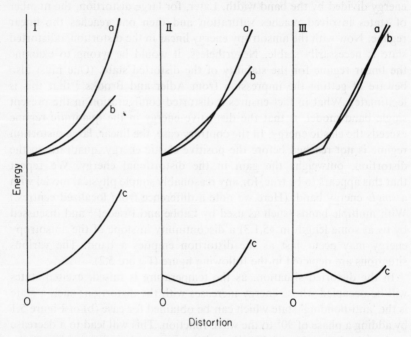

Figure 5.2 Contributions to the total energy: (a) elastic energy, the
same in all three cases, (b) the *negative* of the distortive energy,
(c) the total energy. The cases I and II are for a single band, III is for
multiple bands, one of which gets emptied where the break occurs.

for tunnelling will be of the order the electronic transport-relaxation
time.

There are a number of specific cases which exhibit a descent in symmetry in an originally cubic solid due to the interaction of the Fermi-
surface with the Brillouin-zone boundaries. We shall now describe a few
of these cases, emphasizing that we are entering a domain of speculations.

The first example, vanadium metal, shows anomalous behaviour in
electrical resistivity, in the thermal coefficient of expansion,[15] in the
magnetic susceptibility[16] and in other physical properties. This occurs
at a temperature between 200 and 250 °K, depending on the purity of the
material. The suggestion,[17] that this was due to a transition from the
b.c.c. to a b.c. tetragonal phase as the temperature is lowered, was put

aside for a while in favour of a magnetic transition, for want of crystallographic confirmation. Subsequently Suzuki and Miyahara[18] found by X-ray diffraction an orthorhombic distortion near 230 °K (Figure 5.3). The authors interpreted this hysteresis-free distortion as the J-T-E acting on the Γ_{25}-state (having the symmetries of a t_{2g}-triplet) which lies just above the Fermi-level in the APW calculations of Mattheiss.[19] We have already stated our reservations as to what the degeneracy at a single high-symmetry point can achieve and we reiterate that one should be really talking about a pseudo J-T-E near Γ (the origin of reciprocal space).

Concluding this topic, we call attention to Figure 5.3 once again. It appears that for about 8 degrees below the transition the phase is tetragonal. It is indeed expected that the complete removal of a threefold degeneracy

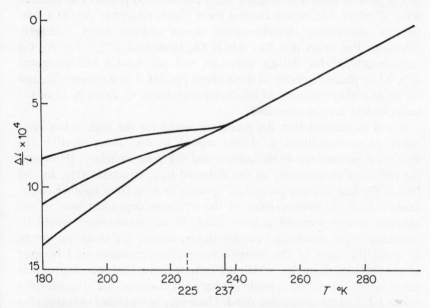

Figure 5.3 Orthorhombic distortion in vanadium. The relative changes $\Delta l/l$ in the three cubic axes are shown as function of temperature. (From Suzuki and Miyahara.)[18]

in a collective J-T-E may take place in two stages, first when a double degeneracy remains and secondly when this also is removed.

The elements As, Sb and Bi crystallize in what is essentially a simple cubic lattice with a slight rhombohedral distortion. There are indications[20] that the distortion can be explained (or described) in terms of a transition from a high-symmetry metallic phase to a phase having a reduced symmetry and an energy gap. We shall further discuss these elements in §5.1.6.

Alloys of In with Pb, Sn and Tl are tetragonally distorted. Extrapolating the c/a vs. electron-concentration curve we obtain a faint indication of the cubic structure which can be thought of as the 'origin' of the tetragonally distorted alloys. [Reference 21, also V. Heine on p. 103 of Reference 22. There is also a statement by V. Heine on p. 106 of this very useful book about the distorted structure of α- and β-Hg, GaII and In from an apparently f.c. cubic parentage, and a question by L. Kaufman on p. 113 about the nearly cubic gapless origin of such diamond-type semiconductors as Sn, Ge, Si; both of which indicate an affinity to the idea of J-T-E as expounded here.]

Upon cooling, the noble metals Cu, Ag or Au in 50–50 alloys with Zn or Cd go over from a disordered b.c.c. (so-called β) phase to an ordered b.c.c. β' phase. On further cooling there occurs (except in AgZn) a continuous, martensitic transformation to an ordered lower symmetry structure. For brass (Cu–Zn) this is f.c. tetragonal (β''),[23] for Ag–Cd orthorhombic,[24] for Au–Zn uncertain and for Au–Cd f.c. tetragonal (β'').[25] For phase diagrams in these alloys pp. 244–5 in Reference 22, and for an alternative account of the transitions due to C. Zener p. 25 of the same source, may be consulted.

It will be realized that the preceding model for the high → low symmetry phase transitions is closely related to the 'Jones-effect':[26] the attraction between the Fermi-surface and the zone-boundary. [However, the reduction in symmetry of the distorted f.c.c. structures (Hg, In and Ga) in the last but one paragraph appears to arise from another cause, namely from the enhancement of the structure dependent part in the cohesive energy (through a term which in our terminology would be quadratic in the distortional coordinate) by causing the structural weight to avoid the zeros of the energy-wave-number characteristic.] In spite of the previous examples, which are only tentative and speculative, it is fair to say in summary that we do not have any clear-cut case of a distortion due to J-T-E in the conduction band. There may be a number of reasons for this. One is that band gaps tend to be rather low compared to, say, the Fermi energy (Reference 27, p. 41). Another is the doubt attached to the Jones-effect, due to the circumstances that the Brillouin zone boundaries appear to be in contact with the non-spherical Fermi-surface even in the noble metals.[28-9]

It is nevertheless undeniable that the Jones-effect is amply documented by experiment in cases when the proximity of the Fermi-surface to the zone-boundary *changes* the lattice parameters in the already non-symmetrical structure (instead of lowering the symmetry as in the previous instances).

King and Massalski[30-2] have found that when the electron concentration is brought up to $\sim 1\cdot 5$ in ζ-hexagonal close-packed alloys of the noble metals with III–V elements there is a continuous change in the c/a-ratio. Actually the volume also changes; this can, however, be corrected for by extrapolation of the volume change in the α-f.c.c. structure which precedes the ζ-phase as the electron concentration is increased.[33] The electron diffraction data of Sato and Toth on $Pd_{3-x}Mn_{1+x}$ [34] and $Au_{3-x}Mg_{1+x}$ [35] alloys were interpreted by them in terms of the attraction between the Fermi surface and a closeby surface of the Brillouin zone boundary. For $x \leq 0$ the alloy structure is based on the occupation of the sites of an f.c.c. lattice in such a manner as to form a one-dimensional super-period four times the original f.c.c. lattice parameter.[34] Upon further addition of Mn $(x > 0)$ the super period stays four; however, now a large and steady tetragonal distortion of the crystal is found to take place, which is maintained by the preference of the Mn to a particular sublattice. The tetragonal distortion ensures the stability of the super period by keeping the Fermi-surface (containing about $1\cdot 2 + 2\cdot 4x$ electrons per formula unit) close to the (110)-type planes of the f.c.c. Brillouin zones. Alternative interpretations are due to Villain[36] and to Pick.[37]

It is instructive from our point of view to formulate the effect in such a manner as to reveal the similarity with the pseudo-J-T-E.

Supposing a one dimensional band, which arises from a parabolic dispersion curve $E(k) = \hbar^2 k^2/(2m)$, we wish to calculate the amount of strain which will ensure that the total energy, including the electronic and elastic energy, will be a minimum. The Fermi momentum $\hbar k_F$ of the linear system *does not coincide* with the distance $\frac{1}{2}\hbar K_0$ of the Brillouin zone edge from the origin in the absence of strain and we shall suppose that $k_F < \frac{1}{2}K_0$.

Since in the distorted structure the shortest reciprocal lattice vector is $K_0 + 2\pi Q/(\hbar a^2)$, the state k will be coupled by the matrix-element V to the state

$$k - K_0 - 2\pi Q/(\hbar a^2).$$

(Figure 5.4). a is the lattice spacing and Q is the distortion, i.e. the strain multiplied by a.

The matrix

$$
\begin{array}{l}
|k\rangle \\
|k - K_0 - 2\pi Q/(\hbar a^2)\rangle
\end{array}
:
\begin{bmatrix}
\hbar^2 k^2/2m & V \\
V & \hbar^2[K_0 + 2\pi Q/(\hbar a^2) - k]^2/2m
\end{bmatrix}
$$

bears a clear similarity to the P-J-T-E (but contains also a quadratic term in Q). From the eigenvalues of this matrix plus an elastic energy term of the form $\frac{1}{2}AQ^2/a^2$ we obtain the energies E. The total energy, equivalent to the free energy at zero temperature, per unit length is

$$E_T = \frac{2}{\pi} \int_0^{k_F} E(k)\, dk$$

It can be shown that the leading term in V goes as $-V$ for large values
of $2mV/(\hbar K_0)^2$ and as $-V^2$ for small values of this ratio. If the total

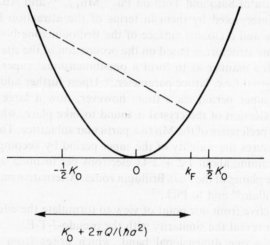

Figure 5.4 Coupling scheme for one-dimensional band. The states between which a matrix element V exists are connected by a broken line.
k_F is the Fermi-momentum, $\frac{1}{2}K_0$ is a zone boundary.

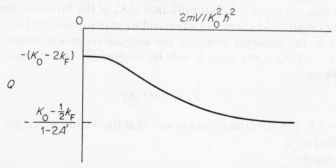

Figure 5.5 The distortion Q as function of the off-diagonal matrix element V divided by the band width $\hbar^2 K_0^2/2m$. A' is the coefficient A in
the elastic energy divided by $(4\pi^2\hbar^2/2ma^2)$.

energy is examined for its minimum as function of Q, one obtains the
strain at equilibrium. This is shown in Figure 5.5 as function of the off-
diagonal matrix element V. The figure shows that if initially there was no

overflow of electrons (i.e. $\frac{1}{2}K_0 > k_F$), then $Q < 0$ and the zone boundary will move inwards. This result was already established by Jones,[26] arguing from more sophisticated models of interaction than ours.

5.1.3 V$_3$Si, *the linear chain model*

In §7.13 we describe the main experimental facts accompanying the cubic-to-tetragonal transition in V$_3$Si and Nb$_3$Sn (at $T_m = 20.5\,°\text{K}$ and $45\,°\text{K}$) and introduce the linear chain model for these structures.

The simplicity of the model has enabled Labbé and Friedel[11] to formulate in a quantitative way the mechanism of the J-T transition in an extended system. In reviewing their theory we shall bear in mind that the model, which was applied to V$_3$Si is very approximative (though the multifarious quantitative agreement with experiment is impressive). We shall therefore concentrate on those qualitative aspects of the model which shed light on the question: What does a band-structure do to the J-T-E (and vice versa)? We should perhaps insert at this point that Anderson and Blount[38] have previously argued from the apparently second order nature of the V$_3$Si transition that in this material something else, other than the tetragonality, must also undergo a change during the transition. This 'something else' may be the polarity or the electronic configuration. This follows since the Landau theory, which is based on a single order parameter, predicts that a cubic $\rightarrow$ tetragonal transition will in general be of the first order ('In general', unless invariants of the third order vanish. We shall return to this qualification in §5.3. Landau's theory was recently reassessed.[39])

Returning now to the picture of the band, which can be seen in Figure 7.26, we assume that at the beginning of the game one of the composite peaks is narrowly filled with electrons or holes. This assumption leads, within the model, to a natural explanation of the high density of electronic states noted in §7.13. At the same time, the narrowness of the filled band promotes the operation of the J-T-E. Let us now see how strains of the type e_θ, e_ε can split a peak. Each component of the peak is associated with one of the vanadium strands, along one of the {001}-directions, and the nearest neighbour overlap of the 3d-functions can be expected to depend exponentially on the distance between neighbours in the strand. A positive (negative e_θ-type strain will then lower (raise) the [100], [010] band peaks by an amount proportional to $e^{-\frac{1}{2}Ke_\theta}$ and will raise (lower) the [001] band by e^{-Ke_θ}. Here K is a constant representing the radial extension of the vanadium wave function. The actual sense of the distortion, and the question whether one or two bands will be lowered, depends on many detailed circumstances. This *a priori* uncertainty in the sign of the distortion is a characteristic feature of a J-T-E in a band (or, more accurately, in a many electron system). It is expected that a very narrow or thinly occupied band will simulate a single electron system in which that spontaneous strain is favoured which leads to maximal stabilization, i.e. to the lowering of the singlet state.

Another distinction between a localized J-T-centre and the J-T-E in a band is that in the former case the distortional energy is linear in the distortional coordinate. In the latter case we find two alternatives. In the first, the net gain in the total energy of the bands, due to the lowering and raising of the sub-bands, is quadratic in the strain e, for small distortion. The distortive tendency is counteracted by other effects, which are lumped together in the positive term $\frac{N}{2}a^3(C_{11} - C_{12})(e_\theta^2 + e_\epsilon^2)$ representing elastic forces in the lattice. (N is the number of unit cells in the crystal, a is the distance between the vanadium atoms in the same chain and C_{11}, C_{12} are elastic constants.)[10] If the elastic forces are weaker than the distortive ones the lattice is quadratically unstable with respect to the distortion. The distortion is then bound to occur. Alternatively, the elastic forces may hold sway for a while, until the upper split sub-band becomes emptied, after which the distortional energies in the bands become *linear* in e and may result in a thermodynamic state of lower energy than the cubic state. For a model like the present one, with a discontinuous density of states, the regime where the strain gradient of the energy is negative starts with a discontinuity in the gradient. For models with more realistic density of states the gradient changes continuously and for sufficiently smoothly varying density of states curves may not become negative at all.

Rather than working through the detailed mathematics of Reference 11 we wish to concentrate on the physical essentials. Perhaps the most important task is to derive the fundamental physical parameter of the problem. In the Introduction (Chapter 1) we stated that the most important single physical parameter in our subject is the P-J-T-ratio,

$$S' = \frac{L^2}{\Delta E \hbar \omega} \tag{5.2}$$

Let us try to interpret the parameters L, ΔE and $\hbar\omega$ in the present context. L, as defined in equation (1.3), equals the change of the energy for a distortion of the size of the zero point amplitude. Supposing for the moment that the spontaneous strain e is of this size, we have

$$L = wKe$$

(Here $2w$ is the width of the band in Figure 7.26 and K appears in the exponential of the overlap). Further

$$\hbar\omega = a^3(C_{11} - C_{12})e^2$$

To interpret ΔE we note that, in the distorted state of the pseudo-J-T-E, ΔE (or the equivalent quantity $2W$ in §3.6.1) gives a measure of the admixture of the state leading to positive distortion into the state of negative distortion and vice versa. (The larger ΔE, the greater the admixture). The quantity $N_e w/3N$ (where N_e is the number of electrons in the upper split band) has a rather similar meaning, since it provides some sort of measure for the number of electrons in the distorted structure whose energy would be raised by the distortion, were they not to relax to lower split bands. Then the ratio in equation (5.2) becomes

$$S' = 3NwK^2/[a^3(C_{11} - C_{12})N_e] \tag{5.3}$$

for which the value 1·1 was found by Labbé and Friedel[11] (Table 5.1). These authors expand the free energy as a Taylor series in the e_θ distortion and find a first order transition. This is of course expected for a one-parameter system undergoing a cubic-tetragonal phase transition. Utilizing the experimental data on V_3Si and estimating some parameters they succeed (with the corrected values of Reference 40) to predict the transition temperature $T_m = 23\cdot2$ °K and a strain amplitude of about $+10^{-3}$, in fair agreement with experiment. Other parameters which appear in the theory take the values in Table 5.1.

Table 5.1 Parameters in V_3Si

$N_e/3N$	K (exponent)	w (half-width)	$\frac{1}{2}(C_{11} - C_{12})a^3$ (elastic energy)	$\Delta E/2$ 0·054 eV	S' [equation (5.3)]
0·036	0·637	4·5 eV	22·5 eV	0·054 eV	1·1

The number of electrons per atom, 0·036, is to be compared with the four available states (d_{xy}, $d_{x^2-y^2}$ for each spin) per atom in the sub-band.

With the value of 1·1 for the P-J-T-E ratio S', the free energy surface in the space of e_θ, e_ϵ appears similar to the lower sheet of the $T \otimes \varepsilon$ potential energy surface in the q_θ, q_ϵ space (Figure 3.24), but there is of course no crossing at the origin. In one of the three equivalent distortions the tetragonal strain e_θ is positive, $e_\epsilon = 0$ and the [100], [010] branches get lowered relative to [001]. At the lowest temperatures only the former branches have filled states, the Fermi surface being 0·003 eV above the [100] [010] density of states peaks and by about that much below the [001] peak. The x and y-oriented chains have an excess $\Delta n = 0\cdot054$ electron per vanadium atom over the z-chains.

As the temperature is raised, excitations to the [001]-branch begin to occur. The preference of [001] for negative e_θ-deformations acts as a feedback to raise additional electrons up to this branch and, ultimately at some temperature, to bring back the crystal to the cubic phase. Here E_F is only 0·0018 eV above the band peak and the density of states at the Fermi surface $n(E_F)$ has the high value 0·036/0·0018 = 20 (eV)$^{-1}$ (including both spins). This explains rather naturally much of the thermal data.[21-3]

The paramagnetic susceptibility increases strongly with decreasing temperature above T_m.[44-5] This behaviour is due to the Pauli paramagnetism,[40] which is known to be proportional to some weighted

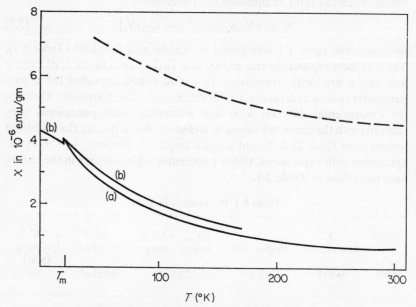

Figure 5.6 Paramagnetic susceptibility in V_3Si.
Broken lines: experiments.[45]
Full lines: Theory (excluding the high frequency part and exchange).
(a) Linear chain model.[40] (b) Dynamic treatment fitted to (a) at $T = T_m$ (E. Pytte, private communication, 1971).

average of the density of states (Reference 10, p. 261, Reference 45–6). This average has a rather similar behaviour to $n(E_F)$, whose main temperature dependence in the cubic phase is due to the strong temperature variation of E_F. This is the explanation[40] for the marked temperature dependence of the paramagnetic susceptibility (Figure 5.6). Labbé[40] has also interpreted the (nearly) temperature-independent 'high-frequency' contribution to the susceptibility and to the Knight-shift.[47] Subsequently, Gossard[47] has observed a difference of 0·588 MHz in the NMR quadrupole splittings of ^{51}V in chains along and transverse to the distortion.

This agrees well with theoretical results if the difference in the occupation number Δn is taken as 0·054, as predicted by Labbé.

In their high resolution measurements of angular correlation arising from positron annihilation of single crystals of V_3Si, Berko and Weger[48] found several peaks spaced at π/a. A planar Fermi surface is implied by this result, in agreement with the independent linear chain model. However, Weger[49] found that interchain interaction also results in plane-like sections of the Fermi-surface situated half way across the Brillouin zone (We refer to the description given in §7.13, Figure 7.27.)

In summary, it can be said in praise of the linear chain model that not only does it reproduce the experimental facts with a good verisimilitude but it also shows—quite ingeniously—how a soluble one-dimensional model can be brought to bear, through the operation of the exclusion-principle, on a three-dimensional problem. From our point of view, it represents the first exact analytical treatment of the J-T-E on a band and this is the reason for the prominence that we have accorded to it. The shortcomings of the solution should also be noted:

(1) Because of the assumed independence of the linear chains an instability in a tetragonal mode implies *a fortiori* an instability to a uniform compression and it is in fact quite difficult to envisage a mechanism which will (within the model of independent chains) stabilize the latter and not the former. Thus supposing that the energy of the conduction electrons depends on the volume of the crystal, but not on its shape, there exists the possibility that at their low density these electrons will promote compression rather than resist it. From the value of 0·06/eV/atom for the conduction-electron density of states which is found, one derives the mean interelectronic radius, r_s, as being 5–6 atomic units. These are rather large values for which the conduction electron system by itself is probably unstable to compression. Some indication of the unstability is provided by the Hartree–Fock approximation (rigorously valid for high densities) which ensures stability to compression only for $r_s < 4·8$.

The opposite view was taken in a paper by Labbé and Barisic,[50] who have ascribed the stability against isotropic compression to the s-electrons. This assumed, they have calculated the contribution of these electrons to the elastic constants by comparison with the experimental data. This does no better than beg the question 'What is the crystal stability due to?' In the last analysis, the final truth lies in the chains not being independent;[49] however, in the present writer's view, it must be shown that this stabilizes the volume sooner than it spoils the distortive mechanism which is the essence of the Labbé–Friedel model.

(2) If the 3d sub-bands differ not only in their widths but also in the position of their baricenters, then we have another mechanism leading

to the splitting of the [100], [010] and [001] branches and to a cooperative tetragonal distortion. This mechanism is not considered by Labbé and Friedel, though it is expected to be comparable to the one studied by them. The splitting of the baricenters by the crystal field of the neighbouring transition metal atoms is presumably several electron volts (as seen from the band structure calculations of L. F. Mattheiss[19]), and is evidently a sensitive function of the V–V distance along a chain. Again a first order transition is anticipated.

(3) It is not clear whether all assumptions of the Labbé–Friedel model are indeed necessary to account for the anomalies in the β–W compounds. Thus a model with interacting chains can have a sharp density-of-state peaks for some values of the crystal momenta.[49] It was also shown[52] that if the Fermi-surface is near to doubly degenerate levels, which split for an e_θ-strain, then a temperature-dependent softening of the elastic constants $C_{11}-C_{12}$ will occur, as observed.

(4) Consideration in the Labbé–Friedel theory of a uniform distortion only represents a simplification which should be improved by including distortional modes with $\mathbf{k} \neq 0$. This was done by Pytte[52a] whose equations of motion give rise to coupled excitations of electron density fluctuations and phonons. Nb_3Sn and V_3Si were shown to differ in that, at some temperature below T_m, two sub-band edges rise above E_F in the former, while in the latter all three sub-bands are partially occupied even at $T = 0$ °K. The paramagnetic susceptibility increases monotonically with T in Nb_3Sn below T_m, but decreases in V_3Si (Figure 5.6).

5.1.4 Super-periodicity

A reduction in the energy of the band electrons can also be achieved in a second way: by reducing the translational symmetry of the crystal through the appearance of a super-periodicity. The theory of this effect, first due to Peierls,[6-7] is easiest to describe in one dimension and so we consider first this case. Then we go over to physically more realistic systems.

Figure 5.7 represents the energy spectrum of a linear system of nearly free electrons. The Fermi momentum k_F is supposed to be well inside the first zone so that a parabolic spectrum, $E = \hbar^2 k^2/2m$, can be regarded as a good first approximation. If now k_F happens to be a submultiple $\pi/(pa)$ of the Brillouin zone boundary π/a, then a p-fold increase of the periodicity of the lattice will have the effect of connecting, through the Fourier-component $V_{k_F} (\equiv V)$ of the potential, a state of wave-number k with one having $k - 2k_F$ as the wave-number. It is expected from the definition of the Fourier expansion of the perturbed potential that V will be proportional to the distortion.

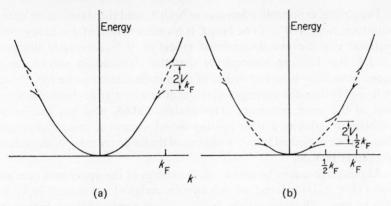

Figure 5.7 Electron energy in distorted lattice.
(a) Lattice distortion of wavelength $2\pi/k_F$.
(b) Lattice distortion of wavelength $4\pi/k_F$.

The energy matrix, appropriate to $k > 0$, is

$$\begin{bmatrix} \hbar^2 k^2/2m & V \\ V & \hbar^2(k - 2k_F)^2/2m \end{bmatrix} \tag{5.3}$$

From the eigenvalues $E(k)$ of this matrix, one can obtain the total energy of all occupied states per unit length of the lattice through the integral

$$E_T = \frac{2}{\pi} \int_0^{k_F} E(k)\, dk$$

Let W denote $\hbar^2 k_F^2/2m$, the width of the occupied part of the uppermost band. The two limiting cases $V/W \ll 1$ and $\gg 1$ are of particular interest. One finds in the former situation

$$E_T \sim 2k_F W/(3\pi) - \frac{k_F V^2}{2\pi W} \log(4W/V) \qquad (V/W \ll 1)$$

i.e. a nearly quadratic dependence on V. An elastic term should of course be added. Since the matrix element V is proportional to the amplitude of the distortion leading to superperiodicity, one can regard the elastic term as quadratic in V. Let this elastic energy per unit length be denoted by $k_F A V^2/(\pi W^2)$. Because of the logarithm in the preceding expression, it is clear that the system will be unstable to the periodic distortion considered. Stability will be attained at

$$V = V_0 = 4W \exp\left(-\tfrac{1}{2} - 2A/W\right),$$

and the associated stabilization energy per unit length will be

$$\frac{k_F V_0^2}{4\pi W} = \frac{4k_F W}{\pi} e^{-1-4A/W}.$$

The strong, exponential decrease of both V_0 and the stabilization energy with increasing A/W is to be noted. It is characteristic of one dimensional systems. For the one dimensional model of V_3Si previously discussed (§5.1.3) this has the consequence that the stabilization energy due to super-periodicity is several orders of magnitude smaller (since for $W \sim w$, $A/W \sim 15$) than the corresponding quantity which arises from the reduction of the point symmetry. This enables Labbé, who has studied the problem in terms of a tight binding model somewhat more complicated than the one presented here,[51] to disregard the tendency to super-periodicity altogether in V_3Si.

At this stage it may be noted that the energy of the uppermost occupied states ($k < k_F$) is reduced not only by a distortion of wavelength $2\pi/k_F$, but also by one with wave-lengths $4\pi/k_F, \ldots$ or a sum of these distortions. This is seen in Figure 5.7 (b). However, the quantity V appearing in the matrix above, equation (5.3), will in these cases be proportional to the second, third, . . . power of the distortion amplitude and will not therefore lead to a thermodynamically stable distorted configuration.

The expansion of the energy in the extreme case $V/W \gg 1$ leads to

$$E_T \sim \tfrac{8}{3}\pi^{-1}k_F W - 2k_F V + k_F A V^2 W^{-2}\pi^{-1}$$

The value of the half band-gap in the stable situation is $V = W^2 A^{-1}\pi^{-1}$ and the stabilization energy per unit length is $k_F W^2 A^{-1}\pi^{-1}$.

We can therefore summarize our conclusions so far in the following. The operative parameter is the ratio of the occupied band width W to the coefficient A in the elastic energy per unit length ($k_F A V^2 W^{-2}\pi^{-1}$). For W/A small, the ratio of the equilibrium band gap to the width W also becomes small: in fact extremely small. When W/A is large, the band gap to band width ratio is also large and a large energy gain results upon distortion. The transitions will take place at a temperature which is of the order of this energy gain.

The preceding one dimensional model can be extended to three dimensions in a number of ways. The appropriate extension depends of course on the physical situation which is studied. We shall consider the metal-to-semiconductor transition in the higher oxides of transition metals, a subject studied theoretically in detail by Adler and Brooks[12] and which will be further discussed by us in §7.12.

The total energy per unit volume is the integral

$$E_T = \frac{2}{(2\pi)^3} \int E(\mathbf{k}) \, d^3\mathbf{k} \tag{5.4}$$

taken over the energies $E(\mathbf{k})$ of all occupied states with either spin. The following forms arise:

$$\tfrac{8}{3}k_F^3(-\alpha V^2 W^{-1} + \tfrac{1}{2}AV^2 W^{-2}) + \text{constant} \qquad (V/W \ll 1)$$

and

$$\tfrac{8}{3}k_F^3(-\beta V + \tfrac{1}{2}AV^2 W^{-2}) + \text{constant} \qquad (V/W \gg 1) \qquad (5.5)$$

Here W is the occupied band width, $2V$ the distortion–induced gap, α and β are numbers and the elastic energy proportional to A has been added. Limiting expressions in these forms were also obtained by Adler and Brooks[12] on the basis of a number of solvable models.

The conditions for the existence of a distortion are for $V/W \ll 1$, that $\alpha > AW^{-1}$, while for $V/W \gg 1$ the form of the expression apparently

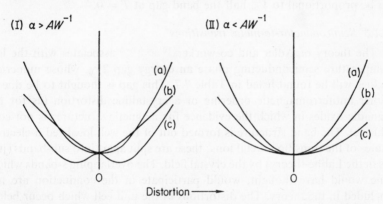

Figure 5.8 The cases of distorted (I) and undistorted (II) equilibrium configurations. (a) is the elastic energy, (b) is the negative of the distortional energy, (c) is the total energy (the free energy at zero temperature) consisting of (a) − (b).

always guarantees a distorted equilibrium configuration. Upon closer investigation of the integral, equation (5.4), over $E(\mathbf{k})$ we find however that, for any reasonable model, the condition $\alpha > AW^{-1}$ is always necessary and only if it is satisfied can one expect the distortion to be given by the second limiting form, equation (5.5). The two situations $\alpha >$ and $< AW^{-1}$ are shown in Figure 5.8.

If the distortion falls in the linear regime, then the half-gap at zero temperature V_0 will take the value $\beta W^2 A^{-1}$ and the energy gain will be

$$\tfrac{8}{3}k_F^3(\tfrac{1}{2}\beta^2 W^2 A^{-1}) = \tfrac{4}{3}k_F^3 \beta V_0$$

Consider now the case of finite temperatures in the narrow band limit where V_0/W is large (note, however, that this implies that $\beta W A^{-1}$ is large). It is expected that, as the temperature is lowered across the transition point, the whole occupied part of the sub-band will experience a sudden drop in energy and a first (rather than second) order transition will occur. Consistently with the abruptness of the transition the latent heat in the transition will be close to the energy gain $\frac{4}{3}k_F^3\beta V_0$ calculated at zero temperature. This quantity is then also equal to the transition temperature times the change in specific entropy or to T_c times the change in specific heat across the transition. This latter quantity is (for a gas of free electrons having small W and large mass) of the order of $(4\pi/3)k_F^3$ (times k, the Boltzmann constant), since all electrons in the band are nearly equally affected by the opening of the gap. It follows therefore that, in the first order transition that is envisaged, the transition temperature is expected to be proportional to V_0, half the band gap at $T = 0$.

5.1.5 Semiconductor-to-metal transitions

The theory of Adler and co-workers[12-3, 53-4] associates with the low temperature semiconducting phase an energy gap $2V_0$, whose numerical values will be found listed in Table 7.7. This gap is thought to be due to either antiferromagnetic ordering or a crystalline distortion (as for the vanadic oxides in which the evidence for magnetic structures is not conclusive). The band structure is formed out of the well localized d-electron states of the transition metal ions; these are split apart into sub-bands (just as in the Labbé-theory) by the crystal field. The overlapping s-bands which, one would have thought, would participate in the conduction are not included in the theory. The distortions in the unit cell which occur below T_c (Table 7.7) are essentially changes in the cation–cation distances. Changes in the dispositions of the oxygen ions are not considered, though such changes may lead to an energy gap even more effectively than the displacement of the cations.

It should be noted that this theory of semiconductor-to-metal transition was preceded by other theories based on magnetic ordering,[55-7] on electron correlation,[58-60] on electron trapping[61-3] or on the formation of strongly bonded cation pairs.[64-5] Subsequently, suggestions involving crystal-field stabilized itinerant d-electrons[66] or a purely displacive type transition[67-8] have also been made. It is outside the reach of our interest to describe any or all of these, or to compare one with another or with the distortional theory[12] of the transition. It has been stated[53] that the distortional theory applies to transitions in VO_2, NbO_2, Mo_9O_{26}, CrS and tetragonal FeS and with less certainty to those in Ti_3O_5, V_3O_5 and V_4O_7, while a combination of the Jahn–Teller mechanism and antiferromagnetism is likely in $V_2O_3(?)$, VO and V_6O_{13}. In the last analysis, it is further experimental evidence, especially of the type which established the distortive nature of the transition in V_3Si (anomalies in the elastic constants, the vanadium NMR, etc.) which will decide between alternative interpretations.

5.1.6 *Lattice instability in semimetals*

In 1961, Mott[69] envisaged phase transitions in these materials to an excitonic phase, i.e. to a phase in which excitons in finite concentration are formed, provided that the band overlap ($-E_G$ in Figure 7.23) is not too large. As is well-known, excitons are elementary excitations which are associated with bound states formed by the Coulomb interactions between electrons in the conduction band and holes in the valence band. The excitons are essentially non-conducting. There can be a direct transition between the semimetallic and non-conducting (also called the semi-conductor phase) phase, or alternatively there may be an indirect transition via a distorted phase.[70]

It is with this distorted phase that we are concerned, in so much as it carries the hall-mark of the Jahn–Teller effect, a distortion or instability due to electronic degeneracy. Not very much is gained by appending the Jahn–Teller label to this or other effects in solids; however, the question may be and has been asked 'why does the J-T-E not operate with the degeneracies in solids?' and the answer is that it does operate under certain circumstances such as the one described in this section.

Coming back to the distorted phase, we note that there may be even a series of distortions as the temperature or the magnitude of the energy gap E_G changes.[71-2] Alternatively there may occur an excitonic- (rather than lattice-) distortion[73] or more accurately a spin density oscillation in the anti-ferromagnetic phase.[72,74] When the lattice distorts, an instability involving the longitudinal optical phonons at $k = 0$ occurs when the two parabolic bands are above each other [case (a) in Figure 7.23], or one involving the mode at $k = \Delta k$, when the conduction band minimum is displaced by this amount from the maximum of the valence band [case (b) in Figure 7.23]. For either mode it is the degeneracy of electronic states which makes the phonon frequency imaginary when the electron–phonon interaction is calculated. This has been done to an arbitrary order in the interaction by Kopaev[73] but by selecting a particular type of diagrams only. Kopaev finds an instability irrespective of the magnitude of the gap. In a more refined treatment (Halperin and Rice, Reference 72) one sees that only below a certain value of $|E_G|$ is instability expected.

There are a number of ways, whereby the lattice stability can be saved. As already noted, a spin-excitonic phase, with $k = 0$ (a) or $k = \Delta k$ (b), may be more stable than the lattice-distorted phase providing that the electron–phonon interaction is insufficiently strong. Alternatively, the lattice may become anisotropic (rather than gain a super-period) which leads, as we know from the general behaviour of the J-T-E, to a stabilized state. This alternative is suggested in Reference 73 for the low temperature semi-metals Bi, As and Sb. Finally, anharmonic interaction between acoustic $\mathbf{k} = 0$ and the unstable optical mode may lead to an isotropic change of lattice dimensions, so that e.g. through a change in E_G the undistorted phase may stay stable.

It is known[75] that even in a normal metal the quasidegeneracy of electronic states (having level separation of the order of the inverse of the linear dimension of the metal) may lead to a phonon instability, essentially of the acoustic modes. This instability is, however, contingent to the coupling strength having a large enough value. The situation considered in this section (semimetals, or in the previous section: narrow-gap semi-conductors) leads *necessarily* to one or other of the instabilities and macroscopic changes discussed above.

5.2 Models for ferroelectrics

The suggestion that the P-J-T-E may be the cause of ferroelectricity in BaTiO$_3$ is due in the first instance to Wigner and then to Jaynes.[76] The idea was further proposed by Liehr[77] and Orgel.[78] Elaborations of the idea, apparently quite independently, were made by Sinha and Sinha[79] and by the present writer[80] who suggested a vibronic mixing between the close-lying 2p and 3s states of the O^{2-} ions. For definiteness we suppose that the triplet t_{1u} at the origin of the Brillouin-zone (the point Γ) is coupled by one (or more) odd vibrations to the singlet a_{1g} (arising for example from an s state) lying at an energy $4W$ higher than t_{1u}. We can assume without much damage to the physical picture that one of these odd modes, denoted by $\mathbf{R} = (X, Y, Z)$ contributes predominantly.

In this and the next sections we abandon the band structure description in favour of the localized states picture. For the sake of simplicity we also replace temporarily the 'rational' coupling coefficients used in this work by the shorter ones employed by Öpik and Pryce.[81] The linear P-J-T matrix is then written as

$$
\begin{array}{c}
|x\rangle \\
\\
|y\rangle \\
\\
|z\rangle \\
\\
|s\rangle
\end{array}
\begin{bmatrix}
-W + A(-2\xi_x \\ \quad + \xi_y + \xi_z) & BQ_\zeta & BQ_\eta & CX \\
\\
BQ_\zeta & -W + A(-2\xi_y \\ \quad + \xi_z + \xi_x) & BQ_\xi & CY \\
\\
BQ_\eta & BQ_\xi & -W + A(-2\xi_z \\ \quad + \xi_x + \xi_y) & CZ \\
\\
CX & CY & CZ & 3W + A'(\xi_x \\ \quad + \xi_y + \xi_z)
\end{bmatrix}
$$

$$\tag{5.6}$$

The vibrational amplitudes are the following combinations of the doubly degenerate modes Q_θ, Q_ϵ and of the isotropic displacements Q_1

$$\left.\begin{aligned}
\xi_x &= \tfrac{1}{2}\sqrt{3}Q_\epsilon - \tfrac{1}{2}Q_\theta + \frac{1}{\sqrt{3}}Q_1 \\[2mm]
\xi_y &= -\tfrac{1}{2}\sqrt{3}Q_\epsilon - \tfrac{1}{2}Q_\theta + \frac{1}{\sqrt{3}}Q_1 \\[2mm]
\xi_z &= Q_\theta + \frac{1}{\sqrt{3}}Q_1
\end{aligned}\right\} \qquad (5.7)$$

To complete the static Hamiltonian we add to equation (5.6) all quadratic terms in the displacements ξ, R and (Q_ξ, Q_η, Q_ζ), as well as all quartic and sixth-order terms in R consistent with cubic symmetry. When one inserts the eigenvalues E_i of the static Hamiltonian in the free energy

$$F = -kT \log \sum_i e^{-E_i/kT}$$

one obtains[80] a remarkable similarity between the resulting expression and the phenomenological free energy of Devonshire[82] for $BaTiO_3$, both as regards the form of the terms and the qualitative temperature dependence. By a suitable choice of parameters one can, in fact, arrange the system to be in the centrosymmetric minimum at high temperature and then, as the temperature is lowered, to go through a series of minima (of the free energy!) which have in turn tetragonal, rhombohedral and orthorhombic symmetries.

The model also accounts for the vanishing of the ferroelectric mode near the transition temperature (§7.12). Since $C^2/(4W)$ is probably the most important term in the electronic polarizability (of O^{2-}, if t_{1u} arises from oxygenic 2p-states) the model naturally incorporates also the elements of Slater's theory[83] for $BaTiO_3$. Actually, the oxygen ions are not at points of cubic symmetry; nevertheless, the matrix of equation (5.6) is correct if an appropriate linear combination of oxygenic orbitals, utilizing the general cubic symmetry, is taken. On the other hand, Jing-Der[84] bases his theory on the D_{4h} point symmetry of each individual O^{2-} arguing that, due to the crystal field and to covalent bonding along Ti–O–Ti, a 3d (e_θ) orbital on the z-oxygen tends to become degenerate with a 2p-level. He supposes that the corresponding P-J-T-E energy difference $4W$ is proportional to $(T - T_c)$ and that this mechanism triggers the phase-transition.

The dominant difficulty in all these theories[76-80,84] is that the energy difference $4W$ must be of the order of k times the Curie temperature ($\sim$400 °K) and an allowed absorption ($t_{1u} \rightarrow a_{1g}$) ought to be clearly observed at such energies. Also it has been shown within the molecular field approximation that for a phase transition to occur the degeneracy of the excited states must be higher than that of the ground state.[85]

Another, ingenious model accounting for the phase transition in $BaTiO_3$ was put forward by Bersuker.[86] Twelve valence electrons of the $BaTiO_3$ unit are supposed to be accommodated within the lowest twelve of the eighteen spin orbitals arising from $3d(t_{2g})$ states on the Ti ion and the six

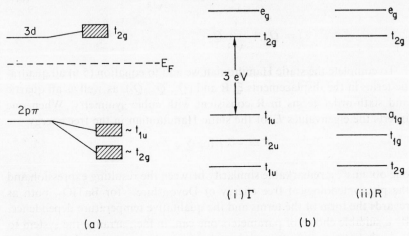

Figure 5.9 (a) The scheme of energy bands in $BaTiO_3$ according to Reference 86.

(b) Computed order of energy levels in $SrTiO_3$ at the symmetry points (i) Γ and (ii) R [1, 1, 1] (Reference 87).

$2p\pi$ orbitals of oxygen. The author of Reference 86 supposes that these latter have strong t_{2g} or t_{1u} characters. (These may arise partly from the point Γ, where the irreducible representations are t_{1u} and t_{2u}, and partly from the corner of the Brillouin-zone, where at $\mathbf{k} = [1, 1, 1]$ the small representations are t_{1g} and t_{2g}). With crystal field and covalency effects included the energy band scheme shown in Figure 5.9 is conceivable. A trigonal distortion of the unit cell, initiated by the displacement of the titanium ion along a body diagonal, may under favourable circumstances further increase the gap $4W$ between the occupied and empty bands. (The author of Reference 86 found that other distortions lead to stationary points of higher energy than the trigonal points). At low temperatures the crystal would then be polarized along one of the body diagonals. As the temperature is raised the system begins to jump the barriers between the minima: first the lowest barrier along the cube edges, then the ones across a face diagonal and finally the ones across the body diagonal. The various phases arise from the averaging of the polarizations.

Here again one must note that the energy gap seems to have escaped observation thus far. Also, available energy band calculations[87] do not agree with the scheme drawn in Figure 5.9.

5.3 Cooperative transitions by localized J-T centres

When J-T-ions are present in solids in finite concentrations a crystallographic phase transition may take place at some temperature T_t (which normally increases with the concentration).[88] Best known are the transitions in spinels and perovskites (§7.12). An example of how the axial ratio in spinels changes with temperature or with the concentration of the Mn^{3+} ions at low temperatures is shown in Figure 5.10. In Appendix VIII experimental data are shown for the phase transitions in spinels doped with various d^n ions. The data pertaining to perovskite type of oxides are less extensive.

The cause and nature of the transition in these oxides are easily understood in broad terms. Qualitatively speaking, what happens in the oxides is that the J-T distortions at neighbouring centres, which we considered in Chapter 3 in isolation, interfere with one another in the solid. At low temperatures an alignment of the distortions takes place giving rise to

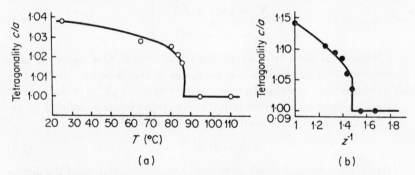

Figure 5.10 The macroscopic tetragonality c/a of spinel lattices with cooperative J-T-distortion. (a) c/a vs. temperature in $Cr(FeCr)O_4$. (From Reference 89.) (b) c/a vs. z^{-1}, where z is the B-site concentration of Mn (the J-T-ion) in $(ZnMn_2O_4)_z(GeCo_2O_4)_{1-z}$. (After Reference 90.)

a macroscopic change in the crystal shape. When we look at the problem from the point of view of the occupation of the degenerate states, the effect becomes similar to the occupation of the up or down spin state through a superexchange mechanism.[91] However, the situation becomes for us further complicated by the fact that the superexchange mechanism does in fact also operate in magnetic oxides and may reinforce, oppose or modify the tendency of the J-T-E.[92] Moreover, the occupation of the degenerate states is not a good quantum number if the dynamic character of the vibrations and the tunnelling induced by these are included.

In addition to spinels and perovskites, cooperative J-T distortions occur in some salts (in $CuSiF_6 \cdot 6H_2O$, at 300 °K),[93] in UO_2 (where magnetic changes are also involved),[92] as well as in higher oxides (§5.1.4), in semi-metals (§5.1.5) and perhaps even in metals and alloys (§5.1.1 and Reference 94). What distinguishes the spinels and perovskites are the relatively large cavity between the oxygen ions which can accommodate the Jahn–Teller ions without interference and the (connected fact of) very narrow impurity bands formed by the J-T-ions in finite concentrations. For these reasons the models of J-T-centres which have localized electronic states and interact through displacements of the anions were very successful in these compounds.

A statistical mechanical treatment was given by Finch, Sinha and Sinha[95] and in an improved form by Wojtowicz[96] and Kanamori.[97] Following the last two, we define for a lattice that is macroscopically tetragonally distorted (along the z-axis) a long-range order parameter s. This gives the number of octahedra N_x, N_y, N_z whose tetragonal axis is along the x, y, z-axes, respectively, according to the relations

$$N_z = \tfrac{1}{3}N(1 + 2s)$$
$$N_x = N_y = \tfrac{1}{3}N(1 - s) \tag{5.8}$$

N is the total number of octahedra in the crystal. One notes the tacit assumptions that the distortion is along one of the cubic axes and that its magnitude is *not* a variable. In terms of the treatment §3.2.3 of the isolated octahedron these assumptions imply a very strong linear coupling $L \gg \hbar\omega$ and the presence of the anisotropy term, equation (3.20)

$$-\beta \cos 3\phi \tag{5.9}$$

Here ϕ represents, through equation (3.6), the orientation of the distortion. In addition, each octahedron is subject to the stress fields of the neighbouring octahedra, which are also distorted. The stress appears in the Hamiltonian in the form, equation (4.1),

$$-(\gamma_\theta \cos \phi + \gamma_\epsilon \sin \phi)q \tag{5.10}$$

The method employed by References 95–7 to evaluate the stress energies γ_θ and γ_ϵ was the mean (or molecular) field approximation.[98] The orientations of the neighbouring octahedra were represented by the average quantities $\langle \cos \phi \rangle$ and $\langle \sin \phi \rangle$, so that equation (5.10) takes the form

$$-\lambda\hbar\omega q_0^2[\langle \cos \phi \rangle \cos \phi + \langle \sin \phi \rangle \sin \phi] \tag{5.11}$$

where q_0, equation (3.15), is the radial displacement of the J-T-centres due to the linear term and the dimensionless quantity λ is defined as follows:

There are four possible types of interactions between a pair of octahedra which have two corners in common. These are shown in the Figure 5.11 together with the labelling of the interaction energies V_{ij}. In the mean field

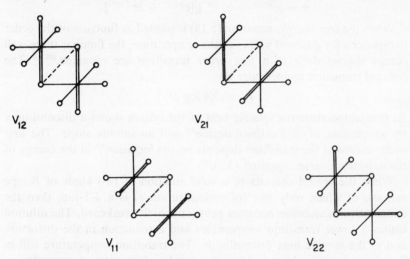

Figure 5.11 Energies of interaction between a pair of distorted octahedra, whose axes of distortion (shown with double lines) have various relative orientations. (After Reference 96.)

approximation the pair interaction energies of all possible relative orientations are replaced by the mean interaction energy of randomly distributed pairs. This defines λ_{spinel} through the relation.

$$\lambda_{\text{spinel}} \hbar \omega q_0^2 = \tfrac{4}{3}(V_{22} - V_{11} + 2V_{21} - 2V_{12}) \qquad (5.12)$$

This quantity is very likely to be positive (although different authors[96,100-1] have assigned different relative magnitudes to V_{ij}).

Since ϕ measures the deviation of distortion from the z-direction, in a z-oriented tetragonal crystal

$$\langle \sin \phi \rangle = 0$$

and $\cos \phi$ is equal to the order parameter s, taking values from -1 to 1.

The number of ways a phase with a given N_x, N_y, N_z, can occur is $N!/(N_x! N_y! N_z!)$ and the free energy is, by equation (5.8),

$$F(s) = F(0) - \tfrac{1}{2}s^2 \lambda \hbar \omega q_0^2 + \tfrac{1}{3}NkT[(1 + 2s) \log (1 + 2s)$$
$$- 2(1 - s) \log (1 - s)] \qquad (5.13)$$

In thermal equilibrium $dF/ds = 0$ and this yields

$$\log \frac{1 + 2s}{1 - s} = 3\lambda\hbar\omega q_0^2 s/(kT) \equiv 3s/T'$$

thus defining a reduced temperature T'. This equation can be rewritten as

$$s = [e^{2s/T'} - e^{-s/T'}]/[e^{2s/T'} + 2e^{-s/T'}]$$

When the free energy, equation (5.13) is plotted as function of the order parameter s for different values of the temperature, the familiar three-well curves characteristic of a first order transition are obtained.[96,98] The reduced transition temperature is

$$T_t' = 3(4 \log 2)^{-1}$$

At this temperature the specific heat of the system shows a discontinuity by an amount of 22·4 cal/mol/degree[96] and an infinite slope. The first order nature of the transition depends on the inclusion[97] in the energy of the anisotropy term, equation (5.9).[99]

When the crystal consists of a solid solution of two kinds of B-type cations, of which only one (of concentration z) is a J-T-ion, then the cooperative mechanism becomes proportionately weakened. The dilution causes a lower transition temperature and a reduction in the distortion and in the specific heat discontinuity. The transition temperature will in fact be proportional to z. Also, the graphs of physical changes during transition when plotted against z^{-1} at constant temperature will have a very similar appearance to the plot against temperature for constant composition. The two parts of Figure 5.10 confirm this prediction of the mean field theory, at least qualitatively.

The Hamiltonian which is the starting point of Kanamori's calculation[97] differs from that given by us in Chapter 3 in that it is written in the lattice waves representation, i.e. in terms of phonons and Pauli-matrices σ_θ, σ_ϵ operating in the function space of Bloch-waves, as well as of macroscopic strain variables. Two cases are investigated by him in detail: the first, ferro-distortive ordering in spinels, we have already discussed. This represents in the lattice wave formalism a distortion along a phonon mode with wave-vector $k = 0$. A condition for such a distortion to occur is that the energy gained through it be larger than the energy gain in any distortion through any other (competing) mode with $k \neq 0$.

The second case treated in Reference 97 is a perovskitic crystal, shown in Figure 7.17, with the composition $\square MnF_3$ (where $\square$ denotes an A-site vacancy). The observed distortion of the crystal indicates that the dominant coupling is to a lattice mode with wave-vector $(\pi, \pi, \pi)/a$. (This is a point on the corner of the first Brillouin-zone of vibrations). This distortion splits the lattice into two interlocking sublattices (designated $+$ and $-$), with nearest neighbour Mn^{3+}-ions belonging to different sublattices. This case has been termed antiferrodistortive.[102]

The coupling between one Mn^{3+} octahedron and the six neighbouring ones can again be represented by a stress term like equation (5.10) or, in the mean field approximation, by equation (5.11). However, there are now three differences. First, the thermal averages $\langle \, \rangle$ appearing in the energy expression, equation (5.11) of one sublattice $(+)$ are in fact averages taken over the other lattice $(-)$, and vice versa. Secondly, the averages $\langle \sin \phi \rangle$ are not zero, although the distorted lattice is macroscopically tetragonal. The tetragonality arises from the following empirical relations that connect the two sublattices.

$$\langle \cos \phi \rangle_+ = \langle \cos \phi \rangle_-$$
$$\langle \sin \phi \rangle_+ = -\langle \sin \phi \rangle_- \qquad (5.14)$$

The structure in Figure 7.20 satisfies these relations.

Thirdly, the coefficient λ appearing in equation (5.11) has not now the meaning given in equation (5.12) and is in fact empirically found to be negative, so that it is better to write $-\lambda = \lambda_{perov} > 0$. An interpretation of this last quantity in terms of the cation–cation force constant was suggested in Reference 102.

Returning now to equation (5.11) and taking its thermal average we realize that, by equation (5.14), the energetically favourable values of the averages are

$$\left.\begin{array}{c} \langle \cos \phi \rangle = 0 \\ \langle \sin \phi \rangle \neq 0 \end{array}\right\} \qquad (5.15)$$

On the other hand, the energy minima of the anisotropy term, equation (5.9), are at $\phi = 0, \pm 2\pi/3$ $(\beta > 0)$. It is clear that these values are not compatible with the averages in equation (5.15) in an ordered arrangement. In reality, a compromise will be found between the purely anti-ferro-distortive and the purely anisotropic tendencies [represented, respectively, by equations (5.15) and (5.9)]. X-ray measurements of the three different distances $l = 2\cdot09$ Å, $m = 1\cdot91$ Å and $s = 1\cdot79$ Å [103] in MnF_3 yield

$$\left| \frac{\langle \cos \phi \rangle}{\langle \sin \phi \rangle} \right| = \frac{2m - l - s}{\sqrt{3}(l - s)} = 0\cdot115$$

showing that the anti-ferrodistortive coupling is quite strong. On the other hand the copper-fluorine distances in $KCuF_3$ are $l = 2\cdot25$ Å, $m = 1\cdot96$ Å and $s = 1\cdot89$ Å,[104–5] so that here

$$\frac{\langle \cos \phi \rangle}{\langle \sin \phi \rangle} = 0\cdot35,$$

indicating stronger anisotropic coupling.

Neither for the spinels nor for the perovskites have we taken into account so far the dynamic nature of the distortions of each octahedron. In fact the treatments of Reference 95–7 correspond to the classical solution of the J-T-E. The quantum mechanical solution will add (even within the mean field approximation) new physical features to the phase transition-problem. These new features, tunnelling and the effect of excited states, will represent additional averaging tendencies whose effect will be quantitative at elevated temperatures (since here thermal excitations already provide a randomizing mechanism) and *qualitative* at zero (or low) temperatures (when classical systems are frozen in).

In the quantum mechanical solutions[100,102] which utilize the molecular field approximation, the strain-energy of equation (5.11) is added to equation (3.21) and the new Hamiltonian H' is used to define the averages in a self-consistent manner, as follows.

$$\left.\begin{aligned}
\langle\cos\phi\rangle &= \mathrm{Tr}\,(\cos\phi\,e^{-H'/kT})/\mathrm{Tr}\,(e^{-H'/kT})\\
\langle\sin\phi\rangle &= \mathrm{Tr}\,(\sin\phi\,e^{-H'/kT})/\mathrm{Tr}\,(e^{-H'/kT})
\end{aligned}\right\} \tag{5.16}$$

These are four self-consistent equations for a perovskite, two for each sublattice. They reduce to two equations after the introduction of the two conditions in equation (5.14). In the spinel case $\langle\sin\phi\rangle = 0$. However, at the transition temperature there are still two equations since in addition to the self-consistency, expressed by equation (5.16), in a first-order phase transition we also require that the free energy of the ordered and disordered phases be equal.

In the cases studied, the transition temperature T_t is of the order of the stress energy $\lambda\hbar\omega q_0^2/k$. (Their ratio is 0·54 for the static case in spinels,[96] dynamic coupling generally reduces this number). Reference to the experimental values of T_t (~1000 °K) for spinels doped at B-sites in Appendix VIII shows the stress energy to be a large quantity. On the other hand we have already noted that the existence of a first order transition, or better the observation of a latent heat, depends on the anisotropy energy, β in equation (5.9), not being much smaller than the stress energy. This puts (with $\alpha \sim 10$ cm^{-1}) the ratio β/α well above 10, so that the tunnelling energy 3Γ, equation (3.28) is very small indeed. In Figures 5.12–5.13 we show some calculated specific heat curves, which indirectly illustrate the dependence of the nature of the transition on the parameter β/α. The tetragonality, c/a, will be shown in Figure 5.15. This is found to be lower in the calculations with dynamic coupling than without. Near zero temperature, where all distortions are aligned, the tetragonality should depend only on the distortion of each centre (and not on the details of the cooperative mechanism). If it is possible to make a guess for c/a from

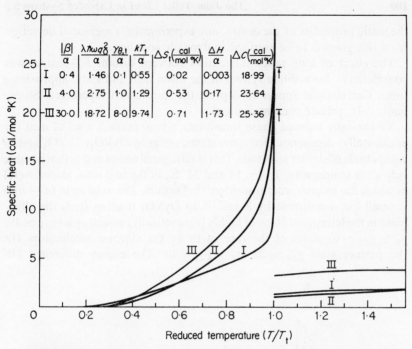

Figure 5.12 Specific heat curves calculated for spinels. The parameters of the three curves are shown in the first two columns of the inset. The results of the self-consistent molecular field calculations[100] are shown in the other columns.

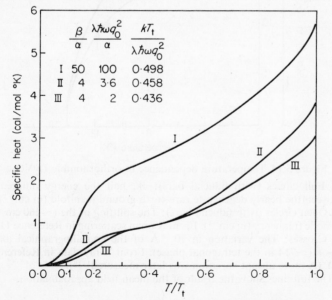

Figure 5.13 Calculated specific heat curves for perovskites.[102]

the static properties of the centre, any experimentally measured deviation from this guess is to be ascribed to dynamic coupling.

The effect of long range interaction between distant J-T-centres was considered[106] for a dilute system in the mean field and static approximations. Correlations appear to depress the transition temperature (Stoneham, 1971, private communication).

Vibronically induced phase transitions, whose cause is a set of near (or accidentally) degenerate electronic states, occur in $DyVO_4$, $TbVO_4$ and in compounds of similar structure. This is tetragonal above and orthorhombic below the temperature T_D (= 14 and 34 °K, in the two salts, respectively) at which the second-order transition[107-9] occurs. The axial ratio $(a - b)/b$ is small but measurable by X-ray.[110] In $DyVO_4$ it arises from the alignment in the tetragonal plane of highly (magnetically) anisotropic ($g_{\perp 1} \simeq 19$, $g_{\parallel} \simeq g_{\perp 2} \simeq 0$) states of the rare earth by the vibronic mechanism. (In the partner state $g_{\perp 2} \simeq 20$, $g_{\parallel} \simeq g_{\perp 1} \simeq 0$). The energy difference $2W$

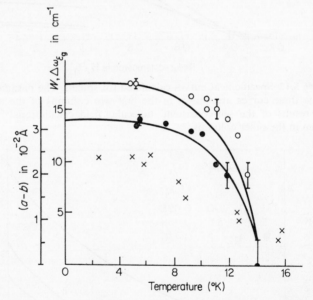

Figure 5.14 Temperature dependence in orthorhombic $DyVO_4$.
(a) Full circles (with fiducial bars): W, half the energy difference between the nearly degenerate rare-earth ground manifold (in cm^{-1}).
(b) Open circles (with fiducial bars): The splitting of the ($\sim$260 cm^{-1}) ε_g mode frequency (in cm^{-1}). From Raman scattering in Reference 111.
(c) Crosses: The variation in 10^{-2} Å of the crystallographical axes $a - b$ ($\simeq$714 in the tetragonal phase). From X-ray data in Reference 110. (A more abrupt variation was later measured[111a]).
The full lines show the result of the mean field approximation.

between these two states (each of which has a Kramers degeneracy) as well as the variation of the ε_g mode frequencies were measured[111] by Raman scattering as a function of temperature below T_D. These, as well as $(a - b)/b$, are expected to be proportional to the order parameter, within the m.f.a. Figure 5.14 shows the results.

The appropriate Hamiltonian of the problem is that in equation (3.82), or more conveniently the equivalent form in equation (7.14), ignoring q_+ and with $q_- \equiv q$. In the static and m.f. approximations, which were used in Reference 111, the self-consistent equation (cf. equation (5.16)) takes the form

$$(4E_{JT}^2 \langle \sigma_\theta \rangle^2 + W^2)^{\frac{1}{2}} = 2E_{JT} \tanh [4E_{JT}^2 \langle \sigma_\theta \rangle^2 + W^2)^{\frac{1}{2}}/kT] \quad (5.17)$$

in which the order parameter $\langle \sigma_\theta \rangle = \langle q \rangle/(L/\sqrt{2\hbar\omega})$ replaces $\langle \cos \phi \rangle$. This has non-zero solutions for $\langle \sigma_\theta \rangle$ if $W < 2E_{JT}$.

A detailed theoretical study of this crystallographical transition is due to Elliott,[112] who first transformed his phonon variables to displaced oscillators and then wrote down the interaction energy between the variables σ_θ on different sites.

J. R. Fletcher (private communication, 1971) considered the corresponding cooperative J-T-problem (i.e., with $W = 0$); however he allowed for the existence of finite macroscopic strains below the transition point T_D.

5.4 Elementary excitations of the vibronic type

Deviations from the completely ordered state described in the previous section should be excitable near zero temperature and should emit energy at somewhat higher temperatures.[113] The deviations, or elementary excitations, in solids are often highly correlated types of motion whose treatment requires advanced calculational techniques. Treatments of this type were undertaken (Reference 113 and H. Thomas, unpublished) for the ordered phase in spinels, subject to strong anisotropic coupling. In this limit, when the distortions of each octahedron are oriented along one of the cubic axes, the theory may be so formulated[106] that neighbouring distortions (Figure 5.11), even if they do not match, are incapable of inducing a reorientation in the octahedra ('there are no spin-flip terms') but they contribute to the relative energy of the distortions. In this formalism the deviations are propagated from site to site because the tunnelling mechanism admixes states representing different orientations of the same site. We have already noted that in spinels the tunnelling is very small and therefore the effect of the propagation of the deviations is rather small. Consequently the energies of the excitations are very nearly equal to the differences of energies shown for tetragonally strained systems in Figure

4.2, which did not include correlation. Knowing the growth of the spontaneous strain with decreasing temperature (Figure 5.15), we can reasonably simply construct an excitation *vs.* temperature curve, such as is

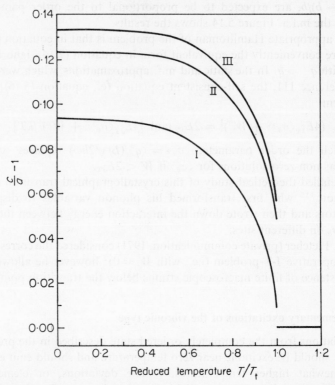

Figure 5.15 The tetragonality in spinels as function of the reduced temperature T/T_t (T_t is the transition temperature). The parameters of the three curves I, II and III are given in Figure 5.12. (From Reference 100.)

shown in Figure 5.16. The excitation energies in this figure are related to the energy levels of the right half of Figure 4.2(a), (b) in the following way:

$$E_a = A_1^{II} - A_1^{I}, \quad E_b = A_1^{II} - B_1^{I}, \quad E_c = B_1^{I} - A_1^{I}$$

Detailed theoretical study[113] of the means of observing these excitations in Mn^{3+}-doped spinels predicted in general rather weak transitions. So far no observations have been made.

Elliott[112] has studied the dynamics of the phase transition for an ensemble of coupled $(A + B) \otimes \beta$ P-J-T systems with application to the distortions in the vanadates (p. 180). As long as the energy difference W or E_{JT} is much less than the phonon frequencies (which appears to be the case in these systems) the phonon modes can be decoupled from a slightly modified form of the electronic excitations. The equations of motion for the latter have been studied[114] and give the result for the energy of elementary excitations at wave-vector $\mathbf{k}$ (assuming periodic boundary conditions):

$$\tfrac{1}{4}[\hbar\omega(\mathbf{k})]^2 = J(0)^2\langle\sigma_\theta\rangle^2 + W[W - J(\mathbf{k})\langle\sigma_\epsilon\rangle]$$

where $J(\mathbf{k})$ is the Fourier transform of the interaction energy between different sites. At the transition temperature, given by equation (5.17) with $\langle\sigma_\theta\rangle = 0$, $\hbar\omega(0) \to 0$, showing typical soft mode behaviour. When coupling to the acoustic modes is included, then it is one of the latter which softens, as also do the corresponding x (adiabatic and isothermal) elastic constants as also do the corresponding x (adiabatic and isothermal) elastic constants.[115].

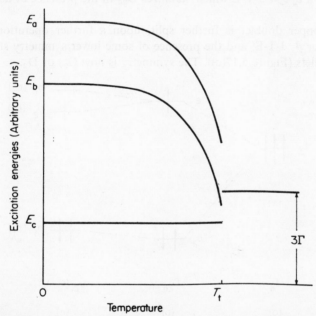

Figure 5.16 Elementary excitations spectrum due to vibronically coupled E states in spinels as function of temperature. For the sake of clarity the size of Γ is increased many-fold over its value in most spinels.[113]

5.5 Few-electron systems

It seems appropriate after elaborating on extended many-body systems to consider some few-electron effects occurring in localized centres.

In a few-electron system, the low symmetry fields which will be discussed in detail in §7.1 can in principle compete with the inter-electronic

Coulomb interaction, though it is unlikely that what we term on p. 218 'random' fields can do so effectively. These fields were, we recall, almost unobservable in a non-J-T situation and came only to the fore in a multiple-well system. The three electron system in T_d symmetry was discussed[116] in terms of a vibronically stabilized distortion with a view of accounting for the low symmetry (C_{2v}) field evident in the electron spin resonance of negatively charged vacancies in diamond[117] and of Ni^- in germanium or Pd^- and Pt^- in silicon.[118-9] In all these cases after the inner shell states are filled three electrons are left over to occupy the six spin-orbitals of t_2 symmetry. The energy level scheme below[116] (Figure 5.17), shows the heights of the one-electron energies first in (a) a static field of T_d symmetry, then (b) when a tetragonal field splits the triplet, placing the singlet lower. We suppose, following Reference 116 that this state of affairs comes about through a $t_2 \otimes \varepsilon$ J-T-E which stabilizes D_{2d} in the presence of a random strain.

The upper doublet is further split, upon a further operation of an $e \otimes \beta_2$ or β_1 J-T-E, and the presence of some lower symmetry strain, to two singlets (Figure 5.17(c)). The symmetry is now C_{2v} or D_2.

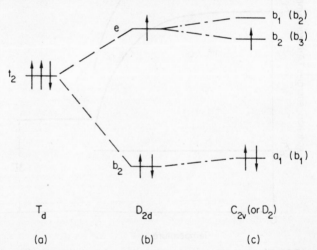

Figure 5.17 Three-electron system in an orbital-triplet manifold, with the D_{2d} distortional energy gain larger than interelectronic Coulomb interaction. The order of the upper levels in C_{2v} (or D_2) is arguable.

With three electrons to be 'placed', as indicated by the arrows in Figure 5.17, the lowest energy configuration has a biaxial symmetry. This presupposes that the Coulomb interaction between various many-electron states, which arise from allocating three arrows among the triplet in various ways, is weaker than the separation in (b). In the opposite case, which is probably more frequent, one gets the following diagrams (Figure 5.18).

Here the level in (a) denotes the sum of one-electron energies of t_2-states in T_d symmetry, in the absence of Coulomb interaction. (b) shows the height of the three-electron terms in T_d with Coulomb interelectronic

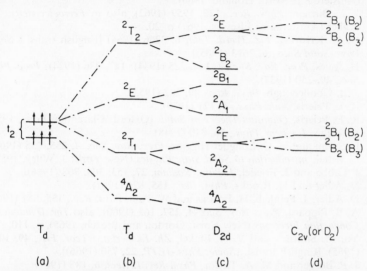

Figure 5.18 Three-electron system, e.g. V^- in Si, in different local symmetries. The level scheme corresponds to the situation where Coulomb interaction dominates the distortive tendencies in the spin-doublet states. Then the P-J-T-E, which would depress a doublet state below the quartet, is ineffective.

interaction. (c) includes the operation of J-T-E on the states of (b) and the effect of a perturbation of D_{2d} symmetry or lower. The ground state in T_d or in D_{2d} symmetry is (in opposition to the previous figure) *orbitally non-degenerate*. When the system is excited to a state which in D_{2d} is degenerate, the J-T-E (reduced by the Ham-effect, Chapter 3) will again operate and a further reduction in symmetry will occur (Figure 5.18(d)).

A distortion to D_{2d}, D_2 or C_{2v} may also arise, starting from the non-degenerate $^4A_2$ ground state in T_d, through an inter-term coupling in the spin *doublets* (a P-J-T-E, which may be represented by reference to Figure 5.18 as $(^2E + {}^2A_2) \otimes \varepsilon$ or $(^2B_1 + {}^2A_2) \otimes \beta_2$ in D_{2d} symmetry, or $(^2T_1 + {}^2E) \otimes \tau_2$ in T_d provided that this coupling is comparable to or larger than the separation between the terms.[120] It is possible that this is the case in silicon in which the configuration interaction is small.[117] More common is the case when the reduced symmetry is the one stabilized by an intra-term J-T-E.

5.6 References

1. H. Jones, *Theory of Brillouin Zones and Electronic States in Crystals* (Amsterdam: North Holland, 1960).
2. J. L. Birman, *Phys. Rev.*, **125**, 1959 (1962), also in *Ferroelectricity*, Ed. Waller (Amsterdam: Elsevier, 1967) p. 20.
3. N. N. Kristofel, *Fiz. Tverd. Tela*, **6**, 3266 (1964) [English transl.: *Soviet Phys.-Solid State*, **6**, 2613 (1965)].
4. H. Jones, *Proc. Roy. Soc. A*, **144**, 225 (1934); **147**, 936 (1934); *Proc. Phys. Soc.*, **49**, 250 (1937).
5. J. B. Goodenough, *Phys. Rev.*, **89**, 282 (1953).
6. R. E. Peierls, *Ann. Phys.*, **4**, 121 (1930).
7. R. E. Peierls, *Quantum Theory of Solids* (Oxford: Clarendon Press, 1956).
8. H. Kuhn, *J. Chem. Phys.*, **16**, 840 (1948).
9. L. Salem and H. C. Longuet-Higgins, *Proc. Roy. Soc. A*, **255**, 435 (1960).
10. C. Kittel, *Introduction to Solid State Physics* (New York: J. Wiley, 1959).
11. J. Labbé and J. Friedel, *J. Phys. Radium*, **27**, 153, 303, 708 (1966).
12. D. Adler and H. Brooks, *Phys. Rev.*, **155**, 826 (1967).
13. D. Adler, J. Feinleib, H. Brooks and W. Paul, *Phys. Rev.*, **155**, 851 (1967).
14. A. B. Pippard, *Proc. Roy. Soc. A*, **257**, 165 (1960); also *The Dynamics of Conduction Electrons* (New York: Gordon and Breach, 1965), p. 110.
15. Yu. M. Smirnov and V. A. Finkel, *Zh. Eksperim. i Teor. Fiz.*, **49**, 1077. (1965). [English transl.: *Soviet Phys.-JETP*, **22**, 750 (1966)].
16. J. P. Burger and M. A. Taylor, *Phys. Rev. Letters*, **6**, 185 (1961).
17. W. Rostoker and A. S. Yamamoto, *Trans. Am. Soc. Metals*, **47**, 1002 (1955).
18. H. Suzuki and S. Miyahara, *J. Phys. Soc. Japan*, **21**, 2735 (1966).
19. L. F. Mattheiss, *Phys. Rev.*, **134**, A970 (1964); **139**, A1893 (1965).
20. M. H. Cohen, L. M. Falicov and S. Golin, *IBM J. Res. Develop.*, **8**, 215 (1964).
21. V. Heine and D. Weaire, *Phys. Rev.*, **152**, 603 (1966).
22. P. S. Rudman, J. Stringer and R. I. Jaffe, Editors, *Phase Stability in Metals and Alloys* (New York: McGraw-Hill, 1967).
23. E. Hornbogen, A. Segemuller and G. Wasserman, *Z. Metallkunde*, **48**, 379 (1957).
24. D. B. Masson and C. S. Barrett, *Trans. AIME*, **212**, 260 (1958).
25. L. C. Chang and T. A. Read, *J. Metals*, **3**, 47 (1951).
26. H. Jones, *Proc. Roy. Soc. A*, **144**, 225 (1934); **147**, 396 (1934), *Proc. Phys. Soc.*, **49**, 250 (1937).
27. P. W. Anderson, *Concepts in Solids* (New York: Benjamin, 1963).
28. A. B. Pippard, *Phil. Trans. Roy. Soc. A*, **250**, 325 (1957).
29. R. W. Morse, A. Myers and C. T. Walker, *Phys. Rev. Letters*, **4**, 605 (1960).
30. H. W. King and T. B. Massalski, *Phil. Mag.*, **6**, 669 (1961).
31. T. B. Massalski and H. W. King, *Acta Met.*, **8**, 677 (1961); also in *The Fermi Surface* (London: J. Wiley, 1960) p. 290.
32. T. B. Massalski, *Acta Met.*, **5**, 541 (1957).
33. W. B. Pearson, *A Handbook of Lattice Spacings and Structures in Metals and Alloys* (Oxford: Pergamon, 1958).

34. H. Sato and R. S. Toth, *Phys. Rev.*, **127**, 469 (1962); **139**, A1581 (1965); *Solid State Commun.*, **2**, 249 (1964).
35. H. Sato and R. S. Toth, *J. Phys. Chem. Solids*, **29**, 2015 (1968).
36. J. Villain, *J. Phys. Rad.*, **23**, 861 (1962).
37. R. Pick, *J. Phys. Rad.*, **24**, 123, 233 (1963).
38. P. W. Anderson and E. I. Blount, *Phys. Rev. Letters*, **14**, 217 (1965).
39. A. P. and M. F. Cracknell and B. L. Davies, *Phys. Stat. Sol.*, **39**, 463 (1970).
40. J. Labbé, *Phys. Rev.*, **158**, 647 (1967).
41. F. J. Morin and J. P. Maita, *Phys. Rev.*, **129**, 1115 (1963).
42. J. E. Kunzler, J. P. Maita, H. J. Levinstein and E. J. Ryder, *Phys. Rev.*, **143**, 390 (1966).
43. J. Bonnaret, J. Hallais, S. Barisic and J. Labbé, *J. Physique*, **30**, 701 (1969).
44. A. M. Clogston and V. Jaccarino, *Phys. Rev.*, **121**, 1357 (1961).
45. A. M. Clogston, A. C. Gossard, V. Jaccarino and Y. Yafet, *Rev. Mod. Phys.*, **36**, 170 (1964).
46. A. M. Clogston, V. Jaccarino and Y. Yafet, *Phys. Rev.*, **134**, A650 (1964).
47. A. C. Gossard, *Phys. Rev.*, **149**, 246 (1966), *Bull. Am. Phys. Soc.*, **13**, 366 (1968).
48. S. Berko and M. Weger, *Phys. Rev. Letters*, **24**, 55 (1970).
49. M. Weger, *J. Phys. Chem. Solids*, **31**, 1621 (1970).
50. S. Barisic and J. Labbé, *J. Phys. Chem. Solids*, **28**, 2477 (1967).
51. J. Labbé, *J. Physique*, **29**, 195 (1968).
52. M. Rosen, H. Klimker and M. Weger, *Phys. Rev.*, **184**, 466 (1969).
52a E. Pytte, *Phys. Rev. Letters*, **25**, 1176 (1970); *Phys. Rev.*, **B4**, 1094 (1971).
53. D. Adler, *Solid State Phys.*, **21**, 1 (1968); *Rev. Mod. Phys.*, **40**, 714 (1968).
54. J. Feinleib and W. Paul, *Phys. Rev.*, **155**, 841 (1967).
55. J. C. Slater, *Phys. Rev.*, **82**, 538 (1951).
56. F. J. Morin, *Phys. Rev. Letters*, **3**, 34 (1959).
57. J. de Cloizeaux, *J. Phys. Radium*, **20**, 600, 751 (1959).
58. N. F. Mott, *Proc. Phys. Soc.*, **A62**, 416 (1949).
59. J. Hubbard, *Proc. Roy. Soc. A*, **276**, 238 (1963); **277**, 237 (1964); **281**, 401 (1964).
60. W. Kohn, *Phys. Rev.*, **133**, A171 (1964).
61. J. Yamashita and T. Kurosawa, *J. Phys. Soc. Japan*, **15**, 802, 1211 (1960).
62. T. Holstein, *Ann. Phys.*, **8**, 343 (1959).
63. R. R. Heikes and W. D. Johnston, *J. Chem. Phys.*, **26**, 582 (1957).
64. J. B. Goodenough, *Phys. Rev.*, **117**, 1442 (1960); **120**, 67 (1960), *Magnetism and the Chemical Bond* (New York: Interscience, 1963).
65. T. Kawakubo, *J. Phys. Soc. Japan*, **20**, 4, 516 (1965).
66. I. Nebenzahl and M. Weger, *Phys. Rev.*, **184**, 936 (1969).
67. C. N. Berglund and H. J. Guggenheim, *Phys. Rev.*, **185**, 1022 (1969).
68. C. N. Berglund and A. Jayaraman, *Phys. Rev.*, **185**, 1034 (1969).
69. N. F. Mott, *Phil. Mag.*, **6**, 287 (1961).
70. J. de Cloizeaux, *J. Phys. Chem. Solids*, **26**, 259 (1965).
71. W. Kohn, *Phys. Rev. Letters*, **19**, 439 (1967).
72. B. I. Halperin and T. M. Rice, *Solid State Phys.*, **21**, 115 (1968).
73. Yu. V. Kopaev, *Fiz. Tverd. Tela*, **8**, 2633 (1966) [English transl.: *Soviet Phys.-Solid State*, **8**, 2106 (1967)].
74. A. N. Kozlov and L. A. Maksimov, *Zh. Eksperim. Teor. Fiz.*, **48**, 1184 (1965) [English transl.: *Soviet Phys.-JETP*, **21**, 790 (1966)].

75. A. B. Migdal, *Zh. Eksperim. i Teor. Fiz.*, **34**, 1438 (1958) [English transl.: *Soviet Phys.-JETP*, **7**, 996 (1958)].
76. E. T. Jaynes, *Ferroelectricity* (Princeton: University Press, 1953).
77. A. D. Liehr, *J. Phys. Chem.*, **67**, 471 (1963).
78. L. E. Orgel, *Discussions Faraday Soc.*, **26**, 138 (1958).
79. K. P. Sinha and A. P. Sinha, *Indian Journ. Pure Appl. Phys.*, **2**, 91 (1964).
80. R. Englman, *Quarterly Progress Report SSMT Group, MIT*, **47**, 74 (1963) (unpublished); 'Microscopic Theory of Ionic Dielectrics', Israel Atomic Energy Commission, *Report IA-994, A.E.C. Accession No. 14129* (Defence Documentation Centre No. AD 612613) (1964).
81. U. Öpik and M. H. L. Pryce, *Proc. Roy. Soc. A*, **238**, 425 (1957).
82. A. F. Devonshire, *Phil. Mag.*, **40**, 1040 (1949); **42**, 1065 (1951).
83. J. C. Slater, *Phys. Rev.*, **78**, 748 (1950).
84. L. Jing-Der, *Acta Phys. Sinica*, **22**, 186 (1966) [English transl.: *Chinese J. Phys.*, **22**, 149 (1966)].
85. S. Strässler and C. Kittel, *Phys. Rev.*, **139**, A758 (1965).
86. I. B. Bersuker, *Phys. Letters*, **20**, 589 (1966).
87. A. H. Kahn and J. A. Leyendecker, *Phys. Rev.*, **135**, A1321 (1964).
88. J. D. Dunitz and L. E. Orgel, *J. Phys. Chem. Solids*, **3**, 20 (1957).
89. T. Yamadaya, T. Mitui and T. Okada, *J. Phys. Soc. Japan*, **17**, 1897 (1962).
90. H. F. McMurdi, B. M. Sullivan and F. A. Mauer, *J. Res. N.B.S.*, **45**, 35 (1950).
91. P. W. Anderson, *Solid State Phys.*, **14**, 99 (1963).
92. S. J. Allen, Jr., *Phys. Rev.*, **166**, 530 (1968); **167**, 492 (1968).
93. B. D. Bhattacharyya and S. K. Datta, *Ind. J. Phys.*, **41**, 181 (1968).
94. J. S. Griffith and L. E. Orgel, *Nature*, **181**, 170 (1958).
95. G. I. Finch, A. P. B. Sinha and K. P. Sinha, *Proc. Roy. Soc. A*, **242**, 28 (1957).
96. P. J. Wojtowitz, *Phys. Rev.*, **116**, 32 (1959).
97. J. Kanamori, *J. Appl. Phys.*, **31**, 145 (1960).
98. R. Brout, *Phase Transitions* (New York: Benjamin, 1965) §2.2.
99. C. Haas, *Phys. Rev.*, **140**, A863 (1965), *J. Phys. Chem.*, **26**, 1225 (1965).
100. R. Englman and B. Halperin, *Phys. Rev.*, **B2**, 75 (1970).
101. P. Novák, *J. Phys. Chem. Solids*, **30**, 2357 (1969); **31**, 125 (1970).
102. B. Halperin and R. Englman, *Phys. Rev.*, **B3**, 1698 (1971).
103. M. A. Hepworth and K. H. Jack, *Acta. Cryst.*, **10**, 345 (1957).
104. A. Okazaki, Y. Suemune and T. Fuchikami, *J. Phys. Soc. Japan*, **14**, 1823 (1959).
105. A. Okazaki and Y. Suemune, *J. Phys. Soc. Japan*, **16**, 176 (1961).
106. A. M. Stoneham and R. Bullough, *J. Phys. C.: Solid State Phys.*, **3**, L195 (1970).
107. A. H. Cooke, C. J. Ellis, K. A. Gehring, M. J. M. Leask, D. B. Martin, B. M. Wanklyn, M. R. Wells and R. L. White, *Solid State Commun.*, **8**, 689 (1970).
108. A. H. Cooke, D. M. Martin and M. R. Wells, *Solid State Commun.*, **9**, 519 (1971).
109. C. J. Ellis, K. A. Gehring, M. J. M. Leask and R. L. White, *J. de Physique*, **32**, CI-488 (1971).
110. F. Sayetat, J. X. Boucherle, M. Belakhovsky, A. Kallel, F. Tcheou and H. Fuess, *Phys. Letters*, **34A**, 301 (1971).

111. R. T. Harley, W. Hayes and S. R. P. Smith in *Proc. Int. Conf. on Light Scattering in Solids*, 1971 (Paris: Flammarion, 1971); *Solid State Commun.*, **9**, 515 (1971).

111a J. B. Forsyth and C. F. Sampson, *Phys. Letters*, **36A**, 223 (1971).

112. R. J. Elliott in *Proc. Int. Conf. on Light Scattering in Solids* (Paris: Flammarion, 1971).

113. B. Halperin and R. Englman, *Solid State Commun.*, **8**, 1555 (1970).

114. R. Brout, K. A. Müller and H. Thomas, *Solid State Commun.*, **4**, 507 (1966).

115. R. J. Elliott, S. R. P. Smith and A. P. Young (private communication, 1971).

116. M. D. Sturge, *Solid State Physics*, **20**, 91 (1967).

117. G. D. Watkins in *Effects de Rayonnements sur les Semiconducteurs*, Ed. Baruch (Paris: Dunod, (1965) p. 97.

118. G. W. Ludwig and H. H. Woodbury, *Phys. Rev.*, **113**, 1014 (1959).

119. H. H. Woodbury and G. W. Ludwig, *Phys. Rev.*, **126**, 466 (1962).

120. A. M. Stoneham and M. Lannoo, *J. Phys. Chem. Solids*, **30**, 1769 (1968).

6 Fluctuations in Jahn–Teller systems (relaxation)

Outline

We examine here time-dependent disturbances on a vibronic system which cause transitions between the eigenstates of the system. The vibronic coupling of degenerate electron states to the vibrations opens up new channels for relaxation between the electronic states: however, for very strong coupling the system is prone to distortion and relaxation is impeded.

6.1 The reorientation process

A strongly coupled Jahn–Teller centre which has taken up a distorted configuration will in the course of time reorient to another distorted position. Such a reorientation should be understood as a real physical transition, involving a generally small but finite energy change. It is commonly held that the reorientation is brought about by the interaction of the J-T centre with the phonons of the lattice which then take up or release the energy necessary for energy conservation. The reorientation rate, denoted by τ^{-1}, will be calculated in this section. τ^{-1} is a strongly varying function of the temperature (its measured value is known in one case to spread from 10^9 to 10^{-6} sec^{-1})[1] and at a given temperature it is slower the deeper and more separated the potential wells are. This behaviour was already noted in the Introduction: it is in line with physical intuition and agrees with the mental picture associated with the term 'frozen-in-state'.

To derive the expression for the reorientation rate we require that the system be initially oriented. This may appear a trivial point, however, we recall that with the vibronic doublet eigenfunctions, equation (3.16) and (3.18) of the $E \otimes \varepsilon$ linear coupling,

$$\langle \cos \phi \rangle = 0$$

so that there is no orientation in the zero-order eigenstates and *a fortiori* no reorientation. We therefore consider the action upon the system of a small stress which ensures an initial orientation. The need for the inclusion of small strains in the consideration of relaxation processes is due in the first place to Sussmann.[2] This strain may be very small, as when due to e.g. random static strains (whose magnitude is of the order of 10^{-4}–10^{-5} in the commonly studied crystals), in fact, so small that its actual size, though not its presence, is immaterial to the effect. Alternatively, its size may determine the relaxation rate, as found by Sussmann[2] for O^{2-} centres in alkali halides.[3] In yet other cases the strain may be externally applied, as by uniaxial compression. While it appears possible to discuss the reorientation process through the correlation function $\langle \cos \phi \cos \phi(t) \rangle$, without the employment of strain, the approach which is based on a strain is more straightforward and, because of the existence of ubiquitous random irregularities, more physical.

The theory of reorientation between equivalent wells, which was developed[2,4] for nondegenerate systems, is directly applicable to vibronically coupled degenerate systems. Nevertheless, fluctuations in a *weakly coupled* dynamic system which is poorly localized, are better described as transitions between degenerate states split apart by strain rather than as a reorientation process. Here the fluctuations are analogous to those which lead to paramagnetic relaxation,[5] in which the degenerate states are split by a constant magnetic field.

We consider first a direct process, namely a transition involving one lattice phonon whose frequency ω_k (k = wave vector) matches the energy difference δ of two strain split vibronic states in two wells. Let the matrix element for a transition due to the phonon k (from the well l to another well u) be written as $\langle l|H_k'|u \rangle$. The transition or jump frequency is given by

$$\nu_{lu} = \frac{2\pi}{\hbar} \sum_k |\langle l|H_k'|u \rangle|^2 \delta(\omega_k, \delta/\hbar) \tag{6.1}$$

where the summation (which in practice reduces to an integration) is over all phonons which satisfy the energy conservation. In the summand δ is the Krönecker delta.

The jump frequencies take different forms depending on the mechanisms by which the phonons admix the states of l and u. Following Sussmann's treatment[2] we differentiate between perturbations whose effect on the two wells is symmetric or anti-symmetric with respect to the two wells.

The following results are obtained.[2]

$$\nu_{lu} = \frac{(dU/de)^2}{\pi} \frac{\delta^3}{\rho \hbar^4 c^5} \frac{1}{e^{\delta/kT} - 1} \qquad \text{Symmetric perturbation} \qquad (6.2)$$

$$\nu_{lu} = \frac{(dU/de)^2}{\pi} \frac{\delta(3\Gamma)^2}{\rho \hbar^4 c^5} \frac{1}{e^{\delta/kT} - 1} \qquad \text{Antisymmetric perturbation} \qquad (6.3)$$

Here δ is the strain splitting of the wells. 3Γ is the tunnelling energy between the wells. (This definition follows the usage of 3Γ in §3.2.4.) ρ is the mass density of the crystal. The sound velocity c is defined as the following average of the longitudinal and transverse sound velocities, c_l and c_t [4]

$$\frac{1}{c^5} = \frac{1}{5}\left(\frac{2}{c_l^5} + \frac{3}{c_t^5}\right) \qquad (6.4)$$

The definition of (dU/de) raises some questions. It is intended to be an average over all phonons of the energy change per unit strain, caused by the strain field of the phonons. The numerical constants which appear in equations (6.2) and (6.3) are appropriate to a Debye-model of lattice-vibrations. In other models or for different physical systems different numerical constants are required. Keeping these constants fixed, we lend somewhat different meanings to the same symbol (dU/de) in different conditions. We shall presently rewrite (dU/de) in terms of the parameters of the localized J-T centre; but this procedure is subject to the difficulties mentioned here.

We now write down the rate equation[2,4] from which the reorientation rate τ^{-1} is derivable. Suppose there are w wells, and these are separated by the strain into w_u upper and w_l lower wells, so that

$$w_u + w_l = w.$$

Let the probability of occupation of the upper well be N_u and that for the lower well be N_l. The assumption is made here that only two states, one in each well and separated by the energy δ, 'count'; rate equations involving a larger number of vibronic states are the subject of §6.6.

The change of N_u with time t is given by

$$\frac{dN_u}{dt} = -w_u N_u \nu_{ul} + w_l N_l \nu_{lu} \qquad (6.5)$$

Using now the condition for detailed balance

$$\nu_{lu}/\nu_{ul} = e^{-\delta/kT},$$

and the sum over all probabilities

$$w_u N_u + w_l N_l = 1,$$

the reorientation rate is immediately derivable.

Examples: For a pair of wells we obtain from equation (6.5)

$$\tau^{-1} = \nu_{1u}\tfrac{1}{2}\,(e^{\delta/kT} + 1); \tag{6.6}$$

for three wells split into a lower well and two higher ones

$$\tau^{-1} = \nu_{1u}\tfrac{1}{3}\,(e^{\delta/kT} + 2); \tag{6.7}$$

for four wells of which one is lowered

$$\tau^{-1} = \nu_{1u}\tfrac{1}{4}\,(e^{\delta/kT} + 3) \tag{6.8}$$

but for four wells split into two and two we regain equation (6.6).

For all cases of equations (6.6–8), near the absolute zero of temperature τ tends to a constant value representing the emission of one quantum from the upper well. At higher temperatures ($T > \delta/k$),

$$\frac{1}{\tau} \simeq BT \tag{6.9}$$

where B is a constant; however, at even higher temperatures an activation (or Orbach) process[6-7] of the form

$$\tau^{-1} = \nu\,e^{-\Delta/kT} \tag{6.10}$$

becomes effective. Here ν is a frequency factor roughly of the order of 10^{11}–$10^{14}\ \mathrm{sec}^{-1}$ and Δ is the energy of an intermediate, excited state to which real transitions occur from the initial state. There are also Raman processes,[8-9] in which the reorienting system scatters rather than absorbs a phonon. These yield terms in τ^{-1} proportional to T^3 and T^5.

The use of the results in equations (6.2–3) in cases of vibronic coupling will now be explained:

In a linear $E \otimes \varepsilon$ case, whose two vibronic ground states levels are separated (by the strain e) to an extent of $qLe(M\omega D^2/\hbar)^{1/2}$ [equation (4.3)] it is this quantity which has to be inserted for δ in equation (6.2), while for dU/de we put $\tfrac{1}{2}qL(M\omega D^2/\hbar)^{1/2}$ (Reference 10).

There is no tunnelling between the components of the vibronic doublet, so that the factor denoted by 3Γ in equation (6.3) vanishes. Thus, the antisymmetric process is inactive. Similarly for $T \otimes \varepsilon$ coupling, 3Γ vanishes since the vibronic states of equation (3.57) are orthogonal. For (dU/de) in equation (6.2) we have to substitute the coefficient $K^{(v)}(T_2)G(\zeta)$ from equation (3.60), appropriate for a local strain of type τ_2, and of unit amplitude.

The strong coupling case of $E \otimes \varepsilon$ was analysed by Williams et al.[11] in connection with the decay of electron spin echos in $La_2Mg_3(NO_2)_{12} \cdot 24H_2O$ containing divalent copper.[12] In equation (6.3) the identification of

(dU/de) with $2^{-1}qL(M\omega D^2/\hbar)^{1/2}$ is appropriate, while 3Γ is the separation between the lowest E- and A-states. With an estimate of 0.5 cm^{-1} for the average strain energy δ, this quantity effectively disappears from equation (6.3) for temperatures above $T = 1.3$ °K (this was the lower limit of the temperature range in Reference 12) and τ^{-1} becomes proportional to T, as in equation (6.9). From the value of 10^3–$10^{4.5}$ (sec. degrees)$^{-1}$ extracted from the measurements for the slope B in equation (6.9), a very small tunnelling ($<10^{-1}$ cm^{-1}) and the stabilization energy $E_{JT} \sim 1000$ cm^{-1} were estimated. Remarkably, τ^{-1} as a linear function did not extrapolate at $T = 0$ to zero, as required by the theory, but rather tended to a value of about 10^4–10^5 sec^{-1}. This may indicate some reorientation process which involves tunnelling only.

The activation process, equation (6.10), was observed by Sturge and collaborators[13] through the attenuation of an acoustic wave of a hundred megacycle/sec frequency in Al_2O_3 doped by doubly degenerate Ni^{3+} ions. The reorientation rate deduced from the loss data led these authors to derive the value of $2\beta = 90$ cm^{-1} for the height of the barriers between the three minima. Near 10 °K, τ from the acoustic data confirms that from ESR broadening to within a factor of 1–3. At lower temperatures, below 6 °K, the linear relation between τ^{-1} and T leads to a value of $3\Gamma = 1$ cm^{-1} (Reference 10).

For both $Al_2O_3 : Ni^{3+}$ and $La_2Mg_3(NO_2)_{12} \cdot 24H_2O : Cu^2$ the electron-spin relaxation is significantly quickened by mechanisms involving vibronic coupling. However, these mechanisms become frozen-in for strong vibronic coupling as is apparent from the dependence in equation (6.3) of the jump frequency on the tunnelling rate 3Γ, which decreases as the coupling strength grows. Thus there is no contradiction between the relaxation rate results and the resistance of the strongly coupled J-T-E to distortion (Chapter 1).

6.2 Spin-lattice relaxation times, T_1

In addition to the normal mechanisms[5] whereby the spin states of an orbital singlet relax, there appear to be several alternatives which are peculiar to multiple well systems. These were summarized by Ham (Reference 10, §11.6) for orbital doublets, and his results are paraphrased here. The importance of these multiple well mechanisms is made evident by a significant shortening of the relaxation times in the aforementioned cases below that occurring in a Tutton salt,[14-5] where the e-orbitals of Cu^{2+} are split apart by the tetragonal crystal field, and of the correlation times in regular octahedrally ligated complexes compared to essentially tetragonal ones.[15a]

As shown in detail in Reference 11 for the case of strong non-linear coupling, if the applied field H is not along a cubic axis the localization of the orbital states in one well aligns, through a spin–orbit coupling mechanism, the electronic spin away from the direction of H. Therefore in the course of a jump to another well the spin may become reversed too.

The expression for the spin-lattice relaxation time T_1 is given by

$$T_1^{-1} = \frac{2}{3} \frac{(g_{\parallel}^{\iota} - g_{\perp}^{\iota})^2}{g_1} (l^2 m^2 + n^2 l^2 + m^2 n^2) \tau^{-1} \tag{6.11}$$

where $g_{\parallel}^{\iota}$ and $g_{\perp}^{\iota}$ ($\iota = \theta$ or ϵ) are the principal values of the g-tensor shown in equation (3.41), g_1 is the average in equation (3.43) and l, m, n are the direction cosines of the magnetic field. The presence of τ^{-1} in equation (6.11) expresses the fact that in the relaxation process the jump accompanied by a spin reversal is followed, or preceded, by a jump with unchanged spin.

In another mechanism[10] the simultaneous reversal of electronic and nuclear spins occurs. Unlike equation (6.11), this mechanism is effective also when the magnetic field is along a cubic axis. There exists also the possibility of relaxation through the operation of spin–orbit coupling within the lowest three states (the doublet and a nearby singlet). This coupling is made allowed by a trigonal field arising from phonons of the appropriate symmetry.[16]

For molecules in liquids there exists the relaxation theory of McConnell[17] which is based on the bodily rotational motion of molecules. It was pointed out by Ham,[10] that essentially the same mechanisms are involved whether the molecule tumbles, as in a liquid, or whether the complex reorients between equivalent distortions, except that in the former case we should interpret τ as the correlation time for tumbling and we should average over all the possible orientations of the magnetic field in equation (6.11). In apparent contrast to this result, the hyperfine structure in the low temperature ESR spectrum of C_5H_5 is explicable by the rapid rotation rate ($\sim 10^7 \text{ sec}^{-1}$ above $70\,^{\circ}\text{K}$) of the molecules but not by a rapid reorientation of the distortion.[18]

6.3 Classification of reorientation times in solids

To derive systematically (and this turns out to be, group-theoretically) the effect of phonon waves on the relaxation process of a J-T centre we have to proceed in two stages. First, we must find out what kind of disturbance is caused by the lattice wave locally, at the centre. This means that we must resolve the stresses and forces due to the wave to irreducible representation of the point group of the centre. This being done, we consider each irreducible representation separately, noting, in particular, whether or not it is capable of producing a reorientation of the distortion and if yes, what relaxation time will it give rise to. We shall see that, because in general a number of different kinds of stresses (belonging to different irreducible representations of the point group) are effective, a number of different relaxation times will be associated with each distortion. We recall that the reorientations we are concerned with here are between equivalent minima in the potential hypersurface of the static J-T problem.

The first stage in the programme is easily carried out in principle, though never yet fully accomplished in practice. It is known[19–20] that each phonon-wave forms a basis for an irreducible representation of the space-group in the crystal. If this representation is 'compatible' with a particular irreducible representation of the point group of the centre, which will be either the point group at the origin in k-space or a subgroup thereof, then this mode, i.e. a disturbance belonging to this particular representation, is engendered locally by the wave. Compatibility tables for points in k-space of high symmetry are available in Heine's book[21] or in 'BSW'.[22] General points in k-space possess representations which are compatible with all irreducible representations of Γ (the origin in k-space), so there is no danger that a relaxation mode will stay inactive. On the other hand, the quantitative measure of the activity of the mode depends on factors which, as we have said earlier, have not been calculated.

For the second part of this problem we have to enumerate all types (i.e. all irreducible representations of the point group) of reorientation modes for each static J-T-E situation. It turns out that the task is rather easier than might have been, in that the enumeration depends only on the two factors: on the (point) symmetry group of the centre and on the group of the reduced symmetry, i.e. that subgroup to which the symmetry group becomes reduced in the course of the static distortion. We are thus enabled to utilize the lists of Nowick and Heller (Reference 23–4, containing references to earlier works). The lists of relaxation modes for the groups O and T_d which follow in Table 6.1 specify the static J-T distortions which lead to a reduced symmetry and to a removal (partial or complete) of the degeneracy. A suggested terminology for the distortions is included, although the essential information is, as already said, fully contained in the original and reduced symmetry groups. The meaning of the various symbols for the static distortions (ε^0, ε^{00}, etc.) is given in the tables of the electric-dipole transition selection rules (Appendix VII).

But not only is the list of reorientation modes a property of the groups, so is also the reorientation time of each mode as function of the jump frequencies. Expressions for the reorientation rates at high temperatures are shown in the last column of Table 6.1. There is agreement between these expressions and those in equations (6.6–8) for $kT \gg \delta$. The jump frequencies v_{ij} are defined with reference to the following figures, which are taken, with some slight changes, from Reference 24 (Figure 6.1).

These figures show stereographic projections of the points of intersection of an axis of the elastic dipole caused by the distortion with a reference sphere. The term elastic dipole is discussed in detail by Nowick and Heller (Reference 23, §§2.2–2.3); for our purposes it is sufficient to note that the strain tensor (being a second order symmetric tensor) associated with a localized distortion can be brought to a diagonal form

$$\begin{pmatrix} e_1 & 0 & 0 \\ 0 & e_2 & 0 \\ 0 & 0 & e_3 \end{pmatrix}$$

by a suitable rotation of the coordinate axes.

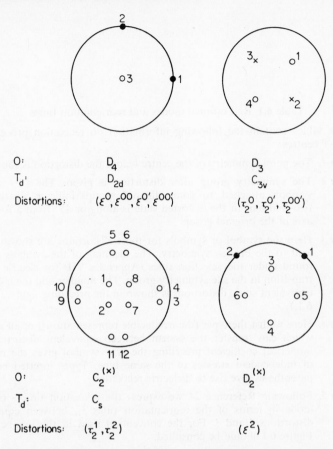

Figure 6.1 Identification of different orientations of J-T distortions. Orientation of a principal axis is shown stereographically. Underneath each drawing the site symmetry group, the distortional mode(s) and the broken symmetry group are marked. (The figures are discussed in the text). (After References 23–4.)

The three numbers e_1, e_2, e_3 define the ellipsoid characterizing the elastic dipole. Specifying the direction of *one* or *two* of the principal axes of the tensor with respect to the original axes of the undistorted lattice gives us fairly full indication of the orientation of the distortion (not, however, of the sense of direction of the defect). The figures give all possible equivalent orientations of the J-T distortions.

As an example we turn to the cubic system of crystals. The legends to the drawings of Figure 6.1 describe the point group of the J-T centre before the distortion, then the point group of the distorted centre. Lastly, we give our terms for the distortion. The first drawing shows intersections of a major axis of the distortions in three equivalent configurations with the reference sphere. Of

Table 6.1 Relaxational modes and reorientation times

These tables contain the following information on relaxation processes in static J-T centres:

Column 1. The point symmetry of the centre before the distortion is shown.

Column 2. The symmetry group after distortion is given. The (+) or (×) symbols attached to some digonal groups indicate whether the two-fold axis in the distorted state is along or 45° from a four-fold axis of the original group.

Column 3. Here the symbol or symbols for those distortions are shown which lead to the broken symmetry. The meaning of the symbols can be found under the selection rules (Appendix VII) for electric dipole transition in the site symmetry group. The number and the nature of equivalent static distortions is shown in the preceding figure. (Figure 6.1.)

Column 4. Here we list the types (the irreducible representations) of all stresses which can transfer the system between equivalent distortions. A numerical coefficient preceding the type symbol gives the number of independent stresses of the same type. Types inserted between parentheses give rise to dielectric relaxation.

Column 5. Following Reference 24 we express the relaxation time τ of each mode in terms of the reorientation rates ν_{ij} between equivalent distortions i and j. For the convention used in the numbers i, j, Figure 6.1 should be consulted.

The numbers 1–6 appearing under $D_2^{(+)}$, $C_{2v}^{(+)}$ which are not shown in Figure 6.1, refer to the six possible ways of arranging the 3 unequal orthorhombic axes of the stress tensor of the distorted centre along the cubic axes of the site. Listed in order of size, the six ways are:

$$1{:}(x, y, z); \quad 2{:}(x, z, y); \quad 3{:}(y, z, x); \quad 4{:}(y, x, z); \quad 5{:}(z, x, y); \quad 6{:}(z, y, x)$$

Relaxation modes and reorientation times—(contd.)

Site symmetry	Reduced symmetry	Distortion	Types of active stress	τ^{-1}
O	D_4	$\varepsilon^0, \varepsilon^{00}, \varepsilon^{0'}, \varepsilon^{00'}$	E, (T₁)	$3\nu_{12}$, $(4\nu_{12} + 2\nu_{14})$
	D_3	$\tau_2^0, \tau_2^{0'}, \tau_2^{00'}$	T₂, (T₁)	$4\nu_{12}$, $(4\nu_{12} + 2\nu_{15} + 2\nu_{16})$
	$D_2^{(+)}$	ε^1	2E, (T₁)	$(\nu_{12} + 3\nu_{13} + \nu_{14} + \nu_{16}) \pm [(\nu_{12} - \nu_{14})^2 + (\nu_{12} - \nu_{16})^2]^{\frac{1}{2}}/\sqrt{2}$, $(4\nu_{12} + 2\nu_{14})$
	$D_2^{(\times)}$	ε^2	E, T₂, (T₁)	$6\nu_{13}$, $2\nu_{12} + 4\nu_{13}$, $(4\nu_{12} + 2\nu_{14})$
	$C_2^{(\times)}$	τ_2^0, τ_2^2	E, 2T₂, (T₁)	$6\nu_{13} + 3\nu_{15} + 3\nu_{16}$, $(\nu_{17} + 3\nu_{12} + 5\nu_{13} + \frac{3}{2}\nu_{15} + \frac{3}{2}\nu_{16})$ $\pm [(\nu_{12} - \nu_{13} - \nu_{17} + \frac{1}{2}\nu_{15} + \frac{1}{2}\nu_{16})^2 + 2(\nu_{15} - \nu_{16})^2]^{\frac{1}{2}}$
				$(2\nu_{12} + 2\nu_{13} + 6\nu_{14} + 2\nu_{17})$
T_d	D_{2d}	$\varepsilon^0, \varepsilon^{00}, \varepsilon^{0'}, \varepsilon^{00'}$	E, (T₁)	$3\nu_{12}$, $(4\nu_{12} + 2\nu_{14})$
	C_{3v}	$\tau_2^0, \tau_2^{0'}, \tau_2^{00'}$	T₂, T₂	$4\nu_{12}$, $(4\nu_{12})$
	$D_2^{(+)}$	ε^1	2E, (T₂)	$(\nu_{12} + 3\nu_{13} + \nu_{14} + \nu_{16}) \pm [(\nu_{12} - \nu_{14})^2 + (\nu_{14} - \nu_{16})^2]^{1/2}/\sqrt{2}$
	$C_{2v}^{(+)}$	ε^2	2E, (T₂)	$(4\nu_{12} + 2\nu_{14})$
	C_s	τ_2^0, τ_2^2	E, 2T₂, (2T₂)	$6\nu_{13} + 3\nu_{15} + 3\nu_{16}$, $(\nu_{17} + 3\nu_{12} + 5\nu_{13} + \frac{3}{2}\nu_{15} + \frac{3}{2}\nu_{16})$ $\pm [(\nu_{12} - \nu_{13} - \nu_{17} + \frac{1}{2}\nu_{15} + \frac{3}{2}\nu_{16})^2 + 2(\nu_{15} - \nu_{18})^2]^{\frac{1}{2}}$,
				(the same again)

these intersections two (depicted by full circles) lie in the plane of the paper and one (depicted by an open circle ○) lies above the paper. The next drawing shows two intersections above (○) and two below (×) the plane of the paper. In both figures, specification of the one major axis suffices to determine the orientation of the deformation.

Not so, however, for some of the other drawings. Thus for distorted centres belonging to an orthorhombic or monoclinic class we need more information. However, we must bear in mind, that an axis of rotation in the distorted centre (if there remains one) coincides with an original axis. The point of intersection denotes that of an additional axis with the reference sphere. Similarly a plane of reflection remaining in the distorted centre necessarily coincides with one in the original site and the intersection provides additional information. The drawings therefore effectively show the directions of the principal axes of the distortion (or of the elastic dipole). A further aid is the usage of (+) and (×) symbols added to the group symbols of orthorhombic and monoclinic systems. These distinguish whether in the cubic systems an axis of the elastic dipole is along (+) or at 45° (×) to one of the original cubic axis.

From the point of view of relaxation processes, we recall that the numbers in the figures denote distortions between which reorientations can take place. It must be clearly noted that they label not the positions of atoms in a distortion but the orientations of the distortion. As an example, number 1 in a trigonal distortion for a cubic site denotes a distortion along the [111] body diagonal of the cube and 2 denotes a distortion along $[-1\,1\,-1]$. While in this particular case all jump frequencies are equal, $v_{12} = v_{13} = \ldots$, this is not so for other cases. Thus for distortions symbolized by ε^1 or ε^2, $v_{12} \neq v_{13}$.

The figures are drawn for 'elastic dipoles', which are even under inversion in the origin, so that only the orientation of the distortion need be given, not its algebraic sense. While the case of even-type distortions occurs commonly in

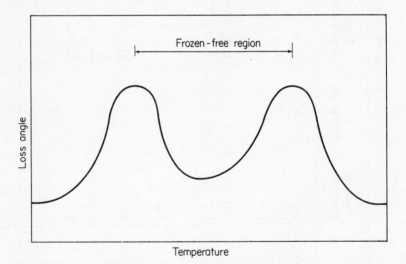

Figure 6.2 Observation of 'frozen-free split' in the loss angle of an applied wave. (After Reference 24.)

the J-T-E, other cases, where the sense of the distortion does matter, are also known (e.g. a distortion in O or T). In this case, following Nowick (Reference 24) we differentiate between reorientation processes caused by stress ('anelastic relaxation') and by an electric field ('dielectric relaxation'). The latter can only occur for polar distortions.

It is possible to draw some qualitative conclusions from Table 6.1 even without detailed knowledge of the jump frequencies. One significant deduction is the very existence of multiple relaxation times. Thus we have the possibility of a 'frozen-free split' in the temperature dependence of, say, a loss angle. The hypothesized outcome of an experiment is shown in Figure 6.2. The two peaks correspond to temperatures such that $\omega \tau_1 \sim 1$ and $\omega \tau_2 \sim 1$, respectively, where ω is the frequency of the absorbed wave and τ_1, τ_2 are two relaxation times of rather different magnitudes. In the gap (indicated by the arrows) between the maxima, the distortion is frozen with respect to one (the slower) relaxation and free with respect to the other, the faster. E.g. for the $C_{2v}^{(+)}$ type distortion in a T_d site the ν_{12} reorientation frequency may conceivably be much larger than ν_{13}. (It helps to look up Figure 6.1 which explains the numbering). Then at a temperature region such that $\tau \cdot 2\nu_{12} \gg 1$, but $\tau \cdot 6\nu_{13} \ll 1$ the distortion will appear, as having, because of the averaging effect of the T_2-type relaxation, a higher effective symmetry than it originally had. In the present example ($C_{2v}^{(+)}$ in T_d) a tetragonal symmetry will be seen. In an actual situation, occurring for a negatively charged vacancy in silicon and also having a T_d site symmetry and a broken symmetry of $C_{2v}^{(+)}$, the dielectric relaxation of type T_2 appears to be faster than the two anelastic relaxations of type E. (Reference 25, p. 174.)

As a last task, the jump frequencies ν_{ij} of the system have to be expressed in terms of the jump frequencies of the individual atoms or ions. For reorientations of a J-T distortion this task has not yet been carried out, though Nowick[24] has solved some examples of elastic dipoles formed by externally introduced defects.

6.4 The line shape

The effects of jumps between potential wells on the ESR line shapes of J-T centres have been considered by Bersuker,[26] O'Brien,[27] Englman and Horn[28] and in a more quantitative fashion by Hudson,[29] Hartmann–Boutron[30] and by Blume and Tjon.[31-2] The physical motivation in the first four references was the coalescence of the ESR lines of Cu^{2+} in several salts as the temperature is raised to somewhere near 20 °K[33-6] or higher,[37] while for References 30–2 the temperature dependence of the quadrupole splitting in the Mössbauer spectrum of Fe^{2+} in some spinels[38-9] and in MgO[40-1] served as a stimulus. In each case the jumps first broaden the distinct frequencies associated with different wells and then, with an increasing reorientation rate τ^{-1} between the wells, the jumps smear out the frequencies until their average is observed. This average is then motionally narrowed.[42-7]

In the works of Hudson,[29] of Hartmann–Boutron[30] and of Blume and Tjon[31-2] an explicit expression is sought for the line shape, which is essentially the absorption intensity regarded as function of the photon frequency ω. The formal expression for the line shape

$$I(\omega) = \frac{2}{\Gamma} \int_0^\infty dt \; e^{i\omega t - \frac{1}{2}\Gamma t} \langle M(0)M(t) \rangle$$

includes the life times $1/\Gamma$ of the states and thermal averages involving the transition operator $M(t)$ in the Heisenberg representation. This operator is given in terms of the time-dependent Hamiltonian $H(t)$ by

$$M(t) = \exp\left(i \int_0^t H(t')\, dt' \right) M(0) \exp\left(-i \int_0^t H(t')\, dt' \right),$$

where the Hamiltonian includes the immediate physical system and the stochastic perturbers but not the electromagnetic field.

To glean an understanding of the theory of Blume and Tjon we consider a system which moves in a three well potential. Jumps between pairs of wells occur at the rate $(3\tau)^{-1}$. Blume and Tjon[31] were concerned with

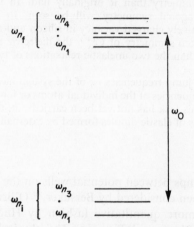

Figure 6.3 Labelling of energy levels in a stochastic system.

transitions to the upper (final) multiplet of the Fe^{2+} nucleus $I = 3/2$, whose levels split up by the quadrupole field when the system is in anyone of the wells. We can consider, more generally, an excited state multiplet which has energy levels $\hbar\omega_{n_f}^{I}$ (or $\hbar\omega_{n_f}^{II}$ or $\hbar\omega_{n_f}^{III}$) if the system is in the first well (or the second or the third). The index n_f (f for final) runs over all upper states and the levels are counted from a common mean (Figure 6.3). Since the wells are equivalent, the sets $\omega_{n_f}^{I}$, $\omega_{n_f}^{II}$, $\omega_{n_f}^{III}$ run over the same values but the states to which the energies belong differ.

Carrying on the discussion now in more formal terms, we split up the Hamiltonian, exclusive of the transition operator M, into a constant part

H_0 (eigenvalue $\hbar\omega_0$) and a part depending on time due to stochastic processes. This part will be written H^{I}, H^{II} or H^{III} depending on which well the system is momentarily in. $\hbar\omega_{n_f}^{\mathrm{I}}$, etc., are the energy levels of H^{I}, etc. (Figure 6.3). In general, this latter part will not commute at different times, i.e.

$$[H^{\mathrm{I}}, H^{\mathrm{II}}] \neq 0, \qquad \text{though } [H^{\mathrm{I}}, H_0] = 0.$$

Specifically, in the model of Reference 31, the commutator of the quadrupole moment operators,

$$[(I_x^2 - \tfrac{1}{3}I^2), (I_y^2 - \tfrac{1}{3}I^2)] \neq 0.$$

This type of stochastic perturbation is called non-adiabatic. If it is suddenly switched on, the system will make a jump from one state to another.

The detailed results of Tjon and Blume[32] are a property of their specific, two-level model ($n = 1, 2$), whose distinction is its solubility for non-adiabatic perturbations. Qualitatively speaking, the initial effect of the jumps on the transition frequencies is that these become shifted inwards from their positions at $\omega = \omega_0 + \omega_{n_f} - \omega_{n_i}$ in the frozen-in phase. There is also some broadening. As the reorientation rate exceeds a certain value (roughly as $\tau\omega_{n_f} \leqslant 1$) the jumps contribute to the broadening but not to the shifts. Ultimately, for $\tau\omega_{n_f} \ll 1$, when only the average of H^{I}, H^{II} and H^{III} is felt, the lines coalesce and become narrowed.

When the Hamiltonians at different times commute, the role of the stochastic process is not to cause transitions between states but rather to modulate the energy of a fixed state. Such 'adiabatic' processes were studied in References 29, 31–2, 42–6. Let us regard the changes in the energy as arising from a random Markovian process and suppose that the eigen-frequencies of the same state are, respectively, $\omega_0 + \omega_2$, $\omega_0 - \tfrac{1}{2}\omega_2$, $\omega_0 - \tfrac{1}{2}\omega_2$ when the system is in the first, second, third well.

The theory outlined, e.g. by Abragam (Reference 47, p. 450), is then applicable and one finds for the normalized line-shape the formula

$$I(\omega) = (\pi\omega_2)^{-1} \operatorname{Re} \frac{iyx - 3}{3iy - (y - 1)(y + \tfrac{1}{2})x} \tag{6.12}$$

where

$$x = \tau\omega_2$$

$$y = (\omega - \omega_0 + i\Gamma/2)/\omega_2.$$

For slow jumping ($x \gg 1$) and for long lifetimes two Lorentzian shapes are found:

$$\frac{1}{3\pi} \frac{2(\tau^{-1} + \tfrac{1}{2}\Gamma)}{(\omega - \omega_0 - \omega_2)^2 + 4(\tau^{-1} + \tfrac{1}{2}\Gamma)^2}, \frac{1}{3\pi} \frac{2(\tau^{-1} + \tfrac{1}{2}\Gamma)}{(\omega - \omega_0 + \tfrac{1}{2}\omega_2)^2 + (\tau^{-1} + \tfrac{1}{2}\Gamma)^2}$$

For fast jumping ($x \ll 1$) the lines coalesce into a single Lorentzian centred in $\omega = \omega_0$ and having a motionally narrowed width of $\tau\omega_2^2/6$. Then

$$I(\omega) \simeq \frac{1}{\pi} \frac{(\frac{1}{6}\tau\omega_2^2 + \frac{1}{2}\Gamma)}{(\omega - \omega_0)^2 + (\frac{1}{6}\tau\omega_2^2 + \frac{1}{2}\Gamma)^2}$$

A historically famous example for this type of broadening is the coalescence of the g-values of Cu^{2+} in several salts.[33-7] With the constant magnetic field along, e.g., the [001]-cubic axis, one sees at low temperatures an ESR spectrum with a g-value equal to $2 + |8\lambda/\Delta|$ corresponding to a tetragonal distortion along [001], and a spectrum of double intensity with $g \simeq 2 + |2\lambda/\Delta|$ due to distortions along [010] and [100]. Referring to the g-values in §3.2.5 equation (3.41) and to Table 3.2 we recognize that for $\beta > 0$ this spectrum is due to an elongated structure of the octahedron, with the ϵ-state of the ^2E-doublet lowest. When the temperature is raised, characteristically to 10 °K–50 °K, the previous spectra are replaced after a temperature region of coexistence by an isotropic line at $g = g_1 = \frac{1}{3}(g_\parallel^\epsilon + 2g_\perp^\epsilon) = 2 + |4\lambda/\Delta|$. In addition to these there exist also spectra of the 'third-type', which exhibit cubic symmetry. These have been discussed earlier in §3.2.5 under 'Ham-effect', and do not concern us here since they do not arise from conditions on which this section is based, namely, deep wells into which the system is thrown by random strains.

The transition from the tetragonal to the isotropic spectra has been interpreted by O'Brien[27] in terms of the adiabatic line shape theory which led to equation (6.12). In this equation we substitute

$$\hbar\omega_2 = \beta H g_2 h(T) \tag{6.13}$$

where g_2 is given in equation (3.43) and $h(T)$ is a function of temperature which will be explained now. According to O'Brien, there are two causes for the broadening of the anisotropic lines and for their coalescence at the mean frequency $\beta H g_1$: First, the increase in the reorientation rate τ^{-1} at elevated temperatures as seen in equations (6.2–3), (6.6–8), and (6.10). Secondly, the thermal occupation of higher lying vibronic states for which the tunnelling rate is higher and $|g_\parallel^\epsilon - g_\perp^\epsilon|$ smaller than in the vibronic ground states. Indeed for states which lie at the level of the barrier, i.e. near 2β, one expects $|g_\parallel^\epsilon - g_\perp^\epsilon| \sim 0$. We have included this effect in equation (6.13) by adding the factor $h(T)$ decreasing with temperature to the anisotropic frequency. Clearly, this is only an approximate description since in reality, because of thermal excitation, there will be a number of independent lines rather than a pair of lines of diminishing separation.

The mechanism envisaged by Bersuker[26] is also based on an enhanced tunnelling energy for higher vibronic states. In this theory the temperature dependent average of the tunnelling energies has to compete with the anisotropic Zeeman energy. If the latter dominates a cubic (anisotropic) spectrum is obtained; if the former, one finds one frequency which tends to the isotropic value and also other anisotropic frequencies which gradually lose their intensity as the temperature is raised. Broadening and relaxation due to reorientations are not considered by Bersuker, the reason being apparently that in this theory the centres are not oriented, since random strains are disregarded.

An alternative explanation[28] for the anisotropic–isotropic transition was based on the linear J-T-E. An argument was given that as the temperature is lowered the *effective* J-T coupling strength in the crystal increases and the rotational states of equation (3.16) with different j become near-degenerate [equation (3.17)]. Under this condition the magnetic field dominates, as in the Bersuker theory, and an anisotropic spectrum is found. At higher temperatures, when the effective J-T coupling is weak, most of the rotational states are non-degenerate and are good eigenstates, not intermixed by the anisotropic Zeeman energy. This results in isotropic lines. In this theory too, the capacity of random strains to orient the centres was ignored.

The hyperfine structure obeys similar line-shape formulae as the Zeeman-transitions. The values of the anisotropic energy entering equation (6.12) are now, for the hyperfine structure on the metal ion

$$\hbar \omega_2 = \beta H A_2$$

(or $\beta H A_2 h(T)$, if an 'average' temperature dependent anisotropic frequency is preferred). A_2 was defined in equation (3.45), in analogy to g_2.

6.5 The line shape (some more examples)

A modified theory of thermal broadening and coalescence was applied by Hartmann-Boutron[30] to the quadrupole splitting in the Mössbauer spectrum of Fe^{2+} in tetrahedral sites of $FeCr_2O_4$. This spinel undergoes a permanent distortion (probably due to a cooperative J-T-E) at $T_t = 135\ °K$[40] and in the low-temperature, tetragonal phase the quadrupole field acting on the nucleus is due to the vibronic ground state A_1^I of Figure 4.2. With increase of temperature one finds experimentally (Figure 6.4) a decrease in the quadrupole splitting as well as some broadening. This goes on even after T_t is exceeded. Then coalescence occurs and at room temperature the spectrum has but one narrow peak.

The model of Reference 30 incorporates two effects: inter-well jumps whose frequency has a strong temperature dependence [different from those in equations (6.2–3) or (6.6–10)] and thermal excitations to the

8

states B_1^I, A_1^{II}, . . . in Figure 4.2. In these excited states the quadrupole splitting is different from (and even opposite to) that in the ground vibronic state A_1^I. In the calculation of Reference 30 the main effect of the

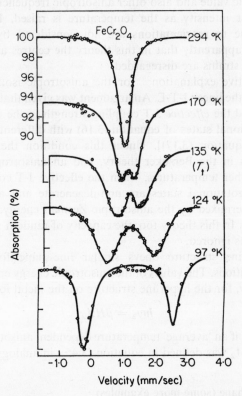

Figure 6.4 Mössbauer spectrum of tetragonal Fe^{2+} in $FeCr_2O_4$ at different temperatures. (From Reference 39.) The values of the velocities are relative to stainless steel.

excitations is to reduce the anisotropic frequency by the factor $h(T)$ in equation (6.13), whose form is written by Hartmann-Boutron as

$$h(T) = \tanh(\Delta/2kT),$$

Δ being a characteristic excitation energy. To reproduce the experimental spectra of Bachella et al.[38] taken at 95 °K and 140 °K, Hartman-Boutron chose $\Delta = 80$ cm^{-1} and $\tau^{-1} = 5 \times 10^6$ sec^{-1} at 140 °K. In actuality, to keep the mathematics tractable, she adopted a heuristic two-well model as well as the static, frozen-in approximation for the vibronic states. This is unlikely to be reliable for the excited states B_1^I, etc. Moreover, the temperature dependence of the quadrupole splitting below the transition temperature which was ascribed to relaxation effects is probably not due to these but to the crystallographic distortion which continues even below T_t.[39]

It is unpleasant to record that one of the most challenging effects in the line shapes of degenerate systems, the anomalous broadening of the ESR lines in $C_6H_6^-$, coronene$^-$, etc.[48] has not yet been satisfactorily explained. Interpretations, whether they emphasized the degeneracy[49] or the removal of the degeneracy by the solvent-radical interaction,[50] whether based on weak[49] or on strong[48,51] coupling, turned out to be insufficient in some respect. Some of the criticisms are noted in Reference 51.

On the other hand, the ESR line shapes[52-3] of impurities in some fluorite type crystals follow the predictions of the theory which is based on weak linear $E \otimes \varepsilon$ coupling and which includes the effect of random strains e.[10,54] From Figure 4.5 we can see that the anisotropy of the spectrum predicted with these assumptions has a broad distribution between the extrema. (On the other hand, in a model for strong anisotropic coupling $(3\Gamma$ small) only a small fraction (given by 3Γ divided by the characteristic strain energy) of intensity would fall outside the width $2\tau^{-1}$ of the anisotropic lines and the line shapes would be more symmetric. This was not observed in References 52–3.) In addition to the line shape, the growth of the isotropic spectra (Figure 4.5(c)) and the form of the spin-Hamiltonian are reported[52-3] to be in accord with Ham's predictions.[10] These advances in the relaxation mechanism in solids give cause for hope that the problem of broadening in hydrocarbon radicals will also be shortly solved.

Strong temperature variations were observed in the g-value and the hyperfine structure constant of Cu^{2+} in CaO[55] (Figure 6.5). No anisotropy–isotropy transition was observed in this case, however only the spectra with $H\|\{111\}$ yielded accurate enough values and in this orientation the isotropic line alone appears [equation (3.42)]. It was suggested[56] that the temperature shift of the lines arises from the coalescence of two systems of lines originally due to the vibronic ground E and excited singlet states, respectively. At a first sight at equation (3.42) the two systems of lines should be identically situated for $H\|\{111\}$ (since for this orientation the square brackets vanish), but it must be realized that there are operators, 'off-diagonal operators', which were neglected (as noted in deriving equation (4.9)) and which introduce a difference between g or A of the doublet and of the singlet. This refinement was first noticed by O'Brien, as quoted in Reference 54.

Adding the hyperfine interaction terms to the spin-Hamiltonian, equation (3.34) and (4.8), we identify the 'off-diagonal operators' as the components along x' and y' of the electronic and nuclear spins S and I. [The system of co-ordinates (x', y', z') is defined by the direction of the magnetic field, $[111] = z'$ and by two perpendicular directions, x' and y'.]

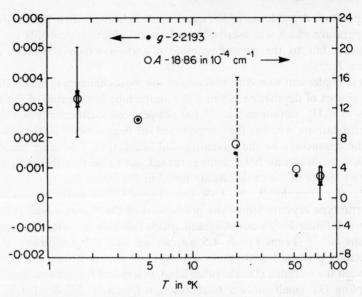

Figure 6.5 ESR data for $CaO:Cu^{2+}$ versus temperature. (From Reference 56, based on the results of Reference 55.) The expected errors are shown at $T = 1\cdot7\,°K$ and $77\,°K$ by single and double vertical lines for the g-factor and the hyperfine constant A, respectively. In the centre the broken line shows how large is the error when the magnetic field is not along [111] (but in fact along [100]).

The coefficients G in the spin-Hamiltonian equation (3.34) take the form:

$$G(A_1) = g_1\beta HS_{z'} + A_1(S_{z'}I_{z'} + S_{x'}I_{x'} + S_{y'}I_{y'})$$

$$G(E_\theta) = -\gamma_\theta q_0 + \tfrac{1}{2}\sqrt{2}g_2\beta HS_{x'}$$
$$+ \tfrac{1}{2}\sqrt{2}A_2\{S_{z'}I_{x'} + S_{x'}I_{z'} + \tfrac{1}{2}\sqrt{2}(S_{x'}I_{x'} - S_{y'}I_{y'})\} \quad (6.14)$$

$$G(E_\epsilon) = -\gamma_\epsilon q_0 - \tfrac{1}{2}\sqrt{2}g_2\beta HS_{y'}$$
$$+ \tfrac{1}{2}\sqrt{2}A_2\{S_{z'}I_{y'} + S_{y'}I_{z'} - \tfrac{1}{2}\sqrt{2}(S_{x'}I_{y'} + S_{y'}I_{x'})\}$$

(The strain energies γ_θ and γ_ϵ were introduced in equation (4.1), q_0 in equation (3.15).)

The off-diagonal operators in the Hamiltonian cause a spin-reversal and will therefore contribute to the g-value only in the second order. In fact the second order correction by these operators to the g-value of the doublet is to a good approximation

$$g_E - g_1 = \tfrac{1}{4}q^2 g_2^2 \left[\frac{2}{g_1} + \frac{1}{g_1 + 2\gamma_\theta qq_0/(\beta H)} + \frac{1}{g_1 - 2\gamma_\theta qq_0/(\beta H)} \right] \quad (6.15)$$

The g-value of the excited singlet is still g_1. If the high temperature g-value is indeed the merged average of g_1 and g_E, then the shift of the low and high temperature g-values is

$$\tfrac{1}{3}(g_E - g_1) \simeq \tfrac{1}{6}q^2 g_2/g_1 \simeq 0\cdot0036q^2 \quad (6.16)$$

The last two terms in equation (6.15) are $O(\beta H/\gamma_\theta q_0)$ and therefore rather small. The experimental value to be compared with equation (6.16) can be read off from Figure 6.5. It is $0{\cdot}0040 \pm 0{\cdot}0015$. Quantitative agreement with equation (6.16) is only possible if q is not much below unity, i.e. the J-T coupling is weak [equation (3.50) and Figure 3.10].

The temperature-shift of the hyperfine-coupling is even easier to compute if it is assumed that $|A_2|$ is rather larger than $|A_1|$. Reference to the definitions in equation (3.45) makes this supposition fairly acceptable. Then, barring accidental cancellation we obtain for the *relative* shift

$$\tfrac{1}{3}(A_E - A_1)/A_E \sim \tfrac{1}{3}$$

In Figure 6.5 we see that this relation is reasonably well reproduced by the experiments.

6.6 Vibronic relaxation in optically excited systems

In §7.6 we present a simplified version of an excited system which consists of two neighbouring electronic states, and in §3.6.2 we gave some details of the corresponding vibronic problem and of its solution. In Figure 6.6, we show a one-dimensional configuration-diagram representing

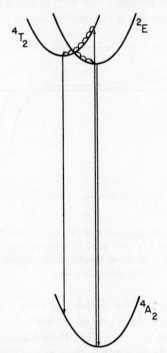

Figure 6.6 Absorption and emission processes typical to a Cr^{3+} compound. Full lines depict radiative transitions, wavy lines show the non-radiative pathways. (From Reference 59.)

a typical Cr^{3+} system in a lattice and the main physical processes occurring therein. We shall now focus our attention on the non-radiative relaxational processes (wavy-lines in Figure 6.6) of the system, since the relative simplicity of the model enables a quantum mechanical and statistical mechanical analysis to be carried through exactly (although at the cost of some stringent assumptions). In a more complex system of multiple well potentials, where the radiation-less processes are no less important, the analogous problem is likely to be uncomputable.

While the vibronic problem is set up in terms of the interaction co-ordinate (the abscissa of Figure 6.6), coupling to the lattice is needed to cause relaxation. It is assumed that the coupling changes the vibrational number in each well from n to $n \pm 1$. Because of vibronic mixing there will be transitions between *all* eigenstates; the probabilities of upward and downward processes obey the requirements of detailed balance. Once these probabilities are known, one can set up and solve rate equations[57] for all vibronic states. From the solutions the time development of the system can be studied. An experimental study of the time variation of the fluorescent emission in emerald was carried out by Pollack.[58] The next figure (Figure 6.7) shows the computed behaviour of a three levels system

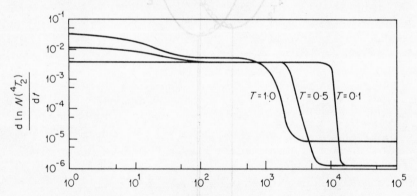

Figure 6.7 Logarithmic derivative of the occupation $N(^4T_2)$ of the lowest vibronic state in the $^4T_2$ potential (shown on the previous figure) as function of the time t, at three temperatures (expressed in units of the vibrational energy). (From Reference 59.)

(such as in Figure 6.6) for five decades of time. The unit of time is the time of a relaxation step along one of the wavy lines in Figure 6.6. (This time is characteristically 10^{-12} sec). The three rate constants, which appear as successive plateaus in Figure 6.7 arise respectively from the following three processes: (A) Thermal activation of the system from the vibronic ground state of $^4T_2$ to the excited vibronic states of $^4T_2$, (B) $^4T_2 \rightarrow {}^2E$

tunnelling, (c) Radiative decay to the $^4A_2$ ground state. There is also a fourth process which acts throughout the time: (d) A radiative depletion of the $^4T_2$ states.

We shall not embark here on a detailed description of the model on which the non-radiative decay mechanism was based.[57] Suffice it to say, that the coupling of the interaction modes with the lattice phonons, acting as a thermal bath, causes the relaxation of the excitation (the wavy lines of Figure 6.6) from every vibronic state but the very lowest.

Relaxation from one electronic state of a molecule to another can also occur through *intramolecular* processes,[60] provided the molecule is sufficiently large, e.g. an aromatic hydrocarbon, and the electronic states are widely separate, say ten vibrational quanta apart. Under such conditions a practically irreversible decay occurs because of the nearly continuous distribution of the vibronic energy levels high above the bottom of the potential. The decay mechanism is vibronic, in that it is the correction term to the Born–Oppenheimer approximation which admixes the two electronic states. An expression can be derived[61] for the relaxation rate which can be formulated analytically in the limiting cases of large and small horizontal shifts of the potentials. It is clear from Reference 61, that the relaxation rate is a (second order) perturbational expression of the Born–Oppenheimer correction term and is therefore, (by the restriction imposed in the Introduction on the non-perturbational nature of the J-T-E) excluded from the purview of our study. The interested reader is referred to the papers quoted.

It was also pointed out in References 61–2 that, apart from electronic relaxation, photochemical rearrangements or distortions can also be envisaged with an intramolecular mechanism—at least for large molecules. Following optical excitation, the molecule can assume, without the assistance of an external agent, a distortion which will persist for a length of time which is infinite on the scale of most physical processes.

The affinity of this situation to that proposed by Jahn and Teller over thirty years ago is evident. On this note we end this chapter.

6.7 References

1. E. L. Elkin and G. D. Watkins, *Phys. Rev.*, **174,** 881 (1968).
2. J. A. Sussmann, *Phys. Kondens. Materie*, **2,** 146 (1964); *J. Phys. Chem. Solids*, **28,** 1643 (1967).
3. W. Känzig, *J. Phys. Chem. Solids*, **23,** 479 (1962).
4. R. Pirc, B. Zeks and P. Gosar, *J. Phys. Chem. Solids*, **27,** 1219 (1966).
5. K. W. H. Stevens, *Repts. Prog. Phys.*, **30,** 189 (1967).
6. C. B. P. Finn, R. Orbach and W. P. Wolf, *Proc. Phys. Soc.*, **A77,** 261 (1961).

7. R. Orbach, *Proc. Roy. Soc. A*, **264**, 458 (1961).
8. I. Waller, *Z. Physik*, **79**, 370 (1932).
9. J. H. Van Vleck, *Phys. Rev.*, **57**, 426 (1940).
10. F. S. Ham, *Jahn–Teller Effects in Electron Paramagnetic Resonance Spectra* in *Electron Paramagnetic Resonance*, ed. Geschwind (New York: Plenum Press, 1971).
11. F. I. B. Williams, D. C. Krupka and D. P. Breen, *Phys. Rev.*, **179**, 255 (1969).
12. D. P. Breen, D. C. Krupka and F. I. B. Williams, *Phys. Rev.*, **179**, 241 (1969).
13. M. D. Sturge, J. T. Krause, E. M. Gyorgy, R. C. Le Craw and F. R. Merritt, *Phys. Rev.*, **155**, 218 (1967).
14. J. C. Gill, *Proc. Phys. Soc.*, **85**, 119 (1965).
15. A. M. Stoneham, *Proc. Phys. Soc.*, **85**, 107 (1965).
15a. M. Noack, G. F. Kokoszka and G. Gordon, *J. Chem. Phys.*, **54**, 1432 (1971).
16. K. P. Lee and D. Walsh, *Phys. Letters*, **27A**, 17 (1968).
17. H. M. McConnell, *J. Chem. Phys.*, **25**, 709 (1956).
18. G. R. Liebling and H. M. McConnell, *J. Chem. Phys.*, **42**, 3931 (1965).
19. E. P. Wigner, *Akad. der Wiss. Göttingen, Mat. Fys. Kl. Nachrichten*, 1930, p. 133 [English transl.: *Symmetry in the Solid State* (New York: Benjamin, 1964) p. 173.]
20. A. A. Maradudin and S. H. Vosko, *Rev. Mod. Phys.*, **40**, 1 (1968).
21. V. Heine, *Group Theory in Quantum Mechanics* (Oxford: Pergamon, 1960).
22. L. P. Bouckaert, R. Smoluchowski and E. P. Wigner, *Phys. Rev.*, **50**, 58 (1936).
23. A. S. Nowick and W. R. Heller, *Adv. Phys.*, **14**, 101 (1965).
24. A. S. Nowick, *Adv. Phys.*, **16**, 1 (1967).
25. M. D. Sturge, *Solid State Phys.*, **20**, 91 (1967).
26. I. B. Bersuker, *Zh. Eksperim. i Teor. Fiz.*, **44**, 1239 (1963) [English transl.: Soviet Phys.—JETP, **17**, 836 (1963)]; *Fiz. Tverd. Tela*, **6**, 436 (1964) [English transl.: *Soviet Phys.–Solid State*, **6**, 347 (1964)].
27. M. C. M. O'Brien in *Paramagnetic Resonance* Vol. I, ed. Low (New York: Academic, 1963) p. 322; *Proc. Roy. Soc. A*, **281**, 323 (1964).
28. R. Englman and D. Horn in *Paramagnetic Resonance* Vol. I, ed. Low (New York: Academic, 1963) p. 329.
29. A. Hudson, *Mol. Phys.*, **10**, 575 (1966).
30. F. Hartmann-Boutron, *J. Phys. Rad.*, **29**, 47 (1968).
31. M. Blume and J. A. Tjon, *Phys. Rev.*, **165**, 446 (1968).
32. J. A. Tjon and M. Blume, *Phys. Rev.*, **165**, 456 (1968).
33. B. Bleaney, R. P. Penrose and B. I. Plumpton, *Proc. Roy. Soc. A*, **198**, 406 (1949).
34. B. Bleaney and D. J. E. Ingram, *Proc. Phys. Soc.*, **A63**, 408 (1950).
35. D. Bijl and A. C. Rose-Innes, *Proc. Phys. Soc.*, **A66**, 954 (1953).
36. B. Bleaney, K. D. Bowers and R. S. Trenam, *Proc. Roy. Soc. A*, **228**, 157 (1955).
37. H. C. Allen, G. F. Kokoszka and R. G. Inskeep, *J. Am. Chem. Soc.*, **86**, 1023 (1964).
38. G. L. Bacchella, P. Imbert, P. Meriel, E. Martel and M. Pinot, *Bull. Soc. Sci. Bretagne*, **39**, 121 (1964).
39. M. Tanaka, T. Tokoro and Y. Aiyama, *J. Phys. Soc. Japan*, **21**, 262, (1966).
40. D. N. Pipkorn and H. R. Leider, *Bull. Am. Phys. Soc.*, **11**, 49 (1966).

41. H. R. Leider and D. N. Pipkorn, *Phys. Rev.*, **165**, 494 (1968).
42. H. S. Gutowsky, D. W. McCall and C. P. Slichter, *J. Chem. Phys.*, **21**, 279 (1953).
43. P. W. Anderson, *J. Phys. Soc. Japan*, **9**, 316 (1954).
44. R. Kubo, *J. Phys. Soc. Japan*, **9**, 935 (1954).
45. R. A. Sack, *Mol. Phys.*, **1**, 163 (1958).
46. P. W. Anderson and P. R. Weiss, *Rev. Mod. Phys.*, **25**, 269 (1963).
47. A. Abragam, *The Principles of the Nuclear Magnetism* (Oxford: Clarendon, 1961).
48. M. G. Townsend and S. I. Weissman, *J. Chem. Phys.*, **32**, 309 (1960).
49. H. M. McConnell, *J. Chem. Phys.*, **34**, 13 (1961).
50. D. Kivelson, *J. Chem. Phys.*, **45**, 751; **45**, 1324 (1966).
51. J. H. Freed and R. G. Kooser, *J. Chem. Phys.*, **49**, 4715 (1968).
52. L. A. Boatner, B. Dischler, J. R. Herrington and T. L. Estle, *Bull. Am. Phys. Soc.*, **14**, 355 (1969).
53. J. R. Herrington, T. L. Estle, L. A. Boatner and B. Dischler, *Phys. Rev. Letters*, **24**, 984 (1970).
54. F. S. Ham, *Phys. Rev.*, **166**, 307 (1968).
55. W. Low and J. T. Suss, *Phys. Letters*, **7**, 310 (1963); Solid State Commun., **2**, 1 (1964).
56. R. Englman, *Phys. Letters*, **31A**, 473 (1970).
57. R. Englman and B. Barnett, *J. Luminescence*, **3**, 37 (1970).
58. S. A. Pollack, *J. Appl. Phys.*, **38**, 5083 (1967).
59. B. Barnett and R. Englman, *J. Luminescence*, **3**, 55 (1970).
60. M. Bixon and J. Jortner, *J. Chem. Phys.*, **48**, 715 (1968); **50**, 3284 (1969).
61. R. Englman and J. Jortner, *Mol. Phys.*, **18**, 145 (1970).
62. J. Jortner, S. A. Rice and R. M. Hochstrasser, *Adv. Photochem.*, **7**, 149 (1969).

7 The physical landscape

In the preceding chapters we illustrated the operation of the J-T-E by reference to physical systems whose characters differ from one to another widely. In this chapter we briefly describe the physical background of the most frequently quoted physical systems. We also derive in each (non-trivial) case the Hamiltonian (H_{JT}) which forms the basis of the mathematics of the J-T-E.

7.1 d^n ions

Transition metal atoms when placed in solvent liquids, or in solids where some ions have large electron affinities, tend to shed their outermost s electrons and to assume a d^n configuration.[1-3] The ground state ionic term is frequently the one with maximum spin multiplicity, in accord with one of Hund's rules, while the orbital part is determined by the competition between two effects of about the same strength (10^4 cm^{-1}). These are the inter-electronic interaction in the ion and the crystal field due to the ligands (the immediate neighbours of the transition metal ion). When this field possesses high symmetry, e.g. an octahedral (also called cubic) or a tetrahedral symmetry, the levels of a single d-electron split into a threefold degenerate manifold t_2 (in T_d or O, t_{2g} in O_h) and a doubly degenerate manifold e (e_g) whose energy separation is traditionally denoted by 10 Dq or Δ. The designation of these one-electron states and their transformation properties in the group are given by

$$t_2: \xi, \eta, \zeta \quad \text{or} \quad |\phi_\xi\rangle, |\phi_\eta\rangle, |\phi_\zeta\rangle \quad \text{transform as } yz, zx, xy$$

$$e: \theta, \epsilon \quad \text{or} \quad |\phi_\theta\rangle, |\phi_\epsilon\rangle \quad \text{transform as } \frac{1}{\sqrt{3}}(2z^2 - x^2 - y^2),$$

$$(x^2 - y^2)$$

(The component ϵ is not to be confused with the representation of the doubly degenerate mode ε.)

Not only are the one-electron states of d split by the crystal field, but so are the many-electron manifolds of the free ion, e.g. ^5D or ^3F; the latter into *terms* (note the terminology!) $^3A_2$, $^3T_2$, $^3T_1$. When the high symmetry crystal field is the dominant perturbation, 'the strong field case', each term has a definite subshell configuration, e.g. a $^3T_1$ term of the vanadium ion V^{3+} will be either t_2^2 *or* et_2. In even more extreme cases of a strong crystal field it can also happen that a configuration with a high number (i.e. > 3) of t_2 electrons (in octahedral surroundings) or of e electrons (in tetrahedral environment) becomes stabilized to such extent that Hund's rules cease to operate and the ground state is one of *low* spin multiplicity. It is more common though that the inter-electron coupling is comparable in magnitude to the crystal field. Then the subshell configuration is not necessarily pure, e.g. a $^3T_1$ term takes the form

$$\cos \alpha |t_2^2\rangle + \sin \alpha |et_2\rangle \tag{7.1}$$

Many of the previous considerations are based on symmetry and group theory and hold perfectly well when the t_2 or e states are not pure ionic states but contain admixtures of a considerable amount of states having t_2 or e symmetry which arise from the ligand orbitals. We thus meet the case of covalent bonding in these complexes, where the d-orbitals play the role of antibonding orbitals.

To the next class of energies (characteristically 100 cm^{-1}) belong three effects, which are particularly important at low temperatures, e.g. in electron spin resonance (ESR) work, and less important in room temperature optical experiments. The first of these is a fixed field of lower symmetry (axial or biaxial) than the large cubic or tetrahedral crystal fields. If present, this field usually arises in solids from the packing of the atoms and in large molecules from the directions of the bonds. Next we have the spin–orbit coupling on the ions. This splits the terms into a number of levels, each characterized by a quantum number J' which is analogous but not quite equivalent to the total angular momentum J of atomic spectroscopy. (More about J' is found on p. 221.) In ESR usually only the lowest J' level is important and this number is shown, together with some other information in the following table. (Table 7.1.)

The last perturbation in the 100 cm^{-1} class is due to the vibrational displacement of the nuclei near the d^n ion. Whether this displacement is to be regarded as a static or a dynamic event depends on the time-scale of the physical effect under consideration. There is no question though that, in a solid as well as in a solvated complex, it is a very useful procedure to consider only the vibrational motion of the nearest neighbours and to

Table 7.1 Ground states of some transition metal ions.

The subscript after the comma represents the usual or likely J' value of the lowest level, whenever this is known to any degree of certainty.

d^n	3d	4d	5d	Octahedral High	Octahedral Low	Tetrahedral High	Tetrahedral Low
d^1	Sc^{2+} Ti^{3+} V^{4+}	Zr^{3+} Tc^{6+}		$^2T_{2,3/2}(t_2)$	$^2T_{2,3/2}(t_2)$	$^2E_{3/2}(e)$	$^2E_{3/2}(e)$
d^2	Ti^{2+} V^{3+}	Nb^{3+} Ru^{6+}	Ta^{3+} Os^{6+}	$^3T_{1,1}(t_2^2)$	$^3T_{1,1}(t_2^2)$	$^3A_{2,1}(e^2)$	$^1E_{3/2}(e^2)$
d^3	V^{2+} Cr^{3+}	Mo^{3+} Rh^{6+}	W^{3+} Ir^{6+}	$^4A_{2,3/2}(t_2^3)$	$^4A_{2,3/2}(t_2^3)$	$^4T_{1,5/2}(t_2e^2)$	$^2E_{3/2}(e^3)$
d^4	Cr^{2+} Mn^{3+}	Tc^{3+} Pd^{6+}	Re^{3+} Pt^{6+}	$^5E_0(t_2^3e)$	$^3T_{1,1}(t_2^4)$	$^5T_{2,3}(t_2^2e^2)$	$^1A_{1,0}(e^4)$
d^5	Mn^{2+} Fe^{3+} Co^{4+}	Ru^{3+}	Os^{3+}	$^6A_{1,5/2}(t_2^3e^2)$	$^2T_{2,3/2}(t_2^5)$	$^6A_{1,5/2}(t_2^3e^2)$	$^2T_2(t_2e^4)$
d^6	Mn^+ Fe^{2+} Co^{3+}	Rh^{3+}	Ir^{3+}	$^5T_{2,1}(t_2^4e^2)$	$^1A_{1,0}(t_2^6)$	$^5E(t_2^3e^3)$	$^1T_{2,1}(t_2^2e^4)$
d^7	Fe^+ Co^{2+} Ni^{3+}	Pd^{3+}	Pt^{3+}	$^4T_{1,3/2}(t_2^5e^2)$	$^2E_{3/2}(t_2^6e)$	$^4A_{2,3/2}(t_2^3e^4)$	$^2E_{3/2}(t_2^3e^4)$
d^8	Ni^{2+}	Pd^{2+} Ag^{3+}	Pt^{2+} Au^{3+}	$^3A_{2,1}(t_2^6e^2)$	$^3A_{2,1}(t_2^6e^2)$	$^3T_{1,0}(t_2^4e^4)$	$^1T_{2,1}(t_2^4e^4)$
d^9	Ni^+ Cu^{2+}	Ag^{2+}	Pt^+ Au^{2+}	$^2E_{3/2}(t_2^6e^3)$	$^2E_{3/2}(t_2^6e^3)$	$^2T_2(t_2^5e^4)$	$^2T_2(t_2^5e^4)$

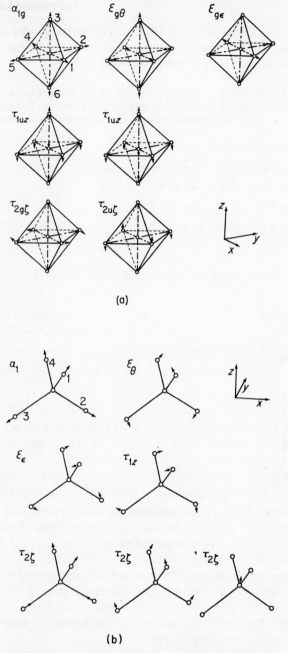

Figure 7.1 Vibrational modes in octahedral and tetrahedral centres. Only the z-component of the threefold modes is shown.

(a) An octahedral MO_6 molecule. The coordinate system, the numbering and the normal modes are shown.

(b) A tetrahedral MO_4 molecule. The coordinate system, the numbering and the normal modes are shown.

disregard, as a start at any rate, the rest. The normal modes of the vibrations in the most commonly met cases, the octahedral and tetrahedral coordinations, are shown in the figures on p. 217.

Table 7.2 Definition of normal coordinates of simple centres in terms of the cartesian coordinates of the ions. The numbering of the ions follows Figure 2.1. Only the z (or ζ) components of triply degenerate modes are written out.

O_h: (Even modes only)

$$Q(\alpha_{1g}) = (X_1 - X_4 + Y_2 - Y_5 + Z_3 - Z_6)/\sqrt{6}$$

$$Q(\varepsilon_{g\theta}) = (2Z_3 - 2Z_6 - X_1 + X_4 - Y_2 + Y_5)/\sqrt{12}$$

$$Q(\varepsilon_{g\epsilon}) = (X_1 - X_4 - Y_2 + Y_5)/2$$

$$Q(\tau_{2g\zeta}) = (Y_1 - Y_4 + X_2 - X_5)/2$$

T_d:

$$Q(\alpha_1) = -[(-X_1 - X_2 + X_3 + X_4) + (Y_1 - Y_2 + Y_3 - Y_4) \\ + (-Z_1 + Z_2 - Z_3 + Z_4)]/\sqrt{12}$$

$$Q(\varepsilon_\theta) = [(-X_1 - X_2 + X_3 + X_4) + (Y_1 - Y_2 + Y_3 - Y_4) \\ + 2(Z_1 - Z_2 - Z_3 + Z_4)]/\sqrt{24}$$

$$Q(\varepsilon_\epsilon) = [(-X_1 - X_2 + X_3 + X_4) + (-Y_1 + Y_2 - Y_3 + Y_4)]/\sqrt{8}$$

$$Q(\tau_{1z}) = [(-X_1 + X_2 - X_3 + X_4) + (-Y_1 - Y_2 + Y_3 + Y_4)]/\sqrt{8}$$

$$Q(\tau_2)_\zeta \begin{cases} = -[(-X_1 + X_2 - X_3 + X_4) + (Y_1 + Y_2 - Y_3 - Y_4) \\ \qquad + (-Z_1 - Z_2 - Z_3 - Z_4)]/\sqrt{12} \\ = [(-X_1 + X_2 - X_3 + X_4) + (Y_1 + Y_2 - Y_3 - Y_4) \\ \qquad + 2(Z_1 + Z_2 + Z_3 + Z_4)]/\sqrt{24} \\ = Z_0 \end{cases}$$

The final class of perturbations is again a low symmetry field, but this time due to randomly placed and oriented, static imperfections or impurities. In solids this displaces the electronic energy levels characteristically by 1 cm^{-1} and would be disregarded except in some special phenomena. It turns out that the strong coupling case of the J-T-E is one of the exceptions where these small perturbations can have important results.

Let us derive now a matrix Hamiltonian describing the first order vibronic coupling when the system is in one of the degenerate electronic states $\phi_\gamma^\Gamma(\mathbf{r})$: Here γ is a component of the degenerate representation Γ of the molecular symmetry group.

Often, but not always, Γ will be an irreducible representation of the group. In other cases, and these are the ones which necessitate an extra bit of sophistication in the derivation, additional degeneracy is induced by the time-reversal symmetry of the Hamiltonian and the resulting equal-energy set is twice as large as the set which forms a basis for the irreducible representation of the group. Appendix III is devoted to these cases.

The potential energy of the system as function of the set of normal coordinates $\{Q\}$ or $\{q\}$ (for terminology see the notes after Table 1.1) is written, up to first order in $\{q\}$, as

$$E(\{q\}) = E(\{0\}) + \sum_i q_i \left(\frac{\partial E}{\partial q_i}\right)_{\{0\}} \qquad (7.2)$$

$\{0\}$ represents the undisplaced configuration of the system. The energy in the previous expression belongs to some electronic eigenstate ϕ^Γ of the molecule. This state is a linear combination of the degenerate states, say,

$$\phi^\Gamma(\mathbf{r}, \{q\}) \equiv \sum_\gamma a_\gamma^\Gamma(q_i)\phi_\gamma^\Gamma(\mathbf{r}) \qquad (7.3)$$

where the coefficients a depend on q_i as indicated. The absence of q_i from $\phi_\gamma^\Gamma(\mathbf{r})$ was discussed in Chapter 2. Then by the Hellmann–Feynman theorem[4-6]

$$E(\{q\}) = E(\{0\}) + \sum_{\Lambda\lambda} q_\lambda^\Lambda \int d\mathbf{r} \phi^{\Gamma*}(\mathbf{r}) \left(\frac{\partial H}{\partial q_\gamma^\Lambda}\right)_{\{0\}} \phi^\Gamma(\mathbf{r}) \qquad (7.4)$$

where $H = H(\mathbf{r}, \{q\})$ is the Hamiltonian of the system and q_λ^Λ is a normal mode, the component λ of the representation Λ. We rewrite equation (7.4).

$$E(\{q_i\}) - E(\{0\}) = \sum_{\Lambda, \lambda\rho\sigma} q_\lambda^\Lambda a_\rho^{\Gamma*} a_\sigma^\Gamma \int d\mathbf{r} \phi_\rho^{\Gamma*} H_\lambda^\Lambda \phi_\sigma^\Gamma$$

$$= \sum_{\Lambda, \lambda\rho\sigma} [\Gamma]^{-\frac{1}{2}} q_\lambda^\Lambda a_\rho^{\Gamma*} a_\sigma^\Gamma \langle\Gamma\|H^\Lambda\|\Gamma\rangle\langle\Gamma\Lambda\Gamma\rho|\Gamma\Lambda\sigma\lambda\rangle \qquad (7.5)$$

in terms of the vector coupling coefficients $\langle\Gamma\Lambda\Gamma'\rho|\Gamma\Lambda\sigma\lambda\rangle$ defined for point groups in References 3 and 7. $\langle\Gamma\|H^\Lambda\|\Gamma\rangle$ is the reduced matrix element and $[\Gamma]$ denotes the dimension of the representation Γ.

From equation (7.5) we can immediately write down the matrix Hamiltonian which arises from the linear vibronic coupling term. The matrix operates on a_γ^Γ regarded as a column vector. The matrix element in the ρth row and σth column is

$$[\Gamma]^{-\frac{1}{2}} \sum_{\Lambda\lambda} q_\lambda^\Lambda \langle\Gamma\|H^\Lambda\|\Gamma\rangle\langle\Gamma\Lambda\Gamma\rho|\Gamma\Lambda\sigma\lambda\rangle \equiv (H_{\mathrm{JT}})_{\rho\sigma} \qquad (7.6)$$

This result can be derived from, e.g. variation of equation (7.5) with respect to the undetermined coefficients a_γ^Γ. (A systematic description of the variational procedure was given by Longuet-Higgins).[8] The complete matrix Hamiltonian includes, in addition to equation (7.6), diagonal matrices of the kinetic and elastic energies as well as higher than linear vibronic terms (matrices), whenever these arise (Chapter 3).

Equation (7.6) is a general expression for the linear vibronic matrix Hamiltonian, which forms the basis of the discussion of the various individual cases of the J-T-E in the preceding chapters. At this point we wish once to generalize it and once to restrict it. By generalization we mean the pseudo-Jahn–Teller effect (P-J-T-E), which was mentioned on p. 3. In this case equation (7.6) goes over to

$$[\Gamma]^{-\frac{1}{2}} \sum_{\Lambda\lambda} q_\lambda^\Lambda \langle \Gamma \| H^\Lambda \| \Gamma' \rangle \langle \Gamma' \Lambda \Gamma \rho | \Gamma' \Lambda \sigma \lambda \rangle \equiv (H_{\mathrm{PJT}})_{\rho\sigma}^{\Gamma\Gamma'} \qquad (7.7)$$

if the vibronic interaction is between the distinct manifolds Γ and Γ'. If we restrict equation (7.4) or (7.6) to single valued representations of the group then more symmetrical results can be obtained in terms of the V-coefficients of the point group.[9] Namely,

$$(H_{\mathrm{JT}})_{\rho\sigma} = \sum_{\Lambda\lambda} q_\lambda^\Lambda \langle \Gamma \| H^\Lambda \| \Gamma \rangle V \begin{pmatrix} \Gamma \Gamma \Lambda \\ \rho\ \sigma\ \lambda \end{pmatrix}, \qquad (7.8)$$

or

$$(H_{\mathrm{PJT}})_{\rho\sigma}^{\Gamma\Gamma'} = \sum_{\Lambda\lambda} q_\lambda^\Lambda \langle \Gamma \| H^\Lambda \| \Gamma' \rangle V \begin{pmatrix} \Gamma \Gamma' \Lambda \\ \rho\ \sigma\ \lambda \end{pmatrix} \qquad (7.9)$$

These results are immediate consequences of the basic relation (Reference 9, equation 2.15)

$$\int \mathrm{d}\mathbf{r} \phi_\rho^{\Gamma*} H_\lambda^\Lambda \phi_\sigma^{\Gamma'} = \langle \Gamma \| H^\Lambda \| \Gamma' \rangle V \begin{pmatrix} \Gamma \Gamma' \Lambda \\ \rho\ \sigma\ \lambda \end{pmatrix} \qquad (7.10)$$

It will be recalled that in equation (1.3) the symbol L was introduced to denote the reduced matrix element $\langle \Gamma \| \partial H/\partial q^\Lambda \| \Gamma' \rangle$ or $\langle \Gamma \| H^\Lambda \| \Gamma' \rangle$.

For double groups the rewriting of equation (7.6) or (7.7) in terms of the V-coefficients requires the introduction of the conjugation matrix U.[10] This is followed up in Appendix III.

7.2 Transition metal hexafluoride molecules

There are two circumstances which warrant the separate consideration of these molecules from the previous d^n systems. One is their special coupling scheme which was established on the basis of absorption spectroscopy in the region of 1,400–20,000 cm^{-1}, particularly by Moffitt et al.[11] The other is the existence of a richly documented and thoroughly studied set of observation on the J-T-E in the vibrational properties of these molecules. This was reviewed by Weinstock and Goodman,[12] whose assignments were confirmed by laser Raman spectroscopy.[12a]

It is generally assumed that the shape of the molecules is a regular octahedron. A dissenting opinion was voiced by Kiseljov,[13] although direct support for the symmetrical configuration is available from electron diffraction measurements.[14-5] On the other hand, the shape of the 'unusual molecule XeF_6' has been subject to some discussion, since a lower symmetry than O_h has been envisaged both in the valence-shell electron-pair repulsion model of Gillespie[16] and in the conjecture of Goodman (1967, unpublished) that a substantial number of XeF_6 molecules are in a thermally excited $^3T_{1u}$ state (which is subject to a Jahn–Teller distortion). However, magnetic deflection studies[17] of molecular beams did not reveal any magnetic moment due to a spin triplet and the electron diffraction[18] data were also explained by a loose but symmetric O_h arrangement. (Note however that Burbank and Bartlett[19] have proposed a non-O_h configuration which also explains the electron diffraction results.)

The spectral assignments[11] for $(4d)^n$ and $(5d)^n$ hexafluorides show that the crystal field of the fluorines dominates over other effects and that the first seven ions in each row of elements have ground state configuration $(t_{2g})^n$ ($n = 0, \ldots, 6$). The next two physical effects, the interelectronic Coulomb repulsion (leading to Russell–Saunders coupling) and the spin–orbit interaction (requiring jj-coupling) are of comparable importance so that, in fact, intermediate coupling scheme is appropriate. Numerical measures of the two effects are given by the values of $3F_2 + 20F_4$ and ζ_d which are 2,400 and 3,400 cm^{-1} in the 4d series and 2,990 and 1280 cm^{-1} in the 5d row. (F_2 and F_4 are Slater integrals[20] and ζ_d is the spin–orbit coupling coefficient for an nd electron). These numbers result from an intuitive fitting of the visible[11] and infrared[12] data.

The hexafluorides are listed in the following table (Table 7.3) according to the number n of t_{2g} electrons in the outermost shell, together with the term symbol for the ground state. For each n we give in the second line of the table the characterizations of the $j'j'$ coupling scheme $(j')^n$, of the Russell–Saunders coupling scheme ($^{2S+1}L'$) and of the total quasi-angular momentum J'. The term $quasi$ and the apostrophes on the angular momenta refer to the use of the 't_{2g}–p equivalence'. (Reference 3, Chapter 9, §5). In this device the states of the two triplets are put in one-to-one correspondence and the matrix elements, in particular of the angular momentum operator, in the t_{2g}-manifold are related to the known matrix elements within the p-manifold.

The E_g and T_{2g} terms for $n = 2$ are very close in OsF_6. In the analogous RuF_6 it turns out that T_{2g} is well stabilized by vibronic coupling. For $n = 3$ the orbital part in IrF_6 is, analogously to Cr^{3+} (Table 7.1) and to a good approximation, a singlet, so that one sees practically no J-T-E. For the 5d-ion Rh the spin–orbit coupling is larger, so that in RhF_6, when the $\zeta_d/(3F_2 + 20F_4)$ ratio is larger, there occurs an interesting case of 'spin-destabilization'. Spin–orbit coupling admixes the orbital singlet with

It is generally assumed that the shape of the molecules is a regular octahedron. A dissenting opinion was voiced by Jøslowitz[?], although direct support for the symmetrical configuration is available from electron diffraction measurements.[?] On the other hand, the shape of the unusual molecule XeF_6 has been subject to some discussion, since a lower symmetry than O_h has been envisaged both in the electron-pair repulsion model of Gillespie[?] and in the continuum. Bartlett 1987, emphasized that a substantial number of XeF_6 molecules in a distorted ... $^1T_{1u}$ state (which is subject to a Jahn–Teller distortion, however magnetic ... detection studies[?] of molecular beams did not reveal any low-lying triplet and the electron diffraction[?] data were also explained by a loosely symmetric O_h arrangement. (Note however that the basis and Bartlett 1987 proposed a model of quantization which also explains the electron quantization.)

The several hexafluorides of the $(4d)^n$ and $(5d)^n$ hexafluorides show that the crystal field of the complex dominates over other effects, and that the first seven ions in each row of elements have a ground state configuration $(t_{2g})^m$, $m = 0, \ldots, 6$. The next two physical facts, the interelectronic Coulomb repulsion (leading to Russell–Saunders coupling) and the spin–orbit interaction require that are of comparable importance so that, in fact, intermediate coupling schemes are appropriate. Numerical measures of the two effects is given by the values of $3P$, $4F$, $20?$, and ξ, which are 2,400 and 1,330 cm and the 4d rows and 2,290 and 1,280 cm in the 5d rows (A_2 and F_2 are States integer and ξ is the spin-orbit screening coefficient for an nd electron). The numbers result from an intuitive fitting of the crystal and infrared.

The hexafluorides ... to the number n of 4d or 5d electrons in the outer shell, together with the term symbol for the ground state. For each entry to the second line of the table the characterizations of the $(j')^n$ ... the scheme $(j')^n$ of the Russell–Saunders coupling scheme $^{2S+1}L'$, J' of the total quasi-angular momentum J', the terms $^{2S+1}L'$ and the superscript on the angular momentum refer to the use of the equivalences (Reference 1, Chapter 9, 55). In this device the angular the two marks are put in one-to-one correspondence and the angular moments, quasi-angular of the angular momentum operator in the t_{2g} manifold are equated to the known matrix elements within the manifold.

The L' and J' terms for $n = 2$ are doublet in OsF_6. In the analogous RuF_6, it turns out that $^3P'$ is weakened/mixed by vibronic coupling. For A_{2u} the orbital part in IrF_6 is analogous to Cr^{3+} (Table 7.1) and to a good approximation a singlet so that one level predictably no 3–1–0. For the 3.2 to 9th the spin-orbit coupling is larger than that in RhF_6, when the $\xi/$3P$_{3}$/$20?$ ratio is larger there occurs an interesting case of 'spin delocalization'. Spin-orbit coupling abolishes the orbital singlet, with

Table 7.3 $(4d)^n$ and $(5d)^n$ hexafluorides
The quasi-atomic schemes in the second line are $(j')^n$, $^{2S+1}L'$, J' (explained in the text)

$n =$	0	1	2	3	4
Quasi-atomic schemes	0, $^1S'$, 0	$\tfrac{3}{2}$, $^2P'$, $\tfrac{3}{2}$	$(\tfrac{3}{2})^2$, $^3P'$, 2	$(\tfrac{3}{2})^3$, $^4S'$, $\tfrac{3}{2}$	$(\tfrac{3}{2})^4$, $^3P'$, 0
4d	$MoF_6(A_{1g})$	$TcF_6(G_{\frac{3}{2}g})$	$RuF_6(T_{2g})$	$RhF_6(G_{\frac{3}{2}g})$	
5d	$WF_6(A_{1g})$	$ReF_6(G_{\frac{3}{2}g})$	$OsF_6(E_g, T_{2g})$	$IrF_6(G_{\frac{3}{2}g} \sim {}^4A_{2g})$	$PtF_6(A_{1g})$

orbitally degenerate states and one observes signs of vibronic splittings and shifts which are larger than in IrF_6, but still outside the accuracy of experiments.

The presence of a moderate strength J-T-E in the hexafluorides is made apparent in the data (a) by the anomalous broadening of combination bands which involve an ε_g-mode, (b) by the lowering of frequencies that is characteristic of electron-vibrational coupling and (c) by the presence of a large number of details in the infra-red spectra. [Effects (a) and (b) are also apparent in the solution spectra of the halides.[21]] The structure of the linear J-T coupling term in the Hamiltonian is again formally given by matrix elements having the form of equation (7.6). An important question, which also forms a major part of the theoretical effort of Reference 12 is how the matrix elements of equation (7.6) (which now refer to many electron states) are derived from the fundamental one-electron vibronic coupling matrix elements. Part of the answer resides in the famous theorem of Wigner–Eckart (Reference 10, Chapter 14) which interrelates matrix elements with different component symbols. To derive the reduced matrix element in a manifold characterized by $(t_{2g})^n e_g^m$ (as in the present case) the method and tables of Appendix V may be of help.

7.3 Heavy metal ion sensitizers

Many singly ionized heavy metals substitute for the positive ion in alkali halides readily and with little deformation of the lattice. They have remarkable optical properties[22] which have been reviewed in various places.[23-5] An explanation was provided by Seitz[26] in terms of transitions between the s^2 and sp configurations of the metal ion, as shown in Figure 7.2.

The upper sp configuration splits up by the spin–orbit coupling (characteristically of the order of several $1,000 \, cm^{-1}$ in the heavier elements of the groups III and IV) to the ionic multiplets shown on the left of Figure 7.2 and to the cubic crystal field terms, shown in the centre, if the cubic environment is also incorporated. This latter is not thought to change the relative level-positions excessively, although the Madelung energy of the crystal raises the levels uniformly by about 5 eV. It was pointed out though,[25] that the s and p functions are bound to undergo a change in the lattice, as is evident from the reduction of the exchange integral to about a third of that in the free ion.

The $A(^1S_0 \rightarrow {}^3P_1)$ and $C(^1S_0 \rightarrow {}^1P_1)$ absorption bands are allowed in electric dipole transitions and are the strongest. The B $(^1S_0 \rightarrow {}^3P_2)$ band is made allowed vibronically through the services of a τ_{2g} or ε_g mode. The occurrence of the J-T-E in the C absorption was noted[27] many years ago.

The theory for emission was evolved by Williams and coworkers in several articles,[28-30] which indeed constituted the basis of the configuration coordinate description (§7.6). At that stage of the theory a single coordinate, the breathing mode of the halogen-octahedron, was regarded as sufficient; however, later experimental work on Raman scattering[31-2] and with optical transitions (which has been referred to in §3.3.4 and §3.3.7) required the inclusion of vibronic coupling with ε_g and τ_{2g}-modes.

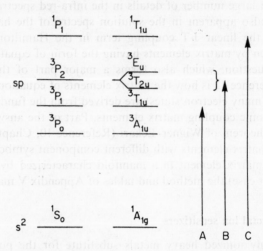

Figure 7.2 sp-states in the Seitz-model[26] in different approximations. First the atomic multiplet levels of the heavy metal are shown (schematically), then the octahedral crystal field terms. The nomenclature for the transitions is indicated.

These also are local modes, arising from the motion of the halogen-octahedron. Their relation to the rest of the modes of the strongly coupled lattice has not yet been elucidated in the literature. We have discussed this in §3.5.2.

The derivation of the linear J-T Hamiltonian follows again the prescription of equation (7.6). The relations between the reduced matrix elements of the various triplet states of Figure 7.2 and the one-electron reduced matrix element are given in Appendix V, following Kamimura and Sugano.[33]

7.4 Defects in diamond-type solids

When an atom is removed by irradiation from a covalent diamond-type lattice, four bonds are broken. The four hybrid orbitals which are left unattached (denoted by a, b, c, d in Figure 7.3) can be combined to form the following singlet and triplet orbitals:

$$|a_1\rangle = \tfrac{1}{2}(a + b + c + d)$$

$$|t_{2\xi}\rangle = \tfrac{1}{2}(a + b - c - d)$$

$$|t_{2\eta}\rangle = \tfrac{1}{2}(-a + b - c + d)$$

$$|t_{2\zeta}\rangle = \tfrac{1}{2}(a - b - c + d)$$

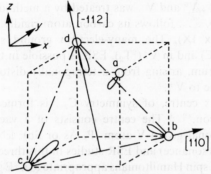

Figure 7.3 The four unattached, 'dangling' orbitals surrounding a vacancy in a diamond-type lattice. The vacancy site is at the centre of the cube. Note that the letters a, b, c, d designate both the orbitals and the site on which the orbitals originate. The C_{3v}-symmetry of a group V atom-vacancy centre is indicated by the dash-dot triangle drawn between the b, c, d-atoms. The a-site may be occupied by P, As or Sb.

In a model where the overlap between the orbitals is taken into account (but not the interelectronic repulsion), the singlet state which has less nodal surfaces than the triplet is lower by several eV. It follows then that in vacancy centres with 3, 4 and 5 electrons (positive, neutral and negatively charged vacancies: V^+, V_0 and V^-) there is a double occupation of the two spin states of a_1 and the triplet is singly, doubly and triply filled. On the other hand, if instead of the vacancy one has a donor then the additional electrons occupy antibonding orbitals whose level scheme is opposite to that of the bonding orbitals. Thus with a nitrogen donor the extra electron enters a triplet state. Indication for this is provided by the trigonal distortion, which is presumably due to a t $\otimes \tau_2$ coupling (§3.3.5), seen in the anisotropic 'low temperature' ($T \leqslant 600\ °K$) ESR spectra of natural and synthetic diamonds.[34-5]

The first strong absorption of what is believed to be a neutral vacancy $(t_2)^2$ in diamond,[36-7] the GR1 band, lies rather lower than expected, at about 2 eV. This result points to the need for inclusion of interelectronic repulsion. Calculations of energy levels[38-9] indeed locate a $^1T_2$ level 2 eV above the ^1E ground state; however, this result, as well as other details of the computed energy level schemes, depends on the nature of the calculations[38-40] rather sensitively. Nevertheless, it is unanimously found that the first excited state in V^0 is the triplet $^3T_1$ and that this lies close enough to the ground state (at a height of a few hundred wave numbers) to be observable by ESR at not too low temperatures.

The derivation of the J-T Hamiltonian follows the group theoretical procedures outlined in §7.1. The static J-T problem for some of the lowest states in V^+, V^0 and V^- was treated by a method[39] in which each of the orbitals a, b, . . . follows its carbon atom rigidly. (The results are listed in Appendix IX). The many-electron ground state is orbitally degenerate in $V^0(^1E)$ and in $V^+(^2T_2)$. ESR is possible in the latter and the so-called Gl spectrum, arising from a tetragonally distorted centre, was interpreted[41] as due to V^+.

A more complex centre, of symmetry C_{3v}, is formed upon electron irradiation of silicon.[42-3] The centre consists of a vacancy and of an adjacent substitutional group V atom (P, As or Sb). ENDOR (Electron Nuclear Double Resonance) and ESR studies in this three-electron system were interpreted by spin Hamiltonians appropriate to a ^2E ground state.[42-3] Although a proper theoretical treatment of the levels is lacking, it is conceivable that the doublet ^2E is the ground state. The physical description of this ground state is that if the impurity atom occupies, say, the a position in Figure 7.3 then two silicon atoms, say b and c, can be considered to form a bonding orbital whose two spin states are occupied, while one unpaired electron is left on d. From the three equivalent states (unpaired electron on b, c or d) two doublet states $^2E_\theta$, $^2E_\epsilon$ (and a singlet $^2A_1$) can be built up which will be coupled to the ε-mode of the silicon triangle. The distorted configuration in which the impurity is on a and the pair of silica b and c pull together (so that the unpaired electron is on d) is labelled by the pair of letters ad.[42-3]

The states formed by excitons bound to defects in diamond type lattices[44] are generally less well localized than the previous vacancy states. Consider for example excitons due to ionized donor and acceptor pairs in GaP. This crystal is a wide gap ($\sim$2·34 eV) semiconductor with a diamond structure, in which the places of the C atoms are occupied by Ga and P alternately. To identify the exciton formed by a hole and an electron the band structure of GaP needs to be known. (In the previously discussed strongly bound vacancy-defect states it was possible to leave the band structure out of considerations). The hole arises from the states near the top of the parabolic valence band which have atomic $p_{3/2}$ character and are therefore fourfold degenerate. As regards the electronic component of the exciton, it turns out that in spite of the complicated form of the conduction band (which possesses equivalent valleys along {100} in reciprocal space) it is possible to regard this component as a state with a twofold Kramers degeneracy.[45] The eightfold degeneracy of the excitons splits into $J = 1$ (A) and $J = 2$ (B) states (where the letter in parenthesis denotes the conventional symbol[45] given to the optical band which comes from the recombination of an exciton in these states). Because of terms of tetrahedral symmetry in the Hamiltonian the five states of $J = 2$ split into E and T_2. The B band is made allowed by admixture of the slightly higher (6 cm^{-1}) A band and below 1·6 °K the B spectrum is much stronger than the 'hot', allowed A band.

An analogous centre is formed when a donor such as Bi, which is isoelectronic with P replaces the latter in GaP. The binding energy of the excitons is small (about 0·1 eV), owing to the near-cancellation of potential and kinetic energies. However, this binding comes from short range forces and therefore the vibronic coupling is strong. The ratio of B- to A-intensities approaches the limit of 1/5 theoretically predicted for very strong coupling.[46]

The C-spectrum in GaP arises from fourfold degenerate exciton states $(G_{3/2})$.[45] It is due to a bound hole and a *pair* of electrons occupying a singlet state. Because of its small binding energy it was suggested that a neutral donor was involved.

In the theoretical treatment of Bir[47] the lattice distortions were represented by elastic deformations. This model may be applicable to diffuse centres or *long*-range forces.

7.5 R-centres in alkali-halides

A cluster of three F-centres (negative ion vacancies which have trapped one electron each) arranged in the shape of an equilateral triangle constitutes an R-centre.[48-9] (Figure 7.4.) The symmetry of the system is C_{3v} and its ground state formed out of the three electrons occupying the available six spin-orbitals is the degenerate ^2E (as in a vacancy-group V atom defect of the previous section). The next level is a quartet, but beyond this there are a number of higher doublets to which optical transitions were observed by Silsbee.[50] All doublets are subject to the J-T-E, whose strength varies from level to level. A preferred direction of polarization was observed after a uniaxial compressive stress was applied. This had a component along the local Y-direction (the crystallographical [−112]-direction) and no component along the local X-direction (the [110] direction). We have seen in §4.7 how this 'stress induced dichroism' yielded information on the vibronic coupling.

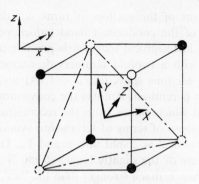

Figure 7.4 The R-centre.

● denotes the alkalis, ○ the halide and ◌ the vacancy. The local X, Y-axes are in the plane of the triangle formed by the three vacancies, the Z-axis is normal to the plane.

7.6 Accidental degeneracies in optical spectra

For a system whose configuration is described by the nuclear coordinates $\{Q\}$ there exists in the Born–Oppenheimer approximation a set of energy surfaces $E_1(\{Q\})$, $E_2(\{Q\})$, . . . for each electronic state 1, 2, . . . The pictorial representations of these are called configurational diagrams. The occurrence of a degeneracy of two electronic states at some point Q_i^0 which is a stationary point for one or both electronic states, is conditional on the following equations.

$$E_1(\{Q^0\}) = E_2(\{Q^0\}), \left(\frac{\partial E_1}{\partial Q_1}\right)_{\{Q^0\}} = 0 \text{ and/or } \left(\frac{\partial E_2}{\partial Q_i}\right)_{\{Q^0\}} = 0$$

In general the number of equations is too great and a solution is not expected to exist except at points of high symmetry. On the other hand, a solution at a point which is not a stationary point on either of the surfaces E_1 and E_2 is expected to be available. Such intersection of the surfaces represents an accidental degeneracy and, provided it is not too high above a minimum, it has important consequences on the optical properties of a number of systems. It can also be shown that such intersection causes the same type of breakdown of the Born–Oppenheimer approximation as the one described in connection with equation (1.1) in the Introduction, except for two circumstances. One, the breakdown will occur at an excited vibrational level, so that ΔE in equation (1.1) represents the separation of this vibrational level from a nearby vibrational level of the other electronic state. Secondly, in some cases, e.g. when the two states have different spin multiplicities, these states would not be coupled by the Born–Oppenheimer correction alone which operates on the space coordinates only, but only in conjunction with spin–orbit coupling. In this case L of equation (1.1) consists of two factors, which come from the Born–Oppenheimer correction and from the spin–orbit coupling.

Since the existence of the degeneracies is meaningful only within the configurational diagram model we shall first make this model more precise:

In its usual sense the term 'configurational diagram' refers to the plot of the potential energies of two or more electronic states, which are partners in optical transitions, as functions of the ionic coordinates. In a solid, or in a fairly big molecule, we should thus be referring to a set of hypersurfaces in a space of extremely high number of dimensions, comparable to the number of normal modes of the system. Actually, we shall regard the potentials as depending only on a few coordinates. These coordinates may be thought of either as those of a local mode which changes in the course of the optical transition or as an interaction coordinate. The term interaction coordinate, which is not a normal coordinate (unfortunately), was introduced in this context by Toyozawa and Inoue.[51] However, a justification for the one-coordinate description, even when many degrees of freedom are involved, was earlier given by Lax.[52] It is not our purpose here to discuss this or alternative justifications[53] for the configurational diagram model, but rather to take its validity for granted and to pursue the consequences.

There are at least four broad areas of research where the configuration diagram description has been extensively used for the optical properties of localized centres. These are: (1) paramagnetic centres,[54-5] (2) phosphors in alkali halides,[29,56-7] (3) colour centres due to vacancies[58-9] and (4) organic molecules (in either the gaseous or the condensed phase).[60-1]

We shall be interested in cases where the potential curves cross each other. Parameters for intersecting configuration diagrams, selected from each of the cases quoted earlier, are shown in the following table and the relevant potential curves appear in the next figure. They give an idea of the variety of the situations which we encounter and are presented here to provide a quantitative framework for the discussions which were given in Chapters 3 and 6 (§§3.6.2 and 6.6).

All the potentials in the figure are assumed to have the same curvature. They differ from one another by the two parameters a, b whose numerical values are given in the Table and which are defined, as the half-separation between potentials in the vertical (a) and horizontal (b) directions. They are given respectively in units of the vibrational quanta ($=\hbar\omega$) and of the zero point motion amplitude $\sqrt{(\hbar/M\omega)}$.

Table 7.4 Configurational coordinate parameters for various systems. The quantity $2a$ is the separation between potentials in the vertical direction and $2b$ the separation in the horizontal direction. $\hbar\omega$ is the vibrational quantum.

System	Upper electronic state	Lower electronic state	Ground electronic state	a (in units of $\hbar\omega$)	b (in units of zero point motion	$\dfrac{\hbar\omega}{(\text{cm}^{-1})}$	Ref.
Emerald	$^4T_2$	^2E	$^4A_2$	0·7	2·7	250	62
Ruby	$^4T_2$	^2E	$^4A_2$	4·6	1·61	250	62
$(Ni \cdot 6H_2O)^{++}$	^1E	$^3T_1$	$^3A_2$	2·15	1·1	300	55
$KCl:Tl^+$	^3P	^1S		141·6	3·83	120	63
F-centre in KCl		^3P	^1S	78·8	5·1	86	59
Anthracene	$^3B_{1u}$	$^3A_{1g}$		2·45	0·86	3000	61

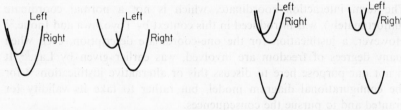

(a) Two level systems   (b) Three level systems

Figure 7.5 Configurational coordinate diagrams for the systems treated in this section and in §3.6.2. Energy is plotted versus the interaction coordinate. (From Reference 64.)

The coupling between different electronic states is an important component of the phenomena that we are discussing. The cause of the coupling may be (among others) spin–orbit coupling, leading to inter-system crossing between states of different spin-multiplicities, or the break-down of the Born–Oppenheimer approximation (vibronic effects). The symbol denoting this coupling will be W. In general W will be an operator, involving vibrational momenta and coordinates (Reference 65 may be viewed for the form of W for vibronic coupling), however, for spin–orbit coupling W will be a simple constant, in fact, the off-diagonal matrix-element linking the electronic states with which the potentials are associated. We shall continue our discussion in terms of the spin–orbit coupling mechanism, since it is simpler to treat than vibronic mixing.

In most cases the physical situation cannot be described by two crossing potential curves between which there exists some form of interaction, since each curve may in reality represent a number of potential curves (e.g. for $^4T_2$ twelve curves, for ^2E four curves), not all of which interact. It is clear, though, that when the relaxation within a set of potentials is fast compared to all other processes in the system then the set can be approximately described by a single curve. Also, when the relaxation from one of the set to all others is very slow then this curve is irrelevant. Problems arise only when relaxation times within the set of potentials are comparable to the crossing or emission times.

The important ratio, equation (1.1) of the Introduction, which shows whether or not perturbation theory is applicable can be expressed in terms of b and W. Suppose that we wish to examine the perturbational correction to the *ground* vibrational level of one electronic state which is contiguous to the nth vibrational level of the other state, the distance between the levels being ΔE. Then equation (1.1) of the Introduction takes the form

$$R' \sim \frac{b^{2n}}{n!} e^{-2b^2} \frac{W^2}{\hbar\omega\Delta E}$$

The correction to the Born–Oppenheimer approximation enters through b and the spin–orbit coupling through W^2.

The Hamiltonian matrix of the intersection can be derived from a somewhat specialized form of the Hamiltonian for the electron–ion system. This can be written, as in equation (2.1),

$$H = T_{el} + T_N + V(q) + V(\mathbf{r}, q)$$

where the successive terms are the electronic and nuclear kinetic energies and the ionic and electronic potentials. The last includes the electron–ion interaction.

Writing $T_{el} + V(\mathbf{r}, q)$ as

$$H_{el}(q = 0) + qV' + H_{s-o}$$

(where the last term is the spin–orbit coupling) and

$$T_N = \tfrac{1}{2}\hbar\omega p^2, \quad V(q) = \tfrac{1}{2}\hbar\omega q^2$$

we obtain the Hamiltonian-matrix[64] for the two states which intersect:

$$\begin{pmatrix} H^- & W \\ W & H^+ \end{pmatrix} \tag{7.11}$$

$$H^\pm = \tfrac{1}{2}\hbar\omega[p^2 + (q \pm b)^2 \pm 2a],$$

where the off-diagonal matrix elements of the spin–orbit coupling yield W, whereas the diagonal matrix elements together with the electronic energy difference constitute $2a\hbar\omega$. The expectation values of V' in the two states differ by $2b\hbar\omega$. p is the vibrational momentum.

7.7 Dimers, trimers, ...

Two weakly linked identical molecules are said to form a dimer or a 'double molecule'. Dimers are a very common (and industrially important) form of aggregation of dyes, however for fundamental work planar, aromatic molecules are preferred, since these have well defined electronic and vibrational properties. The shape of the aggregate may be as in biphenyl (Figure 7.6a) or in two more or less parallel planes above each other (about 3 Å a part) as in a sandwich. (In a sandwich dimer the molecules may be symmetrical, or rotated as in Figure 7.6b.) The formation of trimers, etc., is also possible.

The optical properties of dimers can be markedly different from the monomers of which they are formed, in spite of the weak linkage that exists between the monomers.[66] The theory for this phenomenon is based on electronic as well as vibronic coupling. It was formulated for dimers by Witkowski and Moffitt,[67] Fulton and Gouterman[68] and McRae,[69] and for trimers by Marechal[70] and by Perrin and Gouterman.[71] Recently some of the theoretical predictions were confirmed in anthracene dimers[72-3] and in some cation dimers.[74]

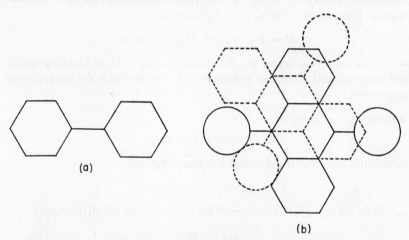

(a)

(b)

Figure 7.6 Structure of some dimers.

(a) The skeleton formed of the carbon atoms in the biphenyl molecule C_6H_5—C_6H_5, viewed from above.

(b) The two molecules (drawn by full and broken lines, respectively) in 9,10 dichloroanthracene, constituting a sandwich dimer. (From Reference 73.)

The vibronic coupling occurs in a state where one of the monomers is excited. Briefly, there are two opposing tendencies: a purely electronic (electrostatic) stabilization of the state in which the excitation resonates between the two monomers and a vibronic effect (due in the first instance to the distortion of that monomer on which the excitation momentarily resides) which reduces, quenches the resonance effect.

The vibronic wave function for the excited state of the dimer is taken to be

$$\psi = \phi_a(\mathbf{r}_A, \mathbf{r}_B)\chi_a(q) + \phi_b(\mathbf{r}_A, \mathbf{r}_B)\chi_b(q) \qquad (7.12)$$

Here

$$\phi_a = \phi_A^T \phi_B^S$$

represents the excitation (state T) on the monomer A, while the monomer B is in its ground state (S). Vice versa for ϕ_b. The wave-functions ϕ_a, ϕ_b are, in general, functions of both the electronic ($\mathbf{r}$) and nuclear (q) coordinates, but we have simplified them by the use of the Born–Oppenheimer approximation, as discussed in Chapter 2, and omitted the dependence on the ionic displacement coordinates. These functions are the solutions of the Hamiltonian for the monomer A (or B)

$$\left[-\tfrac{1}{2}\hbar\omega \frac{\partial^2}{\partial q_A^2} + V(\mathbf{r}_A, q_{A_0}) \right] \phi_A^R = E_R \phi_A^R$$

where R may be S and T, and q_{A_0} represents the coordinates of the nuclei of A in their ground state equilibrium positions. A single coordinate q_A or q_B is assumed for each monomer, for simplicity. The Hamiltonian of the dimer consists, in addition to the monomeric Hamiltonians, also of an interaction term $V^{AB}(\mathbf{r}_A, \mathbf{r}_B; q_A, q_B)$. Writing the vibrational function as a column vector, we find the following matrix equation.

$$H \begin{pmatrix} \chi_a \\ \chi_b \end{pmatrix} \equiv \left[-\tfrac{1}{2}\hbar\omega \left(\frac{\partial^2}{\partial q_A^2} + \frac{\partial^2}{\partial q_B^2} \right) \mathbf{I} + \begin{pmatrix} V_{aa} & V_{ab} \\ V_{ba} & V_{bb} \end{pmatrix} \right] \begin{pmatrix} \chi_a \\ \chi_b \end{pmatrix} = E \begin{pmatrix} \chi_a \\ \chi_b \end{pmatrix}$$

$$(7.13)$$

where $\mathbf{I}$ is the unit matrix.

The diagonal matrix elements V_{aa} and V_{bb} arise from both the monomeric potential and the interaction V^{AB}; the off-diagonal matrix-elements come from the Coulomb-interaction part of V^{AB}. Their nature is given in greater detail in Reference 75. All matrix-elements depend on the vibrational coordinates, but it is sufficient for many purposes to regard $V_{ab} = V_{ba}$ as a constant, $-W$, and to assume a rather simplified form for V_{aa} and V_{bb}. Disregarding a constant energy term, we suppose that V_{aa} (or V_{bb}) has a quadratic dependence of the form $q_A^2 + q_B^2$ and a linear dependence on q_A (or, for V_{bb}, on q_B) only. Introducing now

$$q_+ \equiv \frac{1}{\sqrt{2}}(q_A + q_B)$$

$$q_- \equiv \frac{1}{\sqrt{2}}(q_A - q_B),$$

we derive

$$\left\{-\left[\tfrac{1}{2}\hbar\omega\left(\frac{\partial^2}{\partial q_+^2} + \frac{\partial^2}{\partial q_-^2} - q_+^2 - q_-^2\right) + \frac{1}{\sqrt{2}}Lq_+\right]\mathbf{I} - \right.$$

$$\left. - \frac{1}{\sqrt{2}}Lq_-\sigma_\theta + W\sigma_\epsilon\right)\begin{pmatrix}\chi_a(q_+, q_-)\\\chi_b(q_+, q_-)\end{pmatrix} = E\begin{pmatrix}\chi_a\\\chi_b\end{pmatrix} \quad (7.14)$$

where

$$\sigma_\theta = \begin{pmatrix}-1 & 0\\0 & 1\end{pmatrix}$$

and $\qquad\qquad\qquad\qquad\qquad\qquad\qquad\qquad\qquad\qquad\quad$ (7.15)

$$\sigma_\epsilon = \begin{pmatrix}0 & 1\\1 & 0\end{pmatrix}$$

The coefficient L has been chosen to conform with the Hamiltonian for the case of $E \otimes \beta$ in equation (7.25).

For a trimer, consisting of monomers A, B and C, the matrix for vibronic coupling can be easily derived by generalizing equation (7.13) and expanding the matrix elements up to the first order in the vibrational coordinates. Thus, a possible form of the expansion is

$$V_{aa} = V^0 + q_A X, \text{ etc.}$$
$$\qquad\qquad\qquad\qquad\qquad\qquad\qquad\qquad\qquad (7.16)$$
$$V_{ab} = -W + (q_A + q_B)Y, \text{ etc.}$$

We now go over to the representation given by

$$\begin{pmatrix}\phi_\theta\\\phi_\epsilon\\\phi_{A_1}\end{pmatrix} = \begin{pmatrix}-\dfrac{1}{\sqrt{6}} & -\dfrac{1}{\sqrt{6}} & \dfrac{2}{\sqrt{6}}\\[2mm]\dfrac{1}{\sqrt{2}} & -\dfrac{1}{\sqrt{2}} & 0\\[2mm]\dfrac{1}{\sqrt{3}} & \dfrac{1}{\sqrt{3}} & \dfrac{1}{\sqrt{3}}\end{pmatrix}\begin{pmatrix}\phi_A\\\phi_B\\\phi_C\end{pmatrix} \quad (7.17)$$

and a set of similar relations connecting the vibrational modes q_A, q_B, q_C to q_θ, q_ϵ, q_0. We obtain in this representation

$$H_{JT} = \begin{pmatrix}W - \tfrac{1}{2}Lq_\theta & \tfrac{1}{2}Lq_\epsilon & L'q_\theta/\sqrt{2}\\[2mm]\tfrac{1}{2}Lq_\epsilon & W + \tfrac{1}{2}Lq_\theta & L'q_\epsilon/\sqrt{2}\\[2mm]L'q_\theta/\sqrt{2} & L'q_\epsilon/\sqrt{2} & -2W\end{pmatrix} \quad (7.18)$$

where

$$L\left(= -\frac{2}{\sqrt{6}}X + 2\frac{2}{\sqrt{6}}Y\right) \quad \text{and} \quad L' \equiv L(E, A)\left(= \frac{2}{\sqrt{6}}X + \frac{2}{\sqrt{6}}Y\right)$$

are two coupling parameters arising from the coefficients in equation (7.16). They are the reduced matrix elements of the linear vibronic coupling, consistent with our earlier uses of L for this purpose. Perrin and Gouterman,[71] who set up the trimer problem in an alternative form, employed a more restricted model in which $L = -L'$. We have disregarded, for the sake of simplicity, terms in the totally symmetric coordinate q_0. This is not always justified since the equilibrium position and the force constant of q_0 can be different in the singlet and in the doublet states.

In the complex representations, which have been used in §3.2.1,

$$\phi_{\pm 1} = \mp i(2)^{-\frac{1}{2}}(\phi_\theta \pm i\phi_\epsilon)$$

and

$$q\,e^{\pm i\phi} = q_\theta \pm i\,q_\epsilon$$

Now the linear Hamiltonian appears in the form

$$
H_{JT} =
\begin{array}{c}
|-1\rangle: \\
|+1\rangle: \\
|A_1\rangle:
\end{array}
\begin{pmatrix}
W & \frac{1}{2}Lq\,e^{-i\phi} & -\frac{i}{2}L'q\,e^{i\phi'} \\
\frac{1}{2}Lq\,e^{i\phi} & W & \frac{i}{2}L'q\,e^{-i\phi'} \\
\frac{i}{2}L'q\,e^{-i\phi'} & -\frac{i}{2}L'q\,e^{i\phi'} & -2W
\end{pmatrix}
\tag{7.19}
$$

where we have put a prime on some of the angles, for later reference. For our present purposes $\phi' = \phi$.

7.8 Conjugated hydrocarbons

7.8.1 *The Hückel approximation*

The following theoretical discussion will be largely within a framework specifically constructed to treat these systems, namely the Hückel approximation of the molecular orbital theory. Some results of the alternative approach, the valence-bond theory will also be mentioned. The latter method, which is based on electron correlation, is not so easy to use for the J-T-E, which operates through a one-electron Hamiltonian, as the molecular orbital treatment. Moreover, its general popularity also appears to be on the wane. (See e.g. Reference 76.) A comparison of the two approaches is provided in simple terms in a classic book[77] treating these systems. The gravest fault of the valence-bond method from our point of view is that it leads to a very deep non-degenerate ground-state of the non-existent cyclobutadiene molecule and to an inoperative P-J-T-E. In contrast, molecular orbital theory predicts the incipient instability of this molecule.

Molecular orbital theory is the subject of books by Streitweiser[78] and Salem.[79] The second contains also a very good presentation of the J-T-E in some conjugated systems. Our purpose here is to consider the effect from the broader view of this account, and to add newer developments and results. The experimental results derive mainly from electron spin-resonance and also from the optical spectra. Essential features of the former are the very small spin–orbit coupling[80-1] (so that the complications of odd-electron systems, Appendix III, for the J-T-E are absent), which causes the g-factor to come close to its free electron value, 2·0023 and the nuclear spin of the proton, which shows up in the hyperfine structure as a delicate probe of the electronic spin density.[82]

The planar form of the molecules supports orbital degeneracies of at most two. The occurrence and nature of these orbitals for the itinerant π-electrons in cyclic hydrocarbons may be visualized within the Hückel MO theory and the semi-empirical PPP[83-4] theories by means of the following graphical scheme:

First, draw the regular polygon skeleton of the cyclic molecule with one of the apexes at the bottom. The level at each apex will then represent in appropriate units the energy of a Hückel MO, and the number of apexes at each level the orbital degeneracy of that level. Counted from zero upwards the integer j' shown on the right of each level will give the number of nodal planes normal to the molecule in the Hückel MO wave functions. (This will also be the quasi-angular momentum of the orbital.)

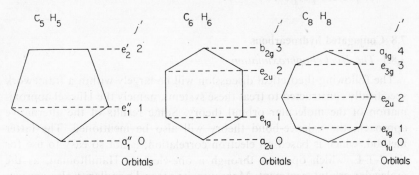

Figure 7.7 Hückel one-electron orbitals and their energy levels in some hydrocarbons. The carbon skeletons are drawn.

We may now fill up the orbitals with π-electrons, two electrons (one of either spin) entering each orbital, starting from the bottom. The σ-orbitals are assumed to be fully occupied. Since each carbon atom provides one π-electron, it is clear that a degenerate ground state is expected for a $C_{2n+1}H_{2n+1}$ molecule, for a molecular ion $(C_{2n+2}H_{2n+2})^+$ or $^-$, or for a molecular ion $(C_{4n}H_{4n})^+$ or $^-$. As representative cases of these systems we take, respectively, cyclopentadienyl (C_5H_5), the benzene ions $(C_6H_6)^\pm$ and the cyclooctatetraene ions $(C_8H_8)^\pm$. The distinction of the type $C_{4n}H_{4n}$ from the rest is due to the result (which may be verified from the symmetric products in Appendix I) that the non-totally symmetric vibrations coupled to the electronic states E_{nu} or E_{ng} of the point group D_{4nh} are the *singly* degenerate modes β_{1g} and β_{2g}, while in the other cases doubly degenerate modes enter the J-T-E. We have seen that this difference has manifold consequences in the behaviour of the two types[85-6] (§3.6.3).

In addition, pseudo J-T-E will occur in excited states of benzene (§7.8.2), in substituted benzene anions (§7.8.3) and in $C_{4n}H_{4n}$-type hydrocarbons (§7.8.5).

The Hückel-type approximation leads for any conjugated molecule to a universal form of the energy W as function of the bond lengths (r_{rs}) in the form

$$W = \sum_{rs} f(r_{rs}) + 2\sum_{rs} p_{rs}\beta(r_{rs}) \qquad (7.20)$$

where the first term is the sum of contributions of the σ-electrons to the bond between the r and s carbon atoms and the second term, due to the π-electrons, is a function of the mobile bond orders p_{rs} (to which we shall soon return) and of the resonance integral $\beta(r_{rs}) = \langle \pi_r | H | \pi_s \rangle$ between the π-orbitals based on the r and s carbon atoms. An explicit form for β is due in the first place to Longuet-Higgins and Salem[87] and, as modified, to Coulson and Golebiewski[88]

$$\beta(r) = \beta_0 \exp\left[-(r - 1\cdot40)/a\right]$$

$$\beta_0 = -8937 \text{ cm}^{-1}, \qquad a = 0\cdot3106 \text{ Å } [87] \qquad (7.21)$$

or $\qquad\qquad \beta_0 = -12{,}867 \text{ cm}^{-1}, \qquad a = 0\cdot3736 \text{ Å } [88]$

where r is expressed in angstroms.

Other models for β, as well as the effect of overlap, are discussed in Reference 78 (§§4.4, 4.5).

The chief distinctive feature of these models, as compared with those employed in the J-T-E of cubic inorganic complexes, is that the higher order vibronic coupling terms follow automatically. Indeed one obtains[89] for the model of Longuet-Higgins and Salem

$$\beta(1\cdot413 + \Delta r) = \beta(1\cdot413) + 27{,}600 \text{ cm}^{-1}(\Delta r) +$$
$$\tfrac{1}{2}88{,}900 \text{ cm}^{-1}(\Delta r)^2 + \ldots$$

(The value $r = 1\cdot413$ Å is close to the average equilibrium distance in C_5H_5 according to the calculations of Reference 90).

The σ-bond energies are given by

$$f(r) = \frac{40}{3}(r - 1\cdot8106) \times 8937 \exp\left[-(r - 1\cdot40)/0\cdot3106\right] \text{ cm}^{-1}\;[87]$$

or

$$f(r) = \frac{40}{3}(r - 1\cdot8376) \times 12{,}867 \exp\left[-(r - 1\cdot40)/0\cdot3736\right] \text{ cm}^{-1}\;[88]$$

The mobile bond orders are measures of the overlap density $\phi_i^*\phi_j$ in the overlap region of π_r and π_s. It is usual to define these for $\phi_i = \phi_j$; however in our case, where the orbitals are degenerate we need an extension of the definition[90] and a corresponding change in notation. In particular, in terms of the LCAO (linear combination of atomic orbitals) $\phi^{(j)} = \sum_r C_r^{(j)}\pi_r$ the bond orders are written as

$$p_{rs}^{(ij)} = \tfrac{1}{2}(C_r^{(i)*}C_s^{(j)} + C_s^{(i)*}C_r^{(j)}) \tag{7.22}$$

To exemplify the general procedure we choose the cyclopentadienyl radical C_5H_5, which has barriers > 50 cm^{-1} between the equivalent distorted minima.[91] The unfilled degenerate set e_1'' consists in the present scheme of the real and imaginary parts of

$$\sqrt{2}\phi^{(1)} = (2/5)^{\frac{1}{2}} \sum_{s=1,\ldots 5} \Pi_s \exp\left[2\pi i(s - 1)/5\right]$$

Properly transforming wave functions ϕ_θ and ϕ_ϵ can be derived from this, as shown in Figure 7.8. The corresponding bond orders $p^{(\theta\theta)}$, $p^{(\epsilon\epsilon)}$, $p^{(\theta\epsilon)}$ follow

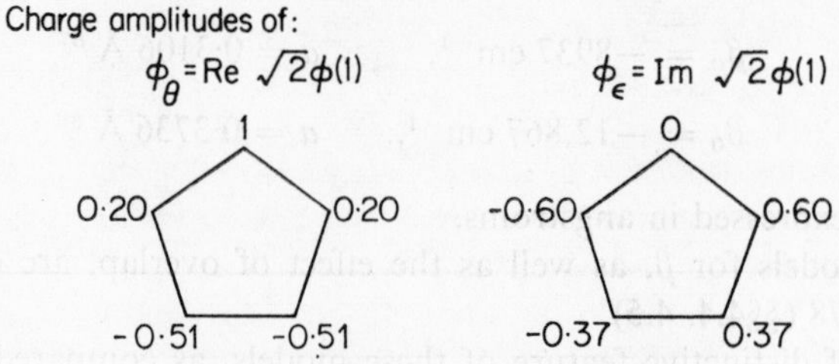

Charge amplitudes of:

$$\phi_\theta = \text{Re } \sqrt{2}\phi^{(1)} \qquad\qquad \phi_\epsilon = \text{Im } \sqrt{2}\phi^{(1)}$$

Figure 7.8 Hückel orbitals in C_5H_5.
The amplitudes at each carbon atom are shown.

from equation (7.22). Instead of using these bond orders we shall work with the more convenient combinations shown in the figure below, since these transform as irreducible representations of D_{5h}.

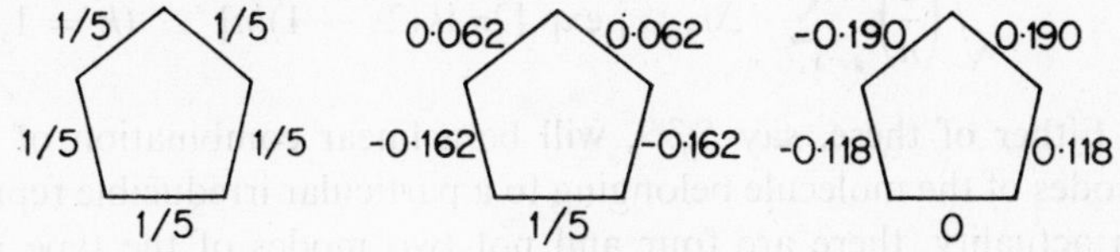

Bond orders: $\frac{1}{2}(\rho_{n,n+1}^{\theta\theta}+\rho_{n,n+1}^{\epsilon\epsilon})$ $\frac{1}{2}(\rho_{n,n+1}^{\theta\theta}-\rho_{n,n+1}^{\epsilon\epsilon})$ $\rho_{n,n+1}^{\theta\epsilon}=\rho_{n,n+1}^{\epsilon\theta}$

Symmetry type: A_1 $E'_{2\theta}$ $E'_{2\epsilon}$

Figure 7.9 Bond orders in C_5H_5. The bond orders p_{rs} were introduced in equation (7.22). Here conveniently transforming linear combinations of p_{rs} are shown. The bond orders in these combinations are shown next to the bonds.

The energy matrix $W^{(ij)}$ operating in the θ, ϵ-manifold is now expanded about the equilibrium bond distances $r_{s,s+1}^0$

$$W^{(ij)} = \sum_{s=1,\ldots 5} \left\{ f(r_{s,s+1})\delta_{(i,j)} + \left[\beta(r_{s,s+1}^0) + \Delta r_{s,s+1} \frac{d\beta(r_{s,s+1}^0)}{dr_{s,s+1}^0} \right. \right.$$
$$\left. \left. + \tfrac{1}{2}(\Delta r_{s,s+1})^2 \frac{d^2\beta(r_{s,s+1}^0)}{dr_{s,s+1}^{02}} \right] p_{s,s+1}^{(ij)} \right\}$$

$\Delta r_{s,s+1} = r_{s,s+1} - r_{s,s+1}^0$, $\delta_{(i,j)}$ is the Krönecker-delta and i, j is (θ, ϵ). The invariance of this expression under the operation of the point group and the transformation properties of the bond order matrices restrict the linear combinations of the extensions Δr which enter the above expression. In fact, as could have been seen immediately from the table for the resolution of symmetric products (Appendix I), only the totally symmetric vibrations and those of type ϵ'_2 enter (in the present, linear approximation). The former represents a change in the dimensions of the molecule and contributes to its stabilization energy; the latter leads (in the static problem) to a set of equivalent minima.

If we stay in the linear approximation of the J-T-E, we can write down the energy-matrix W in the rather more general case of an n-sided cyclic molecule possessing any kind of π-electronic (double) degeneracy. The degenerate set is taken to consist of the real (θ) and imaginary (ϵ) parts of the complex function $\phi^{(k)}$ which itself is such that a $2\pi/n$ rotation increases the phase of $\phi^{(k)}$ by $2\pi k/n$. ϕ_θ and ϕ_ϵ are further specified by their taking the values $(2/n)^{\frac{1}{2}}$ and 0, respectively, on the carbon atom labelled by one. In this representation, the energy-matrix (less the σ-bond energy, which contributes only to the force constants) reads

$$\left(\frac{2}{n}\right)^{\frac{1}{2}} \frac{\partial\beta(r^0)}{\partial r^0} \begin{pmatrix} -S_\theta^{(2k)} & S_\epsilon^{(2k)} \\ S_\epsilon^{(2k)} & S_\theta^{(2k)} \end{pmatrix} \tag{7.23}$$

where the symmetry coordinates $S_\theta^{(2k)}$, $S_\epsilon^{(2k)}$ are respectively the real and imaginary parts of

$$\sqrt{\left(\frac{2}{n}\right)} \sum_{s=1,\dots n} \Delta r_{s,s+1} \exp\left[2\pi \, ik(2s-1)/n\right] \qquad (k=1,2)$$

Either of these, say $S_\theta^{(2k)}$, will be a linear combination of the normal modes of the molecule belonging to a particular irreducible representation. In actuality, there are four and not two modes of the type ϵ_2' in C_5H_5. They are derivable from the above two modes obtained for the C_5 skeleton by supplementing the displacement of each carbon atom by a displacement, once parallel and once antiparallel, of its associated proton. Since the J-T-E provides no immediate cause for a change in the C–H bond-lengths, one would expect those modes to be affected more strongly in which the proton moves along with the carbon.

7.8.2 Excited states in benzene

A P-J-T-E operates between excited orbital doublet states and some nearby singlet of the benzene molecule (Figure 7.10). Evidence for such effect is the deformed structure of the molecule in e.g. the excited $^3B_{1u}$ state, as seen in ESR[92-4] and the phosphorescence spectrum.[95]

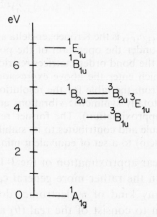

Figure 7.10 Conjectured positions of the excited electronic states of benzene. (After Reference 92.) Strong P-J-T coupling exists between $^3E_{1u}$ and $^3B_{1u}$. Within the doublet $E \otimes \epsilon$ coupling will be weak, since it is due to configuration admixing.[96]

The normal J-T-E in the doublet is weak and is present only thanks to the admixture of excited configurations by Coulomb interactions. Thus, in the Hückel orbital scheme the vibronic matrix element L vanishes for a one-electron operator. The P-J-T-E to the neighbouring $^3B_{1u}$ is strong.[96-7]

The linear J-T Hamiltonian between a doublet and a singlet A_1 or B_1 of the same parity is formally equivalent to that derived for a trimer. (This also follows from the homomorphism between the symmetric group S(3) and the group D_{6h} of benzene). For a singlet A_2 or B_2 the angle ϕ' appearing in the P-J-T Hamiltonian, equation (7.19) is to be amended (in accordance with Table G2.1 of Reference 9)

$$\text{so that } \phi' = \phi - \frac{\pi}{2} \text{ for } A_2 \text{ and } B_2$$

$$(\text{while } \phi' = \phi \qquad \text{for } A_1 \text{ and } B_1)$$

With the addition of second order terms we have the matrix Hamiltonian[92]

$$
\begin{array}{c}
|-1\rangle: \\
\\
\\
|A \text{ or } B\rangle: \\
\\
\\
|+1\rangle:
\end{array}
\left(
\begin{array}{ccc}
W & -\dfrac{i}{2}L'q\,e^{i\phi'} & \dfrac{1}{2}Lq\,e^{-i\phi} \\
 & +\dfrac{i}{4}K'q^2\,e^{-2i\phi'} & -\dfrac{1}{4}Kq^2\,e^{2i\phi} \\
\dfrac{i}{2}L'q\,e^{-i\phi'} & -2W & -\dfrac{i}{2}L'q\,e^{i\phi'} \\
-\dfrac{i}{4}K'q^2\,e^{2i\phi'} & & +\dfrac{i}{4}K'q^2\,e^{-2i\phi} \\
\dfrac{1}{2}Lq\,e^{i\phi} & \dfrac{i}{2}L'q\,e^{-i\phi'} & W \\
-\dfrac{1}{4}Kq^2\,e^{-2i\phi} & -\dfrac{1}{4}K'q^2\,e^{2i\phi'} &
\end{array}
\right)
$$

$$(7.24)$$

7.8.3 *Substituted benzene anions*

According to the results of calculations (collected in Appendix IX) in the J-T stabilized states of the benzene anions the two states ϕ_θ, ϕ_ϵ are virtually degenerate, the barrier height being a mere few tens of cm^{-1}. Substitution of an ethyl (CH_3CH_2) or methyl (CH_3) group at the 1-site (Figure 7.11(a)) splits the nearly degenerate orbitals, raising the symmetric ϕ_θ state by a few hundred wavenumbers. This was established in the first instance by measuring in the fine-structure of the paraxylene negative ion a spin-density of 1/4 at the 2-position. (The spin-density used in the present context is the square of the π-electron wave-function at the position in question on the carbon ring, multiplied by the sign of the spin). If the anion were in the θ-state, one would expect 1/12 for the spin-density. The splitting mentioned earlier is caused by a combination of a number of effects.[98-9] In the order of increasing sophistication one has first an inductive potential effect, which raises the potential of the 4-carbon essentially because the negative end of the dipole on the substituent points toward this carbon. Then there is a covalency-effect which transfers charge between the ring carbons and the substituent, and imparts to CH_n some of the majority spin of the ring. In effect, what happens is that the substituent orbital is a few electron volts below the partially filled π-orbital on the ring; covalent mixing then robs the substituent of the minority spin by transferring it to the ring. The last factor in the splitting is the interatomic exchange which induces on the substituent a spin opposite to the majority ring-spin.

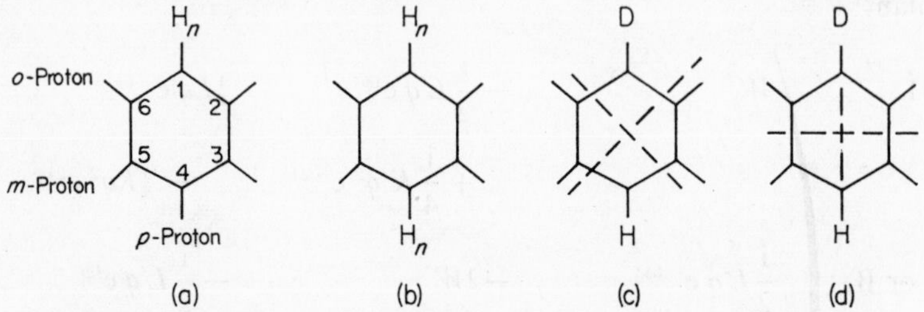

Figure 7.11 (a) Mono-substituted benzene. (b) Di-substituted benzene. (c) Deuterated benzene: High electron density meets low amplitude deuterium vibration. (d) Deuterated benzene: Low electron density meets low amplitude deuterium vibration. In (c) and (d) the broken lines are nodal planes of the electronic wave function.

Calculations of the energy levels of substituted anions, which are based on the first[99] and second[100] mechanism, indeed result, as expected, in the stabilization of the ϵ-state with respect to θ. The state ϵ being antisymmetric with respect to reflection in the symmetry-plane perpendicular to the molecule, there should be zero spin-density at the 4-proton. However, the effects which we have just described lend it a negative (i.e. minority) spin density. ('Lend' it, because the vibronic mechanism takes it back, as we have seen in §3.6.3.) Thus, hyper-conjugation and spin-polarization yield a spin of about -0.063 at a para-proton in toluene$^-$ (compared to 0.26 at sites 2 or 3). In this case, spin-polarization, an intraatomic effect first described by Goeppert-Mayer and Sklar[101] which operates as Hund's rule does in atoms, favours spin alignments. However, most 4-proton spin-densities are found to be experimentally positive, as shown in the table.

Table 7.5 4-proton spin-densities ρ_4 of monosubstituted benzene anions. First row: Observation (from De Boer and Colpa)[99] at about 200 °K. Second row: Calculations (Hobey)[100] for $T = 100$ °K.

	Toluene⁻ $[CH_3C_6H_5]^-$	Ethylbenzene⁻ $[CH_3CH_2C_6H_5]^-$	Isopropylbenzene⁻ $[(CH_3)_2CHC_6H_5]^-$	t-butylbenzene⁻ $[(CH_3)_3CHC_6H_5]^-$
ρ_4 (exp.)	0·020	−0·036	0·045	0·072
ρ_4 (theory)	−0·015	0·032	0·061	0·11

The spin densities in the table are deduced from hyperfine splittings Q_H in ESR and NMR through the McConnell relation $a_4 = Q_H\rho_4$ where $Q_H \sim 24\cdot0$ gauss and ρ_4 is the unpaired spin density on the carbon next to the 4-proton.

7.8.4 Deuterated benzene anions

When the substituent is not a radical but a deuterium atom the spin density of the *para*-proton (H_4) may differ from that of the *meta-* or *ortho*-protons. The difference has now a more subtle cause than the electrical properties of the substituents that we met before; it is the dynamical properties of the deuterium which reduce the symmetry of the benzene anion and render the *para*-proton inequivalent.[102]

Since the asymmetry is very small, arising as it does from the difference in the bonding properties of the deuterium because of its smaller zero point displacement than that of H, the difference in the spin-densities induced by the asymmetry would be extremely small, if the electronic state was an orbital singlet. For an orbital doublet, the components will split in the lower symmetry caused by deuteration and the two states will have very different spin densities as we have seen before. The separation between the doublet will be the difference in the zero point energies of all the vibrations. The physical origin of this difference is the energy difference which occurs when the low amplitude deuterium vibration is brought into contact with states of high and low electron density, respectively. This is illustrated in Figure 7.11(c) and (d).

In alternative approaches[103-4] the *para*-proton spin density depends on the isotopic mass through the resonance energy β, equation (7.21). These have been described in §3.6.3.

7.8.5 Cyclic $C_{4n}H_{4n}$

Cyclic hydrocarbons of the composition $C_{4n}H_{4n}$ belong to the symmetry group D_{4nh} (exactly or, if distorted, approximately). In the Hückel-orbital scheme $4n - 2$ lower lying spin-orbitals of the type a_{1u}, $e_{p,\bar{p}}$

(where $p = 1, 2, \ldots n - 1$ and the second index $\bar{p}$ is g or u according to whether p is odd or even) are first filled in by $4n - 2$ electrons. The remaining electrons are placed in a fourfold degenerate level whose states are the spin-orbitals $^2e_{n,g}$ or $^2e_{n,u}$ (depending on whether n is even or odd). The orbital part of the states is drawn schematically in the next figure (Figure 7.12). With one or three electrons left over, i.e. in the

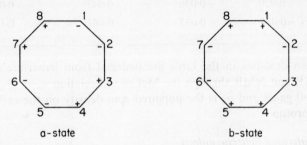

Figure 7.12 Schematic description of the components a and b of an orbital doublet in C_8H_8.

$C_{4n}H_{4n}$ positive or negative ions, vibronic coupling with a β_{1g} or a β_{2g}-mode will stabilize one component or the other or some linear combination of them. This is represented by the linear J-T Hamiltonian

$$- \frac{1}{\sqrt{2}} L(\beta_{1g}) q_{1g} \begin{pmatrix} -1 & 0 \\ 0 & 1 \end{pmatrix} + \frac{1}{\sqrt{2}} L(\beta_{2g}) q_{2g} \begin{pmatrix} 0 & 1 \\ 1 & 0 \end{pmatrix} \qquad (7.25)$$

written out in the a, b representation. (Figure 7.12.)

In a $C_{4n}H_{4n}$-molecule, with two remaining electrons, the fourfold level is split into a triplet $^3A_{2g}$ and three singlets $^1A_{1g}$, $^1B_{1g}$, $^1B_{2g}$ by inter-electronic forces, as shown below.

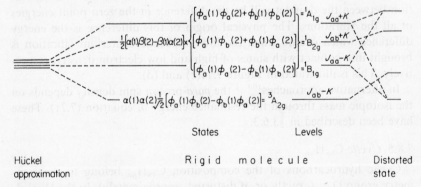

Figure 7.13 Energy levels of a $C_{4n}H_{4n}$ molecule in different approximations. (After Reference 79.)

The exchange part K of the Coulomb energy is positive;[101] in the direct part the diagonal integral J_{aa} $(=J_{bb})$ is generally larger than the non-diagonal integral J_{ab}.[79]

Let us envisage a deformation of the molecule which consists of the shortening of the C_{2n}–C_{2n-1} bonds and the lengthening of the C_{2n+1}–C_{2n} bonds. This deformation will be a linear combination of the C–C stretching β_{1g}-mode and the totally symmetric α_{1g}-mode. The two singlet states $^1A_{1g}$ and $^1B_{1g}$ will be coupled by this deformation. The matrix Hamiltonian is now

$$\begin{matrix} ^1B_{1g}: \\ ^1A_{1g}: \end{matrix} \begin{pmatrix} K & 2^{\frac{1}{2}}Lq_{1g} \\ 2^{\frac{1}{2}}Lq_{1g} & -K \end{pmatrix} \tag{7.26}$$

We have seen in §3.6.1 the shape of the potential curves when equation (7.26) is used.

7.9 Symmetric top molecules

We now give a brief account of the effects of rotational–vibrational interaction in a symmetric top molecule with a threefold axis of symmetry and in a degenerate electronic state. We shall first recall the main spectral consequences of the interaction in the absence of the J-T-E. For a full account, the reader is referred to Herzberg's books. (Reference 105, Chapter IV, §2; Reference 106, Chapter I, §3b, Chapter II, §3b). From our point of view the operation of the rotational–vibrational interaction in the absence of vibronic coupling is best explained with reference to the spectral behaviour of set of levels all with a fixed J. J is the total rotational quantum number including nuclear, vibrational, electronic-orbital and spin angular momenta. For the moment the last two are ignored. K is the component of J in the direction of the symmetric-top axis, taken as the z-direction. In fact, K is an eigenvalue of $\hat{J}_z$. (In this section operators are specified by a caret; the operators for vibrational and electronic orbital angular moments will be distinguished from the total angular momentum operator $\hat{\mathbf{J}}$, by subscripts: $\hat{\mathbf{J}}_\varepsilon$ and $\hat{\mathbf{J}}_e$, respectively). In the absence of rotational–vibrational interaction, the energy levels for the same J (e.g. $J = 6$) and different $K(K = 0, \ldots, 6)$ are spaced at relative heights proportional to K^2. This is shown on the extreme left of the following figure (Figure 7.14), where the vibrational–rotational level degeneracies are also indicated by a number to the left of each level. This number is appropriate for a doubly degenerate vibrational state. In the next two columns, we show the effects of the interaction in its two stages: the Coriolis-interaction and the l-type doubling. The origins of these terms will now be explained.

Coriolis coupling arises from a term in the rotational energy operator having the form

$$-\frac{1}{I_\parallel}\hat{J}_z\hat{J}_{z\varepsilon} - \frac{1}{I_\perp}(\hat{J}_x\hat{J}_{x\varepsilon} + \hat{J}_y\hat{J}_{y\varepsilon}) \qquad (7.27)$$

where $I_\parallel$ and $I_\perp$ are the moments of inertia of the symmetric top molecule along the top axis (the z-direction) and perpendicular to it respectively. [For the meaning of this coupling term p. 381 in Reference 107 may be read and, for a quantum mechanically more rigorous form, Reference 108.]

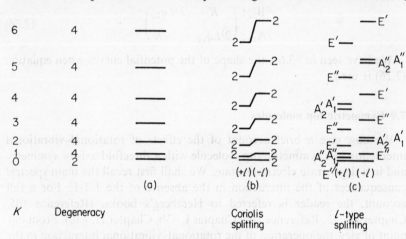

Figure 7.14 Rotational energy levels of a D_{3h} molecule in an E' vibrational state for $J = 6$ in three approximations.

(a) No rotation–vibration interaction.

(b) An interaction term having cylindrical symmetry ($D_{\infty h}$).

(c) Interaction terms of D_{3h}-symmetry.

(The same level scheme obtains for an E' electronic or vibronic state with the interaction in (b), (c) being rotation–electronic, etc.). (After Reference 106.)

The halving of the level degeneracy, which corresponds to an orientation of $J_{z\varepsilon}$ parallel or antiparallel to J_z ($=K$), is called the Coriolis splitting. The resulting levels are identified by two new symbols, $(+l)$, $(-l)$, as shown at the bottom of the middle column (Figure 7.14). Recalling that originally we had a quadratic dependence of the levels on K, we note that we have now two levels at heights proportional to $(K \pm \zeta)^2$. Here ζ is a number of order unity which arises from the expectation value of $\hat{J}_{z\varepsilon}$ in the Coriolis operator term. (The expectation values of $\hat{J}_{x\varepsilon}$, $\hat{J}_{y\varepsilon}$ are zero in a symmetric top eigenstate). A discussion of the quantity ζ

relating to cases where a number of degenerate vibrations are brought into play is given in Reference 109.

It cannot be expected that the Coriolis-coupling term will remove all accidental degeneracy, since its symmetry ($=D_{\infty h}$) is higher than that of the actual symmetry of the molecule (D_{3h}, C_{3h}, etc.). (The situation here is thus entirely analogous to the accidental degeneracy which the reader has met in the linear coupling case of the $E \otimes \varepsilon$ J-T-E.) Therefore a term containing the symmetry of a threefold axis and no more, is needed to lift the remaining degeneracy. An operator expression

$$A[\hat{J}_+^2 \hat{J}_{+\varepsilon} + \hat{J}_-^2 \hat{J}_{-\varepsilon}] \tag{7.28}$$

is of the required form. Here $\hat{J}_\pm$, $\hat{J}_{\pm\varepsilon}$ are the familiar angular momentum raising and lowering operators arising from $\hat{J}_x$, $\hat{J}_y$, etc., and A is a coupling constant. The effect of this term is to separate a pair of accidentally degenerate levels for $K = 1, 2, 4, 5$, etc., into two singlets (A_1'', A_2'' or A_1', A_2') as in the last column of the previous figure (Figure 7.14).

7.10 Spherical top molecules

The basic theory for vibration–rotation coupling in non-degenerate or doubly degenerate states is similar to that for symmetric tops and is described in Reference 105, §IV.3, and Reference 106, §I.3(c). In a threefold degenerate state of a spherical top molecule (such as CH_4) vibration–rotation coupling was treated by Teller[110] and is reviewed in Reference 106 (p. 103). Coriolis coupling adds to the purely rotational energy $BJ(J + 1)$ of the Jth rotational level the quantities

$$-2B\zeta_v J, \quad 0, \quad 2B\zeta_v (J + 1)$$

depending on whether the vibronic angular momentum is parallel, perpendicular or antiparallel to the direction of J. ζ_v is the vibronic Coriolis parameter. For a threefold degenerate vibronic state in O- or T_d-symmetry, $\Gamma_v = T_1$ or T_2 are representations of the vibronic state. ζ_v is defined by the following matrix element of the sum of the electronic l and vibrational m angular momenta.

$$\begin{aligned} ih\zeta_v = ih\zeta_v(\Gamma_v, j) &= \langle \psi_j^{T_1.y} | \hat{l}_z + \hat{m}_z | \psi_j^{T_1.x} \rangle, \quad \Gamma_v = T_1 \\ &= \langle \psi_j^{T_2.\eta} | \hat{l}_z + \hat{m}_z | \psi_j^{T_2.\varepsilon} \rangle, \quad \Gamma_v = T_2 \end{aligned} \right\} \tag{7.29}$$

for the jth state of the species Γ_v.

Higher order vibration–rotation couplings have also been studied,[111-2] but these are several orders of magnitude smaller. They will not be discussed by us.

7.11 Linear molecules

Linear molecules belong to the continuous group $D_{\infty h}$ or $C_{\infty v}$. Their maximum electronic degeneracy is double. The electronic factor in the wave-function can be written for degenerate states as

$$\phi(\mathbf{r}) = f(r)\,e^{i\Lambda\theta} \tag{7.30}$$

in terms of the angle θ which represents the angular whereabouts of the electron with respect to an arbitrary plane containing the linear molecule. $f(r)$ depends on the distance r of the electron from the axis of the linear molecule. For $\Lambda = \pm1, \pm2, \pm3, \ldots$ the state is designated

$$\Pi, \Delta, \Phi, \ldots$$

The linear combinations $\cos\Lambda\theta$ and $\sin\Lambda\theta$ of equation (7.30) yield the even $(+)$ and odd $(-)$ $\Pi, \Delta, \Phi, \ldots$ states.

Other factors in the total wave function are appropriate to the rotation and translation of the molecule as a whole, which will not be treated here. Their role in vibronic systems is discussed by Hougen.[113]

Vibrations which bend the molecule out of the linear position are also doubly degenerate. There may be a number of such modes. However, in a triatomic molecule there is just one such mode, to which the angular frequency $2\pi\nu_2$ is conventionally assigned and which can be given the polar representation

$$q\cos\phi, \qquad q\sin\phi.$$

q is expressed in units of the zero point motion amplitude. The angle ϕ is that subtended between the plane of the instantaneously bent molecule and the arbitrary plane introduced earlier. The vibrational factor in the wave-function has the form

$$\chi_{v,|l|}(q,\phi) = e^{il\phi}\rho_{v,|l|}(q) \tag{7.31}$$

$\rho_{v,|l|}$ is a two-dimensional harmonic oscillator wave-function belonging to level v and rotational state $\pm l$. We have already met this oscillator. [Equation (3.5) and (3.8).]

The first task now is to derive the potential surfaces for the bending vibration when the electron is in the position r, θ. By physical considerations the potential will depend on the *difference* $(\theta - \phi)$ and will be an even function of this argument. Such potential is depicted in the next figure (Figure 7.15).

When the potential is developed as a Fourier series in the angles $\theta - \phi$ it takes the form (for $q \to 0$)

$$V(q, \phi; r, \theta) = a(r)q \cos (\theta - \phi) + b(r)q^2 \cos 2(\theta - \phi) + $$
$$+ c(r)q^3 \cos 3(\theta - \phi) + d(r)q^4 \cos 4(\theta - \phi) + \ldots \quad (7.32)$$

If we calculate the expectation values V^+ and V^- of equation (7.32) within the even $(+)$ and odd $(-)$ states, only even powers of q survive

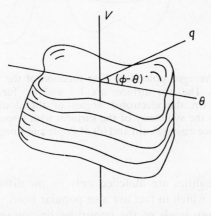

Figure 7.15 Potential surface for the vibration, as function of the coordinates (q, ϕ). The surface is symmetric with respect to reflection in the plane $\phi = \theta$ (where θ is the electronic angle).

the integration over θ and we have essentially the three possibilities which are known as the cases (a), (b) and (c) of Pople and Longuet-Higgins.[114] Figure 7.16 shows these cases. Since V^+ and V^- are even in q there is no logical need for an off-centre minimum (although the possibility for such minima exists, as in cases (b) and (c)). In this respect, then, linear molecules are different[115] from other molecular systems where the first term in q is linear. [Equation (7.2).] When the system is in a Π-state ($|\Lambda| = 1$) the lowest term in V^+ or V^- is quadratic in q, when it is in a Δ-state ($|\Lambda| = 2$) V^+ or V^- starts with q^4 and so on.

7.12 Oxides and other ionic solids

Several oxides serve as convenient host compounds[116] for d^n ions, especially $(3d)^n$, which are accommodated mainly substitutionally. Divalent or trivalent d^n ions are commonly found substituents, e.g., in MgO and Al_2O_3, respectively. The facility of accommodation is probably due to the ionic character of the oxides and the well-defined divalence

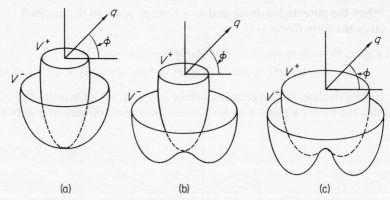

Figure 7.16 Averaged potentials V as function of the vibrational co-ordinates q, ϕ. The two surfaces are V^+ and V^- for even and odd combinations (in the electronic angle) of the doubly degenerate Π-states. Note the symmetry of the surfaces with respect to the plane $\phi = 0$. The three cases (a), (b) and (c) of Pople and Longuet-Higgins[114] are shown.

of O^{2-}. These qualities are matched only by the difluorides having the fluorite structure, which in fact are also popular hosts.

The lower energy levels of the impurities lie presumably in the large energy gap (in MgO about 7·3 eV wide[117]) of the host crystal so that the impurity states can be regarded as localized and the band structure of the host can be ignored with good justification. In contrast, the elastic properties of the host (in general) and its vibrational spectra (in particular) are decisive for many properties of the impurity centre. In the following table (Table 7.6) we quote the reststrahlen wave-numbers of some important ionic compounds. The frequencies of the $(d^n)O_6$ centre and the infra-red frequencies ω_{TO} of the host are not directly identifiable, since the critical points in the phonon spectrum (the points at which the density of phonons states is singular) may be as much characteristic of the localized vibrations of the centre as are the reststrahlen of the host. Yet, Table 7.6 may come to be of use at least in the scaling of the frequencies in similar type crystals.

The frequencies are room temperature values. In the perovskite-type crystal $SrTiO_3$ (Figure 7.17) various interesting phase transitions occur, the one at $T = T_t = 105\cdot5\ °K$ between the cubic and tetragonal modifications.[119] Approaching T_t the lowest optical mode frequency softens in proportion to $(T - T_t)^{\frac{1}{2}}$, as predicted by Cochran[120] and observed in the neutron scattering of Cowley.[121] The microscopic nature of the transition was elucidated through ESR[122] and by elastic neutron diffraction[119] as

Table 7.6 Infrared absorption frequencies ω_{TO} of some host compounds.

Compound	LiF	NaCl	KCl	KBr	MgO	CaO	SrO	MnO	NiO	Al₂O₃	E
ω_{TO} in cm⁻¹	307	164	144	116	394	295	211	268	401	385, 442, 569, 635	⊥c
										400, 583, 654	∥c
Source	a	a	a	a	b	a	b	b	b	c	

Compound	CaF₂	SrF₂	BeF₂	ZnS (cubic)	CdS	CdTe	SrTiO₃	KMgF₃
ω_{TO} in cm⁻¹	257	217	184	273	240	150	86, 176	168, (254?), 299
							265, 544	458
Source	b	b	b	b	b		d	e

a From Reference 118, p. 85.

b From a review by R. Ruppin and R. Englman, *Repts. Progr. Phys.*, **33**, 149 (1970), which contains a bibliography.

c A. S. Barker, Jr., *Phys. Rev.*, **132**, 1474 (1963).

d A. S. Barker, Jr., *Phys. Rev.*, **145**, 391 (1966).

e C. H. Perry and E. F. Young, *J. Appl. Phys.*, **38**, 4616 (1967).

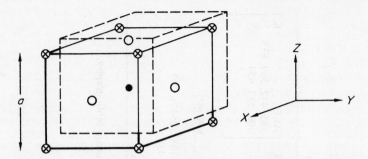

Figure 7.17 The structure of the perovskite-type crystal ABX_3. The broken lines show a cell which contains a formula unit.

⊗ A-ion   ● B-ion   ○ X-ion.

consisting of a rotation (Figure 7.18) of the TiO_6 octahedra about the new tetragonal axis. (The angle of rotation ϕ is about eight degrees at 77 °K.) The macroscopic tetragonality of the crystal is $c/a = 1\cdot0004$, at 77 °K,

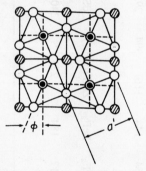

Figure 7.18 Rotation of the TiO_6 octahedra in the tetragonal phase of $SrTiO_3$. (From Reference 119.) A section perpendicular to the tetragonal axis is shown.

whereas in the individual octahedra $c/a' = 1\cdot0001$. There are anomalies in the elastic constants,[123] but the crystal is not polar even in the distorted phase. (A J-T-E in $SrTiO_3$ was discussed in §4.4.)

On the other hand, spontaneous dipole moments are exhibited in other crystals of the perovskite family, by some of them at relatively high temperatures ($PbTiO_3$: 490 °K, $KNbO_3$: 435 °K). The number of ferro-electric perovskites keeps increasing,[124-5] but the best studied of them all is $BaTiO_3$, discovered at the end of the second world war. This crystal undergoes three first-order phase transitions from the cubic to the tetra-gonal (at 393 °K), the orthorhombic (278 °K) and the rhombohedral

(183 °K) phases. The lattice constants of the unit cell are shown in Figure 7.19.

Numerous theoretical approaches have been tried to account for these transitions. (Reference 124, Chapter IV). It is now accepted that the dielectric anomalies are tied to the vanishing of the frequency of a soft vibrational mode of the crystal near the transition.[120] The vanishing of this frequency is a consequence of the anharmonicities in the lattice-vibrational Hamiltonian.[127] The phase transition does not appear to alter

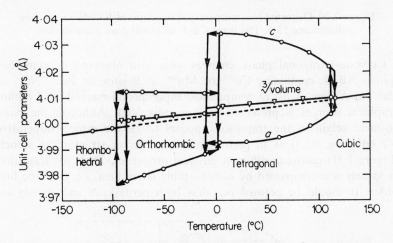

Figure 7.19 Lattice constants of $BaTiO_3$ as function of temperature. (From Reference 126.)

the electronic energies to any significant extent. The energy band structure is probably similar to that in $SrTiO_3$.[128-9] The gap of about 3·5 eV lies between the occupied 2p-states of oxygen and the unoccupied 3d states of Ti. (§5.2 treats the relevance of J-T-E to $BaTiO_3$.)

We meet an evidently different type of phase change in perovskites of the composition ABX_3, in which the octahedrally surrounded B-ion is a transition metal ion having a degenerate electronic state, e.g. Cu^{2+}, Mn^{3+}, Cr^{2+} [130] or Fe^{2+}.[130a] These crystals, e.g. $KCuF_3$,[131] undergo a second order transition from the cubic to the tetragonal phase. In the latter each B-cation has an orthorhombic surrounding and neighbouring B-cations occupy inequivalent lattice points. The structure is seen in Figure 7.20. In MnF_3, where the A-ion is missing, the strain in the direction of the tetragonal axis causes a buckling of the Mn–F–Mn bonds along this axis. Here also the transition appears to be of second order, which is theoretically in accord with the doubling of the unit cell in this transition.

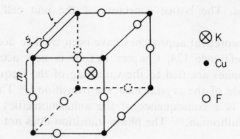

Figure 7.20 Distorted low temperature structure of KCuF$_3$ (according to Reference 131). The three Cu–F distances *l*, *m*, *s* are shown.

Cubic-to-tetragonal phase changes were also observed in numerous spinels AB$_2$O$_4$ containing Cu^{2+} or Mn^{3+} at B sites, or Fe^{2+}, Ni^{2+} or Cu^{2+} at A sites. A list of compounds, experimental methods and bibliographical sources is presented in Appendix VIII. Although the basic physical origin of the transition appears to be the same (cooperative J-T coupling, §5.3) as in perovskites, the different structure of spinels (Figure 7.21) results in different physical properties. Thus, the transition in spinels is accompanied by a measurable latent heat, i.e. is of the first order. It should be pointed out that both perovskites and spinels are

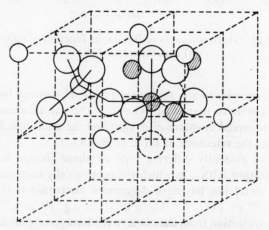

Figure 7.21 Unit cell of a normal spinel AB$_2$O$_4$. Ions situated in two octants are shown, the ions drawn by broken circles belong to other octants.
Large circles: oxygen ions.
Small shaded circles: metal ions B at octahedral sites.
Small open circles: metal ions A at tetrahedral sites.
(From Reference 132.)

prone to distortions, which may be of trigonal, tetragonal or lower symmetries, whose causes (the packing or the magnetic interactions between cations) are unrelated to the distortions here considered.

Crystallographic transitions in $(A + B) \otimes \beta$ P-J-T systems were observed in some rare earth vanadates, arsenates, etc., of tetragonal zircon structure (§5.3).

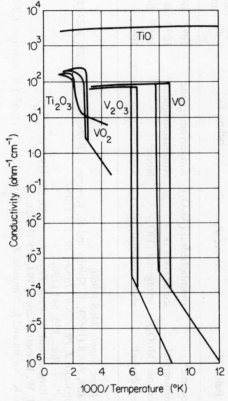

Figure 7.22 Conductivity of some oxides of titanium and vanadium as function of temperature. (After Reference 133.)

Several oxides $M_m O_n$ of transition metals M (with $n/m \leqslant 2$) undergo sudden changes in their structural, conductive, magnetic and optical properties in the 100–500 °K temperature region. The most salient experimental information is still that of Morin whose conductivity vs. temperature curves are reproduced here (Figure 7.22).[133] These show jumps by factors up to 10^6. Some other data for a few selected oxides are collected in Table 7.7. From theoretical considerations concerning the band

Table 7.7 Properties of some titanium and vanadium oxides.*

In the sixth column the relative change in conductivity $\Delta\sigma/\sigma$ at $T = T_c$ is given; ΔS is the entropy change across T_c; V_0 and W are energies introduced in §5.1.4: V_0 is half the energy gap produced distortionally, W is the band width.

Compound	T_c	Crystal Structure Below T_c	Crystal Structure Above T_c	Magnetism below T_c	$\dfrac{\Delta\sigma}{\sigma}$	ΔS in meV/ formula unit/°K	V_0 in eV	W in eV
V_2O_3	150 °K[a]	Monoclinic[b]	rhombohedral (Al_2O_3)[b]	Antiferromagnetic[c]	10^6 [a]	0·3[e]	0·05[e]	0·7[e]
VO	126 °K[a]	Orthorhombic[f]	cubic (NaCl)[f]	Antiferromagnetic[m]	10^6 [a]	0[a]	0·05[f]	
VO_2	340 °K[a]	Monoclinic (MoO_2)[g]	tetragonal (TiO_2)[f]	None[h]	10^4 [k]	0·13[j]	>0·31[l]	
Ti_2O_3	450 °K[a]	Rhombohedral[i]	rhombohedral (Al_2O_3)[i]	Antiferromagnetic[i] (?)	10 [a]	?	>0·03[l]	≫0·05[f]

[a] Reference 133.

[b] E. P. Warekois, J. Appl. Phys. Suppl., 346S (1960).

[c] E. D. Jones, Phys. Rev., 137, A978 (1965); T. Shinjo and K. Kosuge, J. Phys. Soc. Japan, 21, 2622 (1966).

[d] S. Minomura and H. Nagasaki, J. Phys. Soc. Japan, 19, 131 (1964).

[e] J. Feinleib and W. Paul, Phys. Rev., 155, 841 (1967).

[f] D. Adler, J. Feinleib, H. Brooks and W. Paul, Phys. Rev., 155, 851 (1967).

[g] R. Heckingbottom and J. W. Linnett, Nature, 194, 678 (1962).

[h] J. Umeda, H. Kusumoto, K. Narita and E. Yamada, J. Chem. Phys., 42, 1458 (1965); K. Kosuge, J. Phys. Soc., Japan, 22, 55 (1967).

[i] S. C. Abrahams, Phys. Rev., 130, 2230 (1960).

[j] C. N. Berglund and H. J. Guggenheim, Phys. Rev., 185, 1022 (1969); C. N. Berglund and A. Jayaraman, Phys. Rev., 185, 1034 (1969).

[k] H. Sasaki and A. Watanabe, J. Phys. Soc. Japan, 19, 1748 (1969).

[l] Reference 134.

[m] W. W. Warren, G. A. Miranda and W. G. Clark, Bull. Am. Phys. Soc., 12, 1117 (1967).

* Other, mixed oxides are discussed by D. B. McWhan, A. Menth and J. P. Remeika, J. Physique, 32, C1-1079 (1971).

structure of these oxides, the Fermi-level should reside somewhere inside the 3d-bands of M, the oxygenic 2p-bands being fully occupied and the M-4s bands empty. The low temperature non-conducting (or semi-conducting) phase presents an acknowledged mystery in view of the incompletely filled d-bands. The theoretical attempts to overcome this problem and to interpret these transitions were reviewed by Adler,[134] who also summarized the prevalent experimental situation. Since then some doubt has arisen about the occurrence of VO (M. D. Sturge, private communication, 1971).

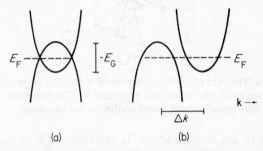

Figure 7.23 A simple model for the energy levels in semimetals as function of the wave-vector **k**. E_F is the Fermi-level and, in (b), Δk is the displacement in k-space between the maximum of the valence band and the minimum of the conduction band. E_G is the negative energy gap between the bands.

It is expected[134-5] that some of the transition metal oxides belong to the class of materials called semimetals. In these there are two overlapping energy-bands. Their band structure is illustrated in a simplified way in Figure 7.23. We have discussed in §5.1.5 the lattice instabilities which are caused by the degeneracy of the overlapping bands.

7.13 V₃Si and isostructural compounds

The crystal structure is shown in the next figure (Figure 7.24). The structure is cubic, with the Si atoms at the centre of the cube and on the corners. The vanadium atoms form three mutually perpendicular sets of chains. Because of the chain-like structure, Weger supposed[136] that the electronic properties would resemble those theoretically predicted for structures with imperfectly screened Coulomb interaction. The high superconducting transition temperature T_c (17·1 °K) in V₃Si, coming near the 'ceiling' of 18 °K postulated at a time by T. H. Geballe et al.,[137] was also thought to confirm Little's hypothesis of the existence of one-dimensional high-temperature superconductors.[138]

Just above the superconducting transition there occur a number of remarkable phenomena, which will be now briefly summarized. The connection of these with the onset of superconducting is a matter of keen interest[139–40] but the issue is not yet settled.

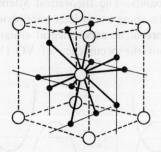

Figure 7.24 The A15 ($\beta - W$) structure of V_3Si. The large circles show the positions of Si, the small dots represent V. The thin lines constitute the three orthogonal sets of linear chains in the structure.

Some three to ten degrees above T_c the cubic V_3Si changes gradually (as far as one can tell, without hysteresis) into the low temperature elongated tetragonal phase.[141] It is apparent from the accompanying

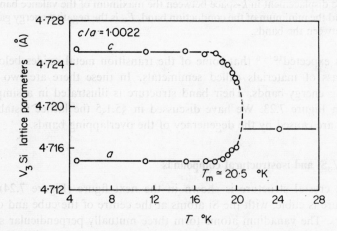

Figure 7.25 The tetragonal lattice parameter c and a of V_3Si vs. T.
(After Reference 141.)

figure (Figure 7.25) that the main part of the deformation takes place above T_c. In contrast, a number of things 'happen' above the martensitic transition temperature T_m, whose relationship to the events below T_m (and above T_c) is not evident, but was in fact elucidated by Labbé and Friedel.

Their theory was reviewed in §5.1.3. The following observations were made:

In V_3Si Testardi and coworkers[142] have found a strong decrease in the velocity of propagation [which is proportional to the elastic constant $\sqrt{(C_{11} - C_{12})}$] of some shear waves, indicating an instability in the e_θ, e_ϵ-strain modes. They also observed a rise in the attenuation of ultrasonics, which may suggest an increase in the electron–phonon coupling. In another A15 compound Nb_3Sn, similar anomalies of the elastic constants were noticed above its superconducting transition temperature.[143-4] It seems now that at T_m ($\simeq$45 °K) $C_{11} - C_{12}$ remains positive and that c/a jumps discontinuously to 0·997.[144a] In V_3Si $c/a > 1$. It was also observed that in V_3Si the electronic specific heat is anomalously large in both the cubic and tetragonal phases[145-7] and the paramagnetic susceptibilities and the Knight shifts were strongly temperature dependent.[148-50] These facts indicate a large and temperature sensitive electronic density of states near the Fermi level E_F. Measurements of the electrical resistance under strain[151] also show signs of the phase transformation. The nuclear relaxation rates in V_3Si and in the analogous compounds V_3Pt, V_3Ge have a peculiar low temperature peak,[152] which can perhaps be ascribed to the density of states anomaly, though this is unlikely (M. Weger, private communication, 1970).

Weger's original suggestion,[136] that the proximity of V atoms along a chain leads to energies which are functions only of the component of the

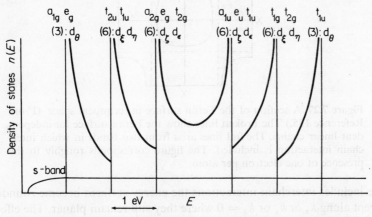

Figure 7.26 Density of state curves of a threefold, orthogonal one-dimensional structure. The states arise from 3d-states tightly bound to vanadium. Above the sub-bands we show the small representations at $\Gamma(k = 0)$, the degeneracies per formula unit and the type of d-orbital for a chain running in the z-direction. (After J. Labbé and J. Friedel, *J. Physique*, **27**, 253 (1966); J. Labbé, *Phys. Rev.*, **158**, 647 (1967).)

wave vector along the chain, yields the density of states curves shown in Figure 7.26. The spreads of the three sub-bands are measures of the overlaps between 3d functions on neighbouring atoms, as follows from the calculation of electronic energies by the tight binding approximation. (Reference 153, Appendix J).

The spreads are different in the three sub-bands, since the overlaps of the different d-orbitals, namely $e_\epsilon(x^2 - y^2)$, $e_\theta(3z^2 - r^2)/\sqrt{3}$, $t_\zeta(yz)$, $t_\zeta(xy)$ are different along a z-oriented chain. The predictions of this simple model were put to a computational test by Mattheiss[154] who found by means of an APW calculation that, quite apart from quantitative disagreements, not even his ordering of the levels at $k = 0$ agrees with that in the linear chain model. A further objectionable feature in the model is that the centre of gravity of all subbands is the same. This disregards any effect of the crystal field on the d-band. Presumably, all these difficulties are reflections of the well-known fact that tight binding approximation is not a good starting point for band structure calculations of metallic alloys.

In Figure 7.27, a section, in the $k_z = 0$ plane, of the Fermi surface is shown (by broken lines) for the model of independent linear chains. If

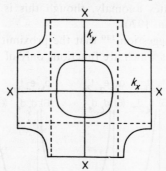

Figure 7.27 A section of the Fermi surface in reciprocal space. (From Reference 155.) The broken lines show the Fermi surface for independent linear chains. The full lines arise from two bands in which interchain interaction is included. The figure corresponds roughly to the presence of one electron per atom.

one includes interchain interactions the energy surfaces become rounded, except along k_x or k_y or $k_z = 0$ where they still remain planar. The effects of such interaction were studied by Weger[155] who found for example, that the high density-of-state portion of Figure 7.26 moves away from the sharp peaks in that figure to other positions intermediate between the sharp peaks. These positions correspond to the point X (Figure 7.27) in reciprocal space.

7.14 References

1. W. Low, *Paramagnetic Resonance in Solids* (New York: Academic Press, 1960).
2. C. J. Ballhausen, *Introduction to Ligand Field Theory* (New York: McGraw-Hill, 1962).
3. J. S. Griffith, *The Theory of Transition Metal Ions* (Cambridge: University Press, 1964).
4. H. Hellmann, *Einführung in die Quanten Chemie* (Leipzig: 1937) p. 285.
5. R. P. Feynman, *Phys. Rev.*, **56**, 340 (1939).
6. W. L. Clinton and B. Rice, *J. Chem. Phys.*, **30**, 542 (1959).
7. G. F. Koster, J. O. Dimmock, R. G. Wheeler and H. Statz, *Properties of the Thirty-two Point Groups* (Cambridge, Mass.: MIT Press, 1963).
8. H. C. Longuet-Higgins, *Adv. in Spectrosc.*, **2**, 429 (1961).
9. J. S. Griffith, *The Irreducible Tensor Method for Molecular Symmetry Groups* (London: Prentice Hall, 1962).
10. U. Fano and G. Racah, *Irreducible Tensorial Sets* (New York: Academic Press, 1959).
11. W. Moffitt, G. L. Goodman, M. Fred and B. Weinstock, *Mol. Phys.*, **2**. 109 (1959).
12. B. Weinstock and G. L. Goodman, *Adv. Chem. Phys.*, **9**, 169 (1965).
12a. H. H. Claassen, G. L. Goodman, J. H. Holloway and H. Selig, *J. Chem. Phys.*, **53**, 341 (1970).
13. A. A. Kiseljov, *J. Phys. B*, **2**, 270 (1969).
14. O. N. Singh and D. K. Rai, *Canad. J. Phys.*, **43**, 378 (1965).
15. M. Kimura, V. Schomaker, D. W. Smith and B. Weinstock, *J. Chem. Phys.*, **48**, 4001 (1968).
16. R. J. Gillespie in *Noble Gas Compounds*, Ed. Hyman (Chicago: University Press, 1963), p. 333.
17. R. F. Code, W. E. Falconer, W. Klemperer and I. Ozier, *J. Chem. Phys.*, **47**, 4955 (1967).
18. L. S. Bartell and R. M. Gavin, *J. Chem. Phys.*, **48**, 2466 (1968).
19. R. D. Burbank and W. Bartlett, *Chem. Commun.*, **1**, 645 (1968).
20. E. U. Condon and G. H. Shortley, *The Theory of Atomic Spectra* (Cambridge: University Press, 1957).
21. L. A. Woodward and M. J. Ware, *Spectrochim. Acta*, **20**, 711 (1964).
22. R. Hilsch, *Z. Physik.*, **44**, 860 (1927); *Z. Angew. Chem.*, **49**, 69 (1936); *Physik. Z.*, **38**, 1031 (1937).
23. D. S. McClure, *Solid State Physics*, **9**, 399 (1959).
24. R. A. Eppler, *Chem. Rev.*, **61**, 523 (1961).
25. C. K. Jorgensen, *Absorption Spectra and Chemical Bonding in Complexes* (Oxford: Pergamon Press, 1962).
26. F. Seitz, *J. Chem. Phys.*, **6**, 150 (1938).
27. P. H. Yuster and C. J. Delbecq, *J. Chem. Phys.*, **21**, 892 (1953).
28. F. E. Williams, *J. Chem. Phys.*, **19**, 457 (1951); *J. Opt. Soc. Am.*, **47**, 869 (1957).
29. F. E. Williams and M. H. Hebb, *Phys. Rev.*, **84**, 1181 (1951).
30. F. E. Williams and P. D. Johnson, *Phys. Rev.*, **113**, 97 (1959).
31. R. T. Harley, J. B. Page and C. T. Walker, *Phys. Rev. Letters*, **23**, 922 (1969).
32. L. C. Kravitz, *Phys. Rev. Letters*, **24**, 884 (1970).

33. H. Kamimura and S. Sugano, *J. Phys. Soc. Japan*, **14**, 1612 (1959).
34. W. V. Smith, P. P. Sorokin, I. L. Gelles and G. J. Lasher, *Phys. Rev.*, **115**, 1546 (1959).
35. L. A. Shulman, I. M. Zaritskii and G. A. Podzyarei, *Fiz. Tverd. Tela*, **8**, 2307 (1966) [English trans. Soviet Phys.–Solid State **8**, 1842 (1967)].
36. E. W. J. Mitchell, *Brit. J. Appl. Phys.*, **8**, 179 (1957).
37. C. D. Clark, R. W. Ditchburn and H. E. Dyer, *Proc. Roy. Soc. A*, **234**, 363 (1956); **237**, 75 (1956).
38. C. A. Coulson and M. J. Kearsley, *Proc. Roy. Soc. A*, **241**, 433 (1952).
39. J. Friedel, M. Lannoo and G. Leman, *Phys. Rev.*, **164**, 1056 (1967).
40. T. Yamaguchi, *J. Phys. Soc. Japan*, **17**, 1359 (1962).
41. G. D. Watkins in *Proceedings of the International Conference on the Physics of Semiconductors*, Paris, 1964, Vol. 3 (New York: Academic Press, 1965) p. 97.
42. G. D. Watkins and J. W. Corbett, *Phys. Rev.*, **138**, A543 (1965).
43. E. L. Elkin and G. D. Watkins, *Phys. Rev.*, **174**, 881 (1968).
44. D. G. Thomas, in *Localized Excitations in Solids*, ed. Wallis (New York: Plenum Press, 1968) p. 239.
45. D. G. Thomas, M. Gershenzon and J. J. Hopfield, *Phys. Rev.*, **131**, 2397 (1963).
46. T. N. Morgan, *Phys. Rev. Letters*, **24**, 887 (1970).
47. G. L. Bir, *Zh. Eksperim i Teor. Fiz.*, **51**, 556 (1966) [English Transl. *Soviet Phys.—JETP*, **24**, 372 (1966)].
48. C. Z. van Doorn, *Philips Res. Repts.*, **12**, 309 (1957), *Philips Res. Repts.*, Supp. No. 4 (1962).
49. H. Pick, *Z. Phys.*, **159**, 69 (1960).
50. R. H. Silsbee, *Phys. Rev.*, **138**, A180 (1965).
51. Y. Toyozawa and M. Inoue, *J. Phys. Soc. Japan*, **21**, 1663 (1966).
52. M. Lax, *J. Chem. Phys.*, **20**, 1752 (1952).
53. E. Mulazzi, G. F. Nardelli and N. Terzi, *Phys. Rev.*, **172**, 847 (1968).
54. R. Englman, *Mol. Phys.*, **3**, 23 (1960).
55. M. H. L. Pryce, G. Agnetta, T. Garofano, M. B. Palma-Vittorello and M. V. Palma, *Phil. Mag.*, (8) **10**, 477 (1964).
56. D. L. Dexter, C. C. Klick and G. A. Russell, *Phys. Rev.*, **100**, 603 (1955).
57. C. C. Klick and J. H. Schulman, *Solid State Physics*, **5**, 97 (1957).
58. F. Seitz, *Rev. Mod. Phys.*, **18**, 384 (1946); **26**, 7 (1954).
59. J. J. Markham, *F-centres in Alkali Halides, Solid State Phys. Suppl.*, **8** (New York: Academic Press, 1966).
60. M. Kasha, *Discussions Faraday Soc.*, **9**, 14 (1950).
61. J. N. Murrell, *The Theory of Electronic Spectra of Organic Molecules*, (London: Methuen, 1963) §§2.1, 14.1.
62. P. Kisliuk and C. A. Moore, *Phys. Rev.*, **160**, 307 (1967).
63. D. Curie, *Luminescence in Crystals* (London: Methuen 1963).
64. R. Englman and B. Barnett, *J. Luminescence*, **3**, 37 (1970).
65. M. Bixon and J. Jortner, *J. Chem. Phys.*, **48**, 715 (1968).
66. W. T. Simpson and D. L. Peterson, *J. Chem. Phys.*, **26**, 588 (1957).
67. A. Witkowski and W. Moffitt, *J. Chem. Phys.*, **33**, 872 (1960).
68. R. L. Fulton and M. Gouterman, *J. Chem. Phys.*, **35**, 1059 (1961); **41**, 2280 (1964).
69. E. G. McRae, *Australian J. Chem.*, **14**, 344 (1965).
70. Y. Marechal, *J. Chem. Phys.*, **44**, 1908 (1966).

71. M. H. Perrin and M. Gouterman, *J. Chem. Phys.*, **46**, 1019 (1967).
72. E. A. Chandross, J. Ferguson and E. G. McRae, *J. Chem. Phys.*, **45**, 3546 (1966).
73. E. A. Chandross and J. Ferguson, *J. Chem. Phys.*, **45**, 3554 (1966).
74. B. Badger and B. Brocklehurst, *Trans. Faraday. Soc.*, **65**, 2576, 2588, 2611 (1969).
75. M. T. Vala, Jr., I. M. Hillier, S. A. Rice and J. Jortner, *J. Chem. Phys.*, **44**, 23 (1966).
76. M. Looyenga, *Mol. Phys.*, **11**, 153 (1966).
77. C. A. Coulson, *Valence* (Oxford: University Press, 1961).
78. A. Streitweiser, Jr., *Molecular Orbital Theory for Organic Chemists* (New York: John Wiley, 1961).
79. L. Salem, *Molecular Orbital Theory of Conjugated Systems* (New York: Benjamin, 1966), chapter 8.
80. D. J. E. Ingram, *Free Radicals as Studied by E.S.R.* (London: Butterworth, 1958) p. 102.
81. S. Basu, *J. Chem. Phys.*, **41**, 1453 (1964).
82. M. G. Townsend and S. I. Weissman, *J. Chem. Phys.*, **32**, 309 (1960).
83. R. Pariser and R. G. Parr, *J. Chem. Phys.*, **21**, 466 (1953); **21**, 767 (1953).
84. J. A. Pople, *Trans. Faraday Soc.*, **49**, 1375 (1953).
85. A. D. McLachlan and L. C. Snyder, *J. Chem. Phys.*, **36**, 1159 (1962).
86. A. Carrington, H. C. Longuet-Higgins, R. E. Moss and P. F. Todd, *Mol. Phys.*, **9**, 187 (1965).
87. H. C. Longuet-Higgins and L. Salem, *Proc. Roy. Soc. A*, **251**, 172 (1959).
88. C. A. Coulson and A. Golebiewski, *Proc. Phys. Soc.*, **78**, 1310 (1961).
89. H. M. McConnell and A. D. McLachlan, *J. Chem. Phys.*, **34**, 1 (1961).
90. W. D. Hobey and A. D. McLachlan, *J. Chem. Phys.*, **33**, 1695 (1960).
91. G. R. Liebling and H. M. McConnell, *J. Chem. Phys.*, **42**, 3931 (1961).
92. J. H. van der Waals, A. M. D. Berghuis and M. S. de Groot, *Mol. Phys.*, **13**, 301 (1967).
93. M. S. de Groot and J. H. van der Waals, *Mol. Phys.*, **6**, 545 (1963).
94. M. S. de Groot, I. A. M. Hesselman and J. H. van der Waals, *Mol. Phys.*, **10**, 91 (1965); **10**, 241 (1966); **16**, 45 (1969); **16**, 61 (1969).
95. G. C. Nieman and D. S. Tinti, *J. Chem. Phys.*, **46**, 1432 (1967).
96. M. H. Perrin, M. Gouterman and C. L. Perrin, *J. Chem. Phys.*, **50**, 4137 (1969).
97. A. D. Liehr, *Z. Naturf.*, **A16**, 641 (1960).
98. M. J. S. Dewar, *Hyper-conjugation* (Chicago: Ronald Press, 1962).
99. E. de Boer and J. P. Colpa, *J. Phys. Chem.*, **71**, 21 (1967).
100. W. D. Hobey, *J. Chem. Phys.*, **43**, 2187 (1965).
101. M. Goeppert-Mayer and A. L. Sklar, *J. Chem. Phys.*, **6**, 645 (1938).
102. B. Sharf and J. Jortner, *Chem. Phys. Letters*, **2**, 68 (1968).
103. M. Karplus, R. G. Lawler and G. K. Fraenkel, *J. Am. Chem. Soc.*, **87**, 5260 (1965).
104. R. G. Lawler and G. K. Fraenkel, *J. Chem. Phys.*, **49**, 1126 (1968).
105. G. Herzberg, *Molecular Spectra and Molecular Structure*, Vol. II (Princeton: Von Nostrand, 1945).
106. G. Herzberg, *Molecular Spectra and Molecular Structure*, Vol. III (Princeton, Van Nostrand, 1966).
107. L. D. Landau and E. M. Lifshitz, *Quantum Mechanics* (Oxford, Pergamon Press, 1958).

108. M. S. Child, *Mol. Phys.*, **5**, 391 (1962).

109. D. R. Boyd and H. C. Longuet-Higgins, *Proc. Roy. Soc. A*, **213**, 55 (1952).

110. E. Teller, Handb. U. Jahrb. *Chem. Physik.*, **9**, 11 (1939).

111. K. T. Hecht, *J. Mol. Spect.*, **5**, 355, 390 (1960).

112. J. Herranz and B. P. Stoicheff, *J. Mol. Spect.*, **10**, 448 (1963).

113. J. T. Hougen, *J. Chem. Phys.*, **36**, 519 (1962).

114. J. A. Pople and H. C. Longuet-Higgins, *Mol. Phys.*, **1**, 372 (1958).

115. H. A. Jahn and E. Teller, *Proc. Roy. Soc. A*, **161**, 220 (1931).

116. R. S. Rubins and W. Low in *Paramagnetic Resonance*, Vol. I, ed. Low (New York: Academic Press, 1963) p. 59.

117. H. P. R. Frederikse in *AIP Handbook*, ed. Gray (New York: McGraw-Hill, 1963) §9–21.

118. M. Born and K. Huang, *Dynamical Theory of Crystal Lattices* (Oxford: Clarendon Press, 1951).

119. B. Alefeld, *Z. Physik.*, **222**, 155 (1969).

120. W. Cochran, *Phys. Rev. Letters*, **3**, 412 (1959); *Adv. in Phys.*, **9**, 387 (1960).

121. R. A. Cowley, *Phys. Rev. Letters*, **9**, 159 (1962).

122. K. A. Müller, W. Berlinger and F. Waldner, *Phys. Rev. Letters*, **21**, 814 (1968).

123. R. O. Bell and G. Rupprecht, *Phys. Rev.*, **129**, 90 (1963).

124. F. Jona and G. Shirane, *Ferroelectric Crystals* (Oxford: Pergamon, (1962)).

125. F. A. Devonshire, *Repts. Prog. Phys.*, **27**, 1 (1964).

126. H. F. Kay and P. Vousden, *Phil. Mag.*, **40**, 1019 (1949).

127. B. D. Silverman, *Phys. Rev.*, **135**, A1569 (1964).

128. A. H. Kahn and A. J. Leyendecker, *Phys. Rev.*, **135**, A121 (1964).

129. E. Simánek and Z. Šroubek, *Phys. Stat. Solidi*, **8**, K47 (1965).

130. J. B. Goodenough, *Magnetism and the Chemical Bond* (New York: Inter-Science, 1963).

130a. J. B. Goodenough, N. Menyuk, K. Dwight and J. Kafalas, *Phys. Rev.*, **B2**, 4640 (1970).

131. A. Okazaki and Y. Suemune, *J. Phys. Soc. Japan*, **16**, 176 (1961).

132. E. W. Gorter, *Philips Res. Repts.*, **9**, 295 (1954).

133. F. J. Morin, *Phys. Rev. Letters*, **3**, 34 (1959).

134. D. Adler, *Solid State Phys.*, **21**, 1 (1969).

135. N. F. Mott, *Rev. Mod. Phys.*, **40**, 677 (1968).

136. M. Weger, *Rev. Mod. Phys.*, **36**, 175 (1964).

137. T. H. Geballe *et al.*, *Physics*, **2**, 293 (1966).

138. W. A. Little, *Phys. Rev.*, **134**, A1416 (1964).

139. T. R. Finlayson, E. R. Vance and W. Rachinger, *Phys. Letters*, **26A**, 474 (1968).

140. J. Labbé, *Phys. Rev.*, **172**, 451 (1968).

141. B. W. Batterman and C. S. Barrett, *Phys. Rev. Letters*, **13**, 390 (1964); *Phys. Rev.*, **145**, 296 (1966).

142. L. R. Testardi, T. B. Bateman, W. A. Reed and V. G. Chirba, *Phys. Rev. Letters*, **15**, 250 (1965).

143. K. R. Keller and J. J. Hanak, *Phys. Letters*, **21**, 263 (1966); *Phys. Rev.*, **154**, 628 (1967).

144. R. W. Cohen, G. D. Cody and J. J. Halloran, *Phys. Rev. Letters*, **19**, 840 (1967).

144a L. J. Vieland, R. W. Cohen and W. Rehwald, *Phys. Rev. Letters*, **26**, 373 (1971).

145. F. J. Morin and J. P. Maita, *Phys. Rev.*, **129**, 1115 (1963).

146. J. E. Kunzler, J. P. Maita, H. J. Levinstein and E. J. Ryder, *Phys. Rev.*, **143**, 390 (1966).

147. J. Bonnaret, J. Hallais, S. Barisic and J. Labbé, *J. Physique*, **30**, 701 (1969).

148. A. M. Clogston and V. Jaccarino, *Phys. Rev.*, **121**, 1351 (1961).

149. A. M. Clogston, A. C. Gossard, V. Jaccarino and Y. Yafet, *Rev. Mod. Phys.*, **36**, 170 (1964); *Phys. Rev. Letters*, **9**, 262 (1962).

150. A. C. Gossard, *Phys. Rev.*, **149**, 246 (1966).

151. M. Weger, (unpublished).

152. B. G. Silbernagel, M. Weger, W. G. Clark and J. H. Wernick, *Phys. Rev.*, **153**, 535 (1967).

153. C. Kittel, *Introduction to Solid State Physics* (New York: John Wiley, 1966).

154. L. F. Mattheiss, *Phys. Rev.*, **138**, A112 (1965).

155. M. Weger, *J. Phys. Chem. Solids*, **31**, 1621 (1970).

Appendix I

Symmetric and anti-symmetric products of representations

Resolution of symmetric products [ ] of single-valued degenerate representations and of anti-symmetric products { } of double-valued degenerate representations in point groups and in continuous axial groups:

The Mulliken notation is used for the single valued representations and Jahn–Teller's[1] or Herzberg's[2] notation for the double-valued representations and for the representations of the icosahedral groups (except that the latter T is kept to designate triply degenerate representations as in the cubic groups). In the continuous axial groups two alternative notations are used.

I_h: $[T_{1g}^2] = [T_{1u}^2] = [T_{2g}^2] = [T_{2u}^2] = A_g + H_g$

$[G_g^2] = [G_u^2] = A_g + G_g + H_g$

$[H_g^2] = [H_u^2] = A_g + G_g + 2H_g$

$\{E_{\frac{1}{2}g}^2\} = \{E_{\frac{1}{2}u}^2\} = \{E_{\frac{3}{2}g}^2\} = \{E_{\frac{3}{2}u}^2\} = A_g$

$\{G_{\frac{3}{2}g}^2\} = \{G_{\frac{3}{2}u}^2\} = A_g + H_g$

$\{I_{\frac{5}{2}g}^2\} = \{I_{\frac{5}{2}u}^2\} = A_g + G_g + 2H_g$

I: The same as in I_h but drop all g or u subscripts. *Note* (and aid to nomenclature in I): The degeneracies of the representations are as follows: A(1), T_1(3), T_2(3), G(4), H(5), $E_{\frac{1}{2}}$(2), $E_{\frac{3}{2}}$(2), $G_{\frac{3}{2}}$(4), $I_{\frac{5}{2}}$(6).

O_h: $\quad [E_g^2] = [E_u^2] \quad = A_{1g} + E_g$

$\qquad [T_{1g}^2] = [T_{1u}^2] = [T_{2g}^2] = [T_{2u}^2] = A_{1g} + E_g + T_{2g}$

$\qquad \{E_{\frac{3}{2}g}^2\} = \{E_{\frac{3}{2}u}^2\} = \{E_{\frac{5}{2}g}^2\} = \{E_{\frac{5}{2}u}^2\} = A_{1g}$

$\qquad \{G_{\frac{3}{2}g}^2\} = \{G_{\frac{3}{2}u}^2\} = A_{1g} + E_g + T_{2g}$

O, T_d: As in O_h, but drop all g or u subscripts.

T_h: $\quad [E_g^2] = [E_u^2] \quad = A_g + E_g$

$\qquad [T_g^2] = [T_u^2] \quad = A_g + E_g + T_g$

$\qquad \{E_{\frac{3}{2}g}^2\} = \{E_{\frac{3}{2}u}^2\} \quad = A_g$

$\qquad \{G_{\frac{3}{2}g}^2\} = \{G_{\frac{3}{2}u}^2\} = A_g + E_g + T_g$

T $\quad$ As for T_h but drop all g or u subscripts.

Note (for the cubic groups) on the degeneracies of the representations in O or T_d: $A_1(1)$, $A_2(1)$, $E(2)$, $T_1(3)$, $T_2(3)$, $E_{\frac{3}{2}}(2)$, $E_{\frac{5}{2}}(2)$, $G_{\frac{3}{2}}(4)$.

$\text{D}_{2p\text{h}}$ (p is an integer)

$\quad [E_{kg}^2] = [E_{ku}^2] = A_{1g} + E_{2k,g} \qquad k < p/2$

$\qquad\qquad\qquad\;\, = A_{1g} + B_{1g} + B_{2g} \quad k = p/2$

$\qquad\qquad\qquad\;\, = A_{1g} + E_{2p-2k,g} \quad\; k > p/2 \quad (k = 1, \ldots, p-1)$

Note: In this and all following groups the anti-symmetric product of every double-valued representation contains the fully symmetric representation, only once. For proof see the concluding paragraph to this appendix.

$\text{D}_{2p+1\,\text{h}}$: $\quad [E_k'^2] = [E_k''^2] \quad = A' + E_{2k}' \qquad\qquad k \leq p/2$

$\qquad\qquad\qquad\qquad\quad = A' + E_{2p+1-2k}' \qquad k > p/2$

$\qquad\qquad\qquad\qquad\qquad\qquad\qquad\qquad\qquad (k = 1, \ldots, p)$

$\text{D}_{2p\text{d}}$: $\quad [E_k^2] \qquad\quad = A_1 + E_{2k} \qquad\qquad k < p$

$\qquad\qquad\qquad\qquad\;\; = A_1 + B_1 + B_2 \qquad k = p$

$\qquad\qquad\qquad\qquad\;\; = A_1 + E_{4p-2k} \qquad\;\; k > p$

$\qquad\qquad\qquad\qquad\qquad\qquad\qquad\qquad (k = 1, \ldots, 2p-1)$

$\text{D}_{2p+1\,\text{d}}$: $\quad [E_{kg}^2] = [E_{ku}^2] = A_{1g} + E_{2k,g} \qquad k < p/2$

$\qquad\qquad\qquad\qquad\qquad = A_{1g} + B_{1g} + B_{2g} \quad k = p/2$

$\qquad\qquad\qquad\qquad\qquad = A_{1g} + E_{2p-2k,g} \quad\; k > p/2$

$\qquad\qquad\qquad\qquad\qquad\qquad\qquad\qquad\qquad (k = 1, \ldots, p-1)$

D_{2p}: $[\mathrm{E}_k^2]$ $= \mathrm{A}_1 + \mathrm{E}_{2k}$ $k < p/2$

                                    $= \mathrm{A}_1 + \mathrm{B}_1 + \mathrm{B}_2$ $k = p/2$

                                    $= \mathrm{A}_1 + \mathrm{E}_{2p-2k}$ $k > p/2$

D_{2p+1}: $[\mathrm{E}_k^2]$ $= \mathrm{A}_1 + \mathrm{E}_{2k}$ $k \le p/2$

                                    $= \mathrm{A}_1 + \mathrm{E}_{2p+1-2k}$ $k > p/2$

$$(k = 1, \ldots, p)$$

C_{2pv}:   The same as D_{2p}

$\mathrm{C}_{2p+1\,v}$:   The same as D_{2p+1}

S_{4p}: $[\mathrm{E}_k^2]$ $= \mathrm{A} + \mathrm{E}_{2k}$ $k < p$

                                    $= \mathrm{A} + 2\mathrm{B}$ $k = p$

                                    $= \mathrm{A} + \mathrm{E}_{4p-2k}$ $k > p$

$$(k = 1, \ldots, 2p - 1)$$

$\mathrm{C}_{(2p+1)1}$: $[\mathrm{E}_{kg}^2] = [\mathrm{E}_{ku}^2]$ $= \mathrm{A}_g + \mathrm{E}_{2k,g}$ $k \le p/2$

$\quad = \mathrm{S}_{2(2p+1)}$ $= \mathrm{A}_g + \mathrm{E}_{2p+1-2k,g}$ $k > p/2$

$$(k = 1, \ldots, p)$$

C_{2ph}: $[\mathrm{E}_{kg}^2] = [\mathrm{E}_{ku}^2]$ $= \mathrm{A}_g + \mathrm{E}_{2k,g}$ $k < p/2$

                                    $= \mathrm{A}_g + 2\mathrm{B}_g$ $k = p/2$

                                    $= \mathrm{A}_g + \mathrm{E}_{2p-2k,g}$ $k > p/2$

$$(k = 1, \ldots, p - 1)$$

$\mathrm{C}_{2p+1\,h}$:   Same as $\mathrm{D}_{2p+1\,h}$

C_{2p}: $[\mathrm{E}_k^2]$ $= \mathrm{A} + \mathrm{E}_{2k}$ $k < p/2$

                                    $= \mathrm{A} + 2\mathrm{B}$ $k = p/2$

                                    $= \mathrm{A} + \mathrm{E}_{2p-2k}$ $k > p/2$

$$(k = 1, \ldots, p)$$

C_{2p+1}: $[\mathrm{E}_k^2]$ $= \mathrm{A} + \mathrm{E}_{2k}$ $k \le p/2$

                                    $= \mathrm{A} + \mathrm{E}_{2p+1-2k}$ $k > p/2$

$$(k = 1, \ldots, p)$$

$\mathrm{D}_{\infty h}$: $\begin{cases} [\mathrm{E}_{\kappa g}^2] = [\mathrm{E}_{\kappa u}^2] & = \mathrm{A}_{1g} + \mathrm{E}_{2\kappa,g} \\ [\Lambda_{\kappa g}^2] = [\Lambda_{\kappa u}^2] & = \Sigma_g^+ + \Lambda_{2\kappa,g} \end{cases}$ $(\kappa = 1, 2, \ldots)$

$\quad$ $(\Lambda_{1g} = \Pi_g, \Lambda_{2g} = \Delta_g,$

$\quad\quad\quad \Lambda_{3g} = \Phi_g,$ etc.$)$

$\mathrm{C}_{\infty v}$: $\begin{cases} [\mathrm{E}_\kappa^2] & = \mathrm{A}_1 + \mathrm{E}_{2\kappa} \\ [\Lambda_\kappa^2] & = \Sigma^+ + \Lambda_{2\kappa} \end{cases}$ $(\kappa = 1, 2, \ldots)$

The result stated above for D_{2ph} follows since the degenerate representations in the groups starting with D_{2ph} are two-fold. Now the representation matrices for the operations of the group (rotation and/or inversion) on a double-valued doubly-degenerate representation are two-dimensional unitary matrices of determinant 1. The most general form of such a matrix is

$$M_{\sigma\rho}(R) = \begin{pmatrix} a & b \\ -b^* & a^* \end{pmatrix}, \qquad |a|^2 + |b|^2 = 1 \qquad (\sigma, \rho = 1, 2)$$

The character of the antisymmetrical product for any group operation R is

$$\tfrac{1}{2}[(\sum_\sigma M_{\sigma\sigma}(R))^2 - \sum_\sigma M_{\sigma\sigma}(R^2)] = \tfrac{1}{2}[(\sum_\sigma M_{\sigma\sigma}(R)]^2 - \sum_{\sigma\sigma'} M_{\sigma\sigma'}(R)M_{\sigma'\sigma}(R)] =$$

$$= aa^* + bb^* = 1 \text{ for all } R$$

A I.1 References

1. H. A. Jahn and E. Teller, *Proc. roy. Soc. A*, **161**, 220 (1937).
2. G. Herzberg, *Molecular Spectra and Molecular Structure*, Vol. 3 (Princeton: Van Nostrand, 1966).

Appendix II

The vibrational modes of cubic systems

Symmetry properties of sets of atoms and reduction of their motion in the cubic groups (based on Reference 1):

Mulliken's notation is used for the irreducible representations and Koster et al.'s[2] for the symmetry elements. (Similar information on other point groups is found in Table I of Reference 1.)

Table A.II.1 shows the Schönflies symbol of each group, in the first column (I), then the species of the translational and rotational motions under (II) and (III). In column IV the various types of equivalent points are designated with a latin letter, alongside with the number of these points in each set, shown in column V. The sixth column VI shows on which symmetry elements are the points to be placed. When a letter is bracketed, this means that this type of points is sufficient to produce the group symmetry in question.

Otherwise, when not bracketed, the type will produce a graph of higher symmetry, which may be found by seeking out the group in which the same letter is bracketed. Thus six identical atoms f are compatible with, e.g., T_h when placed on the two-fold axes of rotation (as the six oxygens in an aquated complex $M \cdot 6H_2O$). However, by themselves they will not produce this group, but rather the higher symmetry group O_h where the letter f is found bracketed. A further set of 12 identical atoms j, say twelve hydrogens, when placed on the reflection planes σ_h will produce T_h. This is indicated by $\boxed{j}$ in column IV of the group, together with another combination $\boxed{i}$ which is also sufficient to yield the group T_h. The last column VII enumerates the species to which the modes of vibrations of each set belong. Before proving the assertion in Chapter 1 ('the Jahn–Teller theorem') that the reduced matrix element L is non-vanishing in each group for

I	II	III	IV	V	VI	VII
O_h	T_{1u}	T_{1g}	a	48	None	$3(A_{1g} + A_{1u} + A_{2g} + A_{2u}) + 6(E_g + E_u) + 9(T_{1g} + T_{1u} + T_{2g} + T_{2u})$
			b	24	σ_h	$2(A_{1g} + A_{2g}) + A_{1u} + A_{2u} + 4E_g + 2E_u + 4(T_{1g} + T_{2g}) + 5(T_{1u} + T_{2u})$
			c	24	σ_d	$2A_{1g} + A_{1u} + A_{2g} + 2A_{2u} + 3E_g + 3E_u + 4T_{1g} + 5T_{1u} + 5T_{2g} + 4T_{2u}$
			d	12	C_2', σ_h, σ_d	$A_{1g} + A_{2g} + A_{2u} + 2E_g + E_u + 2T_{1g} + 3T_{1u} + 2T_{2g} + 2T_{2u}$
			e	8	C_3, σ_d	$A_{1g} + A_{2u} + E_g + E_u + T_{1g} + 2T_{1u} + 2T_{2g} + T_{2u}$
			f	6	C_2, C_4, σ_h, σ_d	$A_{1g} + E_g + T_{1g} + 2T_{1u} + T_{2g} + T_{2u}$
			g	1	All	T_{1u}
O	T_1	T_1	h	24	None	$3A_1 + 3A_2 + 6E + 9T_1 + 9T_2$
			d	12	C_2'	$A_1 + 2A_2 + 3E + 5T_1 + 4T_2$
			e	8	C_2	$A_1 + A_2 + 2E + 3T_1 + 3T_2$
			f	6	C_2, C_4	$A_1 + E + 3T_1 + 2T_2$
			g	1	All	T_1
T_h	T_u	T_g	i	24	None	$3A_g + 3A_u + 3E_g + 3E_u + 9T_g + 9T_u$
			j	12	σ_h	$2A_g + A_u + 2E_g + E_u + 4T_g + 5T_u$
			e	8	C_3	$A_g + A_u + E_g + E_u + 3T_g + 3T_u$
			f	6	C_2, σ_h	$A_g + E_g + 2T_g + 3T_u$
			g	1	All	T_u
T_d	T_2	T_1	k	24	None	$3A_1 + 3A_2 + 6E + 9T_1 + 9T_2$
			l	12	σ_d	$2A_1 + A_2 + 3E + 4T_1 + 5T_2$
			f	6	C_2, σ_d	$A_1 + E + 2T_1 + 3T_2$
			m	4	C_3, σ_d	$A_1 + E + T_1 + 2T_2$
			g	1	All	T_2
T	T	T	n	12	None	$3A + 3E + 9T$
			f	6	C_2	$A + E + 5T$
			m	4	C_3	$A + E + 3T$
			g	1	All	T

each degenerate representation and for at least one non-totally symmetric mode, the rotational and translational modes columns II and III must be subtracted off. This subtraction has not been carried out in the Table, since more than one set of equivalent atoms may occur in any given molecule, while the subtraction has to be carried out only once. In addition to proving the J-T theorem by enumeration for the cubic groups the Table also shows the normal modes of the molecules. Furthermore, since one and the same letter denotes any particular set of equivalent points throughout, the role of each set in different groups can be traced easily.

A proof[3] of the Jahn–Teller theorem which is not enumerative is based on the physical idea that in degenerate electronic states the charge clouds can be chosen to be non-symmetric. Therefore, a non-symmetric electric field can be shown to act on the nuclei of the molecules. Some of the nuclei will then move one way and others in a different way. The proof utilizes a theorem in the induction of representations of finite groups and is not elementary.

A II.1 References

1. H. A. Jahn and E. Teller, *Proc. roy. Soc. A*, **161**, 220 (1937).
2. G. F. Koster, J. O. Dimmock, R. G, Wheeler and H. Statz, *Properties of the 32 Points Groups* (Cambridge, Mass.: M.I.T. Press, 1963).
3. E. Ruch and A. Schönhofer, *Theoret. Chim. Acta*, **3**, 291 (1965).

Appendix III
Derivation of the Jahn–Teller theorem for odd-fermion systems

In Chapter 1 it was claimed that from the point of view of symmetry there is *always*, for *every* non-linear molecule in *any* degenerate electronic state, a linear vibronic coupling term. This claim was expressed by saying that the reduced matrix element L, defined in equation (1.3), does not vanish. In an equivalent way, the non-vanishing of the integrals

$$\int \phi_\rho^{\Gamma'*} H_\lambda^\Lambda \phi_\sigma^\Gamma \, dv \equiv V_{\lambda\sigma\rho} \qquad \text{(A III.1)}$$

introduced in equation (7.10), with $\Gamma = \Gamma'$, is involved. For real functions, when $\phi_\rho^* = \phi_\rho$ this result is conditional upon the well-known group theoretical criterion that the representation Λ of the vibrational coordinate q_λ^Λ be included in the symmetric product $[\Gamma^2]$ of the Γ representation. For cubic groups the information contained in Appendix I and II suffices to prove that for the single-valued representations this condition is met. In other groups Table I of Jahn and Teller's paper is needed. However, not every wave-function or set of wave-functions is real, nor can it always be reduced to a set of real functions by a unitary transformation, and we must reinvestigate the question of the non-vanishing of the above integral for those cases which occur in systems with an odd number of fermions.

The method used in this Appendix utilizes the formalism of tensorial sets[1] and will differ in form, but not in content, from Jahn's original proof[2] for his theorem that *for a double valued representation Γ the anti-symmetric product $\{\Gamma^2\}$ must contain Λ* if the integral is not to vanish identically.

Since the integral is independent of time, it remains invariant under the operation of time-reversal. The time-reversal operator K is defined as the product K_0U of the complex-conjugation operator (K_0) and a unitary matrix U which is such that the representation matrices $M^{\Gamma}_{\gamma'\gamma}(R) \equiv M(R)$ are transformed into their complex conjugates (denoted by a star *) in the manner of

$$UM(R)U^{-1} = M^*(R) \qquad\qquad \text{(A III.1)}$$

[It is necessary to add at this stage that according to the convention of Reference 1, the coefficients a^{Γ}_{γ} in the wave-function of equation (7.3) transform as the components of a standard set and the basic state-vectors ϕ^{Γ}_{γ} transform contrastandardly. Further, $\phi^{\Gamma}_{\gamma}*$ are standard, $a^{\Gamma}_{\gamma}*$ are contrastandard and the state-function ϕ^{Γ} of equation (7.3) is an invariant. The transformation properties of the sets under a group operation R are related to the representation matrices as follows.

$$R\phi^{\Gamma}_{\gamma} = \sum_{\gamma'}\phi^{\Gamma}_{\gamma'}M^{\Gamma}_{\gamma'\gamma}(R).$$

$$Ra^{\Gamma}_{\gamma} = \sum_{\gamma'}M^{\Gamma}_{\gamma\gamma'}(R^{-1})a^{\Gamma}_{\gamma'}.$$

Similarly, since the time-reversal operation is defined for a standard set

$$Ka = K_0Ua$$

then for a contrastandard set the same operation takes the form

$$K\phi = K_0(\phi U^{-1}) = K_0U^*\phi = K_0(K_0UK^{-1})\phi = UK_0^{-1}\phi = UK_0\phi$$

We see now why Wigner[3] has U^* in the definition of K.]

Because of the reality and time-invariance of H_{λ}

$$V_{\lambda\sigma\rho} = KV_{\lambda\sigma\rho} = [\int(U^*\phi)_{\rho}H_{\lambda}(U^*\phi)_{\sigma}\,d\mathbf{r}]^* = \int(U^*\phi)^*_{\sigma}H_{\lambda}(U^*\phi)_{\rho}\,d\mathbf{r}$$

$$= \tfrac{1}{2}\int[\phi^*_{\rho}H_{\lambda}\phi_{\sigma} + U^*_{\sigma\gamma}\phi_{\gamma}H_{\lambda}U_{\rho\nu}\phi_{\nu}]\,d\mathbf{r} \qquad \text{(A III.2)}$$

(summation over repeated suffixes is implied in this and the following formulae. Also, we have suppressed the representation symbols Γ and Λ.)

We have to consider now the matrix U. It is shown in Reference 1 that, for the $2J + 1$ dimensional representation $\Gamma_{(2J+1)}$ of the rotation (and inversion) group, $U_{(2J+1)}$ is unitary and symmetric for $2J + 1 = $ odd and is unitary and antisymmetric for $2J + 1 = $ even. Now the $2J + 1$ dimensional representation is reducible in the molecular symmetry group (M.S.G.). Let us suppose that it contains, for a particular J the representation Γ which is spanned by our degenerate set. A basis of the representation Γ may be derived from the basis for $\Gamma_{(2J+1)}$ in the following way. Denote the basis for the representation $\Gamma_{(2J+1)}$ by $u^{(J)}_{M}$ and apply to this a unitary substitution $Au^{(J)}_{M}$, such that the transformed rotation and inversion matrix $AD^{J}_{MM'}A^{-1}$ will consist of blocks which do not mix under the operations of the M.S.G. One of these blocks will be the representation Γ. The matrix $U'_{(2J+1)}$ appropriate to the new rotation matrix will be $U'_{(2J+1)} = A^* U_{(2J+1)} A$ and this will again be symmetric or anti-symmetric as before, since

$$U'^*_{(2J+1)} U'_{(2J+1)} = A U^*_{(2J+1)} A^{-1} A^* U_{(2J+1)} A^{-1} = A U^*_{(2J+1)} U_{(2J+1)} A^{-1} =$$
$$= (-1)^{2J} A A^{-1} = (-1)^{2J},$$

as before the transformation.

Coming now to the matrix U, it is clear from the reality and time-invariance of the Hamiltonian, that it will mix only states with equal energy. (It may mix states of different and non-equivalent representations if these have a symmetry-inherent degeneracy). Also, since U is part of $U_{(2J+1)}$ it will also be unitary and symmetric or antisymmetric depending on whether it arises from $2J + 1$ odd or even.

Let us use these facts to derive the character (Reference 3) of the integrand in equation (AIII.2). We have to apply to this integrand the operation R^{-1} of the M.S.G. We consider the transformation properties of the factor H_λ and the remaining factors in the integrand separately. Noting that $M_{\alpha\beta}(R^{-1}) = M^{-1}_{\alpha\beta}(R) = M^*_{\beta\alpha}(R)$ the transformation properties of the factor H_λ are given by

$$\chi^\Lambda(R)^* \qquad\qquad (\text{A III.3})$$

where $\chi^\Lambda(R)$ are the characters, i.e. the traces of the transformation matrices of the irreducible representation Λ.

To derive the character of the remaining part we have to manipulate the transformation matrices as follows:

$$\tfrac{1}{2}[M_{\rho\nu}(R) M^*_{\sigma\mu}(R) + U_{\sigma\alpha} M_{\alpha\nu}(R) U^*_{\rho\tau} M^*_{\tau\mu}(R)] \phi^*_\nu H_\lambda \phi_\mu$$
$$= \tfrac{1}{2}[M_{\rho\nu} U_{\sigma\alpha} M_{\alpha\beta} U^{-1}_{\beta\mu} + U_{\sigma\alpha} M_{\alpha\nu} U^*_{\rho\tau} U_{\tau\gamma} M_{\gamma\beta} U^{-1}_{\beta\mu}] \phi^*_\nu H_\lambda \phi_\mu$$
$$= \tfrac{1}{2}[U_{\sigma\alpha} U^{-1}_{\beta\mu}]\{M_{\rho\nu} M_{\alpha\beta} + [(-1)^{-2J} U^{-1}_{\rho\tau} U_{\tau\gamma}] M_{\alpha\nu} M_{\gamma\beta}\} \phi^*_\nu H_\lambda \phi_\mu$$
$$= \tfrac{1}{2}[U_{\sigma\alpha} U^{-1}_{\beta\mu}][M_{\rho\nu} M_{\alpha\beta} + (-1)^{2J} M_{\alpha\nu} M_{\rho\beta}] \phi^*_\nu H_\lambda \phi_\mu$$

Comparing this expression with the expression for $V_{\lambda\sigma\rho}$, we obtain the characters for this factor in the integrand by putting $\nu = \rho$, $\mu = \sigma$ and summing over ρ and σ. We find

$$\tfrac{1}{2}[\chi^{\Gamma^2}(R) + (-1)^{2J} \chi^\Gamma(R^2)]$$

If the integral is not to vanish, these characters must not be orthogonal to the characters, equation (A III.3), of the normal modes. Noting that J is integer for an even number of electrons but is half integer for an odd number of electrons, we find that the integral will not in general vanish, if Λ is contained in the symmetric product $[\Gamma^2]$ for an even number of electrons, or in the antisymmetric product $\{\Gamma^2\}$ for an odd number of electrons.

To obtain physically meaningful results for an odd-fermion system in which the spin–orbit coupling is weak or negligible, one must do two things. First, one must examine direct products like $\Gamma\Gamma'$ for vibronic coupling between the states having representations Γ and Γ, which are split by spin–orbit coupling. Speaking purely formally, a P-J-T-E is involved here. Then, vibronic couplings which are formally allowed but in fact rely on matrix elements between different spin-states must be discounted, since these are weak. In practice this procedure is inconvenient and it is better to restrict oneself to the orbital part of the wave-function alone. This belongs to a single-valued representation, even for an odd fermion system and the symmetric product has to be examined.

More information about the effects of spin-degeneracy, in particular on the infrared spectra of molecules, can be found in Reference 4.

A III.1 References

1. U. Fano and G. Racah, *Irreducible Tensorial Sets* (New York: Academic, 1959).
2. H. A. Jahn, *Proc. roy. Soc. A*, **164**, 117 (1938).
3. E. P. Wigner, *Group Theory and its Applications to the Quantum Mechanics of Atomic Spectra* (New York: Academic, 1959).
4. M. S. Child, *Phil. Trans. roy. Soc. A*, **255**, 31 (1962).

Appendix IV
Classification of higher order couplings

A IV.1 Classification of higher order couplings

The purposes of this Appendix are to introduce coupling terms second and third order in the vibrational coordinates in a systematic way and to define the coefficients of these terms in a consistent, non-arbitrary manner. It will be seen that the arbitrariness is eliminated by agreeing on a certain formal similarity between the definition of the coefficient and the form of the corresponding coupling term. The method developed here[1] is based on the theory of tensorial sets.[2]

A IV.2 Derivation of higher order coupling terms

In order to facilitate algebraic manipulations it is preferable to go over temporarily from the matrix notation of the J-T Hamiltonian to an equivalent Hamiltonian operator. For the linear term, equation (7.8), this operator is written as

$$H(\Lambda) = L(\Lambda) \sum_{\alpha\delta\varepsilon} V \begin{pmatrix} \Lambda & \Gamma & \Gamma \\ \alpha & \delta & \varepsilon \end{pmatrix} q_\alpha^\Lambda |\delta\rangle\langle\varepsilon| \qquad \text{(A IV.1)}$$

and is seen to be an invariant under the operations of the group.

Here Λ and Γ are the irreducible representations of the vibrations and of the degenerate electronic states, respectively. q_α is the dimensionless vibrational coordinate, defined in terms of the normal coordinate Q_α by

$$q_\alpha = \sqrt{\left(\frac{M\omega}{\hbar}\right)} Q_\alpha$$

The coefficients L have the dimensions of energy. They are given by

$$L(\Lambda)V\begin{pmatrix} \Lambda & \Gamma & \Gamma \\ \alpha & \delta & \varepsilon \end{pmatrix} = \langle\delta\left|\frac{\partial H}{\partial q_\alpha}\right|\varepsilon\rangle$$

For the sake of simplicity the formal theory in the present section is given for real sets belonging to single-valued representations. Then the U-matrix (introduced in the previous Appendix) which relates standard and contrastandard sets is simply $U_{v,\rho}^{(\Gamma)} = \delta_{v,\rho}$.

Let us now build up that part in the Hamiltonian which is quadratic in the vibrational coordinates and operates within the electronic manifold. This will consist of a sum of invariants $I(X)$ and their constant coefficients $K(X)$, in the form

$$H(\text{quadratic}) = \sum_X K(X)I(X) \tag{A IV.2}$$

The symbol X, which classifies the invariants, denotes the irreducible representation which is spanned by the vibrational part in the expression for $I(X)$, namely

$$I(X) = \sum_{\alpha\beta\delta\varepsilon} B_{\alpha\beta\delta\varepsilon}q_\alpha q_\beta|\delta\rangle\langle\varepsilon| \tag{A IV.3}$$

That this is a meaningful notation and a sensible classification may be seen by exhibiting the vibrational part of $I(X)$ explicitly, i.e. by rewriting $I(X)$ as

$$I(X) = \sum_{x\delta\varepsilon} V\begin{pmatrix} X & \Gamma & \Gamma \\ x & \delta & \varepsilon \end{pmatrix} \chi\begin{pmatrix} X \\ x \end{pmatrix}|\delta\rangle\langle\varepsilon| \tag{A IV.4}$$

where

$$\chi\begin{pmatrix} X \\ x \end{pmatrix} = \sum_{\alpha\beta} V\begin{pmatrix} \Lambda & \Lambda & X \\ \alpha & \beta & x \end{pmatrix} q_\alpha q_\beta \tag{A IV.5}$$

forms a basis for the x-component of the X-representation. Because q_α and q_β are chosen from one and the same set, the coefficient V must be symmetric under interchange of the indices, α and β. Following the rule of Reference 3 (pp. 13–15) for odd permutations of the columns of V, we see that this requirement amounts to the equation:

$$1 = (-1)^{\Lambda+\Lambda+X} = (-1)^X$$

Thus for the group O, X may be A_1, E, T_2 but not A_2 or T_1.

We can accordingly replace the coefficients B in equation (A IV.3) by the more explicitly characterized coefficients.

$$k\begin{pmatrix} \Lambda & \Gamma \\ \alpha\beta & \delta\varepsilon \end{pmatrix} X = \sum_x V\begin{pmatrix} \Lambda & \Lambda & X \\ \alpha & \beta & x \end{pmatrix} V\begin{pmatrix} X & \Gamma & \Gamma \\ x & \delta & \varepsilon \end{pmatrix} \tag{A IV.6}$$

These coefficients have some orthogonality properties, which are useful for extracting explicit formulae for the coefficients $K(X)$ in terms of the matrix elements of the potential. To derive these, one equates the matrix elements of the quadratic Hamiltonian with the corresponding quantities in the Maclaurin series of the potential.

$$\sum_X K(X) \sum_{\alpha\beta} k \begin{pmatrix} \Lambda & \vdots & \Gamma \\ \alpha\beta & \vdots & \delta\varepsilon \end{pmatrix} X \Big) q_\alpha q_\beta = \tfrac{1}{2} \sum_{\alpha\beta} \langle \delta \Big| \frac{\partial^2 H}{\partial q_\alpha \partial q_\beta} \Big| \varepsilon \rangle q_\alpha q_\beta$$

Because q_α are independent normal coordinates one can further write

$$\sum_X K(X) k \begin{pmatrix} \Lambda & \vdots & \Gamma \\ \alpha\beta & \vdots & \delta\varepsilon \end{pmatrix} X \Big) = \tfrac{1}{2} \langle \delta \Big| \frac{\partial^2 H}{\partial q_\alpha \partial q_\beta} \Big| \varepsilon \rangle$$

The orthogonality relations already referred to are

$$\sum_{\alpha\beta\delta\varepsilon} k \begin{pmatrix} \Lambda & \vdots & \Gamma \\ \alpha\beta & \vdots & \delta\varepsilon \end{pmatrix} X \Big) k \begin{pmatrix} \Lambda & \vdots & \Gamma \\ \alpha\beta & \vdots & \delta\varepsilon \end{pmatrix} X' \Big) =$$

$$= \sum_{\alpha\beta\delta\varepsilon} \sum_{xx'} V \begin{pmatrix} \Lambda & \Lambda & X \\ \alpha & \beta & x \end{pmatrix} V \begin{pmatrix} \Lambda & \Lambda & X' \\ \alpha & \beta & x' \end{pmatrix} V \begin{pmatrix} X & \Gamma & \Gamma \\ x & \delta & \varepsilon \end{pmatrix} V \begin{pmatrix} X' & \Gamma & \Gamma \\ x' & \delta & \varepsilon \end{pmatrix}$$

$$= \sum_{xx'} [X]^{-1} \delta_{XX'} \delta_{xx'} [X]^{-1} \delta_{XX'} \delta_{xx'}$$

$$= \sum_x [X]^{-2} \delta_{XX'} = [X]^{-1} \delta_{XX'}$$

Here $[X]$ denotes the dimension of the representation X.

One finds therefore readily

$$K(X) = \frac{[X]}{2!} \sum_{\alpha\beta\delta\varepsilon} k \begin{pmatrix} \Lambda & \vdots & \Gamma \\ \alpha\beta & \vdots & \delta\varepsilon \end{pmatrix} X \Big) \langle \delta \Big| \frac{\partial^2}{\partial q_\alpha \partial q_\beta} \Big| \varepsilon \rangle \qquad \text{(A IV.7)}$$

On the other hand, the invariant $I(X)$ (equation (A IV.4)) appearing in the Hamiltonian equation (A IV.2), can be rewritten with the aid of equation (A IV.5) to (A IV.6) in the form

$$I(X) = \sum_{\alpha\beta\delta\varepsilon} k \begin{pmatrix} \Lambda & \vdots & \Gamma \\ \alpha\beta & \vdots & \delta\varepsilon \end{pmatrix} X \Big) q_\alpha q_\beta |\delta\rangle\langle\varepsilon| \qquad \text{(A IV.8)}$$

The analogy between equation (A IV.7) and (A IV.8) is evident and is in fact the basis of our choice for the quadratic coupling constants $K(X)$, as in §3.2.1. A list of the invariants $I(X)$ and of the coefficients $K(X)$ for various groups and representations is given in the next section, [A IV.3].

When we continue, in the same manner, to construct invariants cubic in the vibrational coordinates, i.e.

$$I(Y, n) = \sum_{\alpha\beta\gamma\delta\varepsilon} B_{\alpha\beta\gamma\delta\varepsilon} q_\alpha q_\beta q_\gamma |\delta\rangle\langle\varepsilon| \qquad \text{(A IV.9)}$$

we encounter two difficulties which are indicated in the notation $I(Y, n)$. Here, as in the quadratic case, the symbol Y denotes the irreducible representation which is spanned by the vibrational coordinate-part. A further symbol n ($=1, 2, \ldots$) is, however, necessary since now more than one invariant with the same Y may be present. Yet another difficulty is that, while it is easy to make B symmetric in any two of the three indices $\alpha\beta\gamma$, a special procedure is needed to ensure that it is simultaneously symmetric in the three indices. (By 'symmetric in the indices' we mean that, e.g. $B_{124\delta\varepsilon} = B_{421\delta\varepsilon}$.) In fact the problem is analogous to that of forming wave-functions antisymmetric in the coordinates of three electrons and the method of solution is also related.[4]

We now enlarge somewhat the meaning and notation of the function χ introduced in equation (A IV.4) by pretending for the present that the vibrational coordinates arise from three non-equivalent sets, $q_\alpha^{(1)}, q_\alpha^{(2)}, q_\alpha^{(3)}$.

$$\chi_{12}\begin{pmatrix} X \\ x \end{pmatrix} = \sum_{\alpha\beta} q_\alpha^{(1)} q_\beta^{(2)} V \begin{pmatrix} \Lambda & \Lambda & X \\ \alpha & \beta & x \end{pmatrix}$$

This function will be symmetric with respect to the permutations of 1 and 2, if X is chosen so as to be contained in the symmetric product of Λ. The next function

$$\chi_{12,3}\begin{pmatrix} Y \\ y \end{pmatrix}, X \end{pmatrix} = \sum_{x\gamma} V \begin{pmatrix} X & \Lambda & Y \\ x & \gamma & y \end{pmatrix} \chi_{12}\begin{pmatrix} X \\ x \end{pmatrix} q_\gamma^{(3)}$$

$$= \sum_{\alpha\beta\gamma x} q_\alpha^{(1)} q_\beta^{(2)} q_\gamma^{(3)} V \begin{pmatrix} \Lambda & \Lambda & X \\ \alpha & \beta & x \end{pmatrix} V \begin{pmatrix} X & \Lambda & Y \\ x & \gamma & y \end{pmatrix}$$

(A IV.10)

will be symmetric with respect to permutations of 1–2 but not with respect to 1–3 or 2–3. We wish to correct this by seeking coefficients $a(X, n)$, functions of the representations only, such that

$$F_{12,3}\begin{pmatrix} Y \\ y \end{pmatrix}, n \end{pmatrix} = \sum_X a(X, n) \chi_{12,3}\begin{pmatrix} Y \\ y \end{pmatrix}, X \end{pmatrix}$$

(A IV.11)

possesses full permutational symmetry. To determine the coefficients $a(X, n)$ we introduce three functions which are ostensibly symmetric with respect to 2–3.

$$\chi_{23}\begin{pmatrix} X' \\ x' \end{pmatrix} = \sum_{\beta\gamma} q_\beta^{(2)} q_\gamma^{(3)} V \begin{pmatrix} \Lambda & \Lambda & X' \\ \beta & \gamma & x' \end{pmatrix}$$

$$\chi_{1,23}\begin{pmatrix} Y \\ y \end{pmatrix}, X' \end{pmatrix} = \sum_{\alpha'x} \chi_{23}\begin{pmatrix} X' \\ x' \end{pmatrix} q_\alpha^{(1)} V \begin{pmatrix} X' & \Lambda & Y \\ x' & \alpha & y \end{pmatrix}$$

$$= \sum_{\alpha\beta\gamma x'} V \begin{pmatrix} \Lambda & \Lambda & X' \\ \beta & \gamma & x' \end{pmatrix} V \begin{pmatrix} X' & \Lambda & Y \\ x' & \alpha & y \end{pmatrix} q_\alpha^{(1)} q_\beta^{(2)} q_\gamma^{(3)}$$

and

$$F_{1,23}\begin{pmatrix} Y \\ y \end{pmatrix}, n\end{pmatrix} = \sum_{X'} a(X', n)\chi_{1,23}\begin{pmatrix} Y \\ y \end{pmatrix}, X'\end{pmatrix} \qquad \text{(A IV.12)}$$

We now require that

$$F_{1,23}\begin{pmatrix} Y \\ y \end{pmatrix}, n\end{pmatrix} = F_{12,3}\begin{pmatrix} Y \\ y \end{pmatrix}, n\end{pmatrix}$$

Since the vibrational coordinates were supposed to originate from three independent sets, the above equality implies the equality of all the corresponding coefficients in the two expressions, equation (A IV.11) and (A IV.12), i.e.

$$\sum_X a(X, n)\sum_x V\begin{pmatrix} \Lambda & \Lambda & X \\ \alpha & \beta & x \end{pmatrix} V\begin{pmatrix} X & \Lambda & Y \\ x & \gamma & y \end{pmatrix}$$

$$= \sum_{X'} a(X', n)\sum_{x'} V\begin{pmatrix} \Lambda & \Lambda & X' \\ \beta & \gamma & x' \end{pmatrix} V\begin{pmatrix} X' & \Lambda & Y \\ x' & \alpha & y \end{pmatrix}$$

Multiplication by $V\begin{pmatrix} \Lambda & \Lambda & X_0 \\ \alpha & \beta & x_0 \end{pmatrix}$ and summation over α, β yields for the two sides of the previous equation

$$\sum_X a(X, n)\sum_x V\begin{pmatrix} X & \Lambda & Y \\ x & \gamma & y \end{pmatrix} [X_0]^{-1}\delta_{XX_0}\delta_{xx_0}$$

$$= a(X_0, n)V\begin{pmatrix} X_0 & \Lambda & Y \\ x_0 & \gamma & y \end{pmatrix} [X_0]^{-1}$$

$$= \sum_{X'} a(X', n) \sum_{\alpha\beta x'} V\begin{pmatrix} \Lambda & \Lambda & X_0 \\ \alpha & \beta & x_0 \end{pmatrix} V\begin{pmatrix} \Lambda & \Lambda & X' \\ \beta & \gamma & x' \end{pmatrix} V\begin{pmatrix} X' & \Lambda & Y \\ x' & \alpha & y \end{pmatrix}$$

A further multiplication by $V\begin{pmatrix} X_1 & \Lambda & Y \\ x_1 & \gamma & y \end{pmatrix}$ and summation over γ and y yield the following equation for $a(X_0, n)$.

$$[X_0]^{-2}\delta_{X_0X_1}\delta_{x_0x_1}a(X_0, n)$$

$$= \sum_{X'}\left\{ \sum_{\alpha\beta\gamma yx'} V\begin{pmatrix} X_1 & \Lambda & Y \\ x_1 & \gamma & y \end{pmatrix} V\begin{pmatrix} \Lambda & \Lambda & X_0 \\ \alpha & \beta & x_0 \end{pmatrix} V\begin{pmatrix} \Lambda & \Lambda & X' \\ \beta & \gamma & x' \end{pmatrix} \right.$$

$$\left. \times V\begin{pmatrix} X' & \Lambda & Y \\ x' & \alpha & y \end{pmatrix} \right\} a(X', n) \qquad \text{(A IV.13)}$$

$$= [X_0]^{-1}\delta_{X_0X_1}\delta_{x_0x_1}\sum_{X'} W\begin{pmatrix} \Lambda & \Lambda & X_1 \\ \Lambda & Y & X' \end{pmatrix} a(X', n)$$

where the curly brackets in equation (A IV.13) have been rewritten as the recoupling coefficient W by use of the formula (Reference 3, p. 36)

$$\sum_{\alpha\beta\delta\varepsilon\phi} V\begin{pmatrix} a & b & c \\ \alpha & \beta & \gamma \end{pmatrix} V\begin{pmatrix} a & e & f \\ \alpha & \varepsilon & \phi \end{pmatrix} V\begin{pmatrix} b & f & d \\ \beta & \phi & \delta \end{pmatrix} V\begin{pmatrix} c' & d & e \\ \gamma & \delta & \varepsilon \end{pmatrix}$$

$$= [c]^{-1}\delta_{cc'}\delta_{\alpha\alpha'} W\begin{pmatrix} a & b & c \\ d & e & f \end{pmatrix}$$

In conclusion, from equation (A IV.13) we are led to determine the orthonormal eigenvectors $[X]^{-1/2}a(X, n)$ of the symmetric matrix

$$W_{XX'} = [X']^{1/2} W\begin{pmatrix} \Lambda & \Lambda & X \\ \Lambda & Y & X' \end{pmatrix} [X]^{1/2}$$

which have the eigenvalue $+1$. There may be (for any Y) zero, one or more independent eigenvectors a.

Dropping the superscripts (1), (2) and (3) from q the third order invariant equation (A IV.9) may be rewritten

$$I(Y, n) = \sum_{\alpha\beta\gamma\delta\varepsilon} h\left(\begin{matrix} \Lambda & \vdots & \Gamma \\ \alpha\beta\gamma & \vdots & \delta\varepsilon \end{matrix} \middle| Y, n\right) q_\alpha q_\beta q_\gamma |\delta\rangle\langle\varepsilon| \qquad \text{(A IV.14)}$$

where $h = \sum_{X,x,y} a(X, n)V\begin{pmatrix} \Lambda & \Lambda & X \\ \alpha & \beta & x \end{pmatrix} V\begin{pmatrix} X & \Lambda & Y \\ x & \gamma & y \end{pmatrix} V\begin{pmatrix} Y & \Gamma & \Gamma \\ y & \delta & \varepsilon \end{pmatrix}.$

The orthogonality properties of h are derived in the same way as those for k. One finds that

$$\sum_{\alpha\beta\gamma\delta\varepsilon} h\left(\begin{matrix} \Lambda & \vdots & \Gamma \\ \alpha\beta\gamma & \vdots & \delta\varepsilon \end{matrix} \middle| Y, n\right) h\left(\begin{matrix} \Lambda & \vdots & \Gamma \\ \alpha\beta\gamma & \vdots & \delta\varepsilon \end{matrix} \middle| Y', n'\right) = [Y]^{-1}\delta_{YY'}\delta_{nn'}$$

Comparing now the two forms of the cubic Hamiltonian,

$$H(\text{cubic}) = \sum_{Y,n} N(Y, n)I(Y, n)$$

and

$$H(\text{cubic}) = \frac{1}{3!} \sum_{\alpha\beta\gamma} \frac{\partial^3 H}{\partial q_\alpha \partial q_\beta \partial q_\gamma} q_\alpha q_\beta q_\gamma$$

one finds, upon application of the orthogonality properties of h, that

$$N(Y, n) = \frac{[Y]}{3!} \sum_{\alpha\beta\gamma\delta\varepsilon} h\left(\begin{matrix} \Lambda & \vdots & \Gamma \\ \alpha\beta\gamma & \vdots & \delta\varepsilon \end{matrix} \middle| Y, n\right) \langle\delta| \frac{\partial^3 H}{\partial q_\alpha \partial q_\beta \partial q_\gamma} \middle| \varepsilon\rangle$$

$$\text{(A IV.15)}$$

The likeness between equation (A IV.14) and (A IV.15) is again evident.

A IV.3 List of quadratic and cubic invariants for the group O

The invariants will be exhibited in matrix form. A matrix element (δ, ε) is obtained from equation (A IV.8) or (A IV.14) by premultiplying by $\langle\delta|$ and postmultiplying by $|\varepsilon\rangle$.

(The unit matrix in each electronic manifold Γ will be denoted by $\mathbf{I}$.)

(A) *Quadratic invariants*

(1) $\Gamma = E \qquad \Lambda = E$

$$I(A_1) = \tfrac{1}{2}(q_\theta^2 + q_\epsilon^2)\mathbf{I}$$

$$I(E) = \tfrac{1}{4} \begin{pmatrix} q_\theta^2 - q_\epsilon^2 & 2q_\theta q_\epsilon \\ 2q_\theta q_\epsilon & q_\epsilon^2 - q_\theta^2 \end{pmatrix} \begin{matrix} :|E_\theta\rangle \\ :|E_\epsilon\rangle \end{matrix}$$

(2) $\Gamma = T_2 \text{ (or } T_1) \qquad \Lambda = E$

$$I(A_1) = \frac{1}{\sqrt{6}}(q_\theta^2 + q_\epsilon^2)\mathbf{I}$$

$$I(E) = \frac{1}{4\sqrt{3}} \begin{pmatrix} \begin{matrix} -(q_\theta^2 - q_\epsilon^2) \\ -2\sqrt{3}q_\theta q_\epsilon \end{matrix} & 0 & 0 \\ 0 & \begin{matrix} -(q_\theta^2 - q_\epsilon^2) \\ +2\sqrt{3}q_\theta q_\epsilon \end{matrix} & 0 \\ 0 & 0 & 2(q_\theta^2 - q_\epsilon^2) \end{pmatrix} \begin{matrix} :|\xi\rangle \\ \\ :|\eta\rangle \\ \\ :|\zeta\rangle \end{matrix}$$

(3) $\Gamma = T_2 \text{ (or } T_1) \qquad \Lambda = T_2 \text{ (or } T_1)$

$$I(A_1) = \tfrac{1}{3}(q_\xi^2 + q_\eta^2 + q_\zeta^2)\mathbf{I}$$

$$I(E) = \tfrac{1}{6} \begin{pmatrix} 2q_\xi^2 - q_\eta^2 - q_\zeta^2 & 0 & 0 \\ 0 & 2q_\eta^2 - q_\zeta^2 - q_\xi^2 & 0 \\ 0 & 0 & 2q_\zeta^2 - q_\xi^2 - q_\eta^2 \end{pmatrix} \begin{matrix} :|\xi\rangle \\ :|\eta\rangle \\ :|\zeta\rangle \end{matrix}$$

$$I(T_2) = \tfrac{1}{3} \begin{pmatrix} 0 & q_\xi q_\eta & q_\xi q_\zeta \\ q_\xi q_\eta & 0 & q_\eta q_\zeta \\ q_\xi q_\zeta & q_\eta q_\zeta & 0 \end{pmatrix} \begin{matrix} :|\xi\rangle \\ :|\eta\rangle \\ :|\zeta\rangle \end{matrix}$$

(4) $\Gamma = G_{3/2}$　　$\Lambda = E$

$$I(A_1) = \frac{1}{2\sqrt{2}} (q_\theta^2 + q_\epsilon^2)\mathbf{I}$$

For the representation $G_{\frac{3}{2}}$ the basic states of Reference 5 are used.

$$I(E) = \frac{1}{4\sqrt{2}} \begin{pmatrix} -(q_\theta^2 - q_\epsilon^2) & 0 & 4\sqrt{2}q_\theta q_\epsilon & 0 \\ 0 & q_\theta^2 - q_\epsilon^2 & 0 & 4\sqrt{2}q_\theta q_\epsilon \\ 4\sqrt{2}q_\theta q_\epsilon & 0 & q_\theta^2 - q_\epsilon^2 & 0 \\ 0 & 4\sqrt{2}q_\theta q_\epsilon & 0 & -(q_\theta^2 - q_\epsilon^2) \end{pmatrix} \begin{matrix} :|-\frac{3}{2}\rangle \\ :|-\frac{1}{2}\rangle \\ :|\frac{1}{2}\rangle \\ :|\frac{3}{2}\rangle \end{matrix}$$

(5) $\Gamma = G_{\frac{3}{2}}$　　$\Lambda = T_2$ (or T_1)

$$I(A_1) = \frac{1}{2\sqrt{3}}(q_\xi^2 + q_\eta^2 + q_\zeta^2)\mathbf{I}$$

$$I(E) = -\frac{1}{4\sqrt{6}} \begin{pmatrix} -(2q_\zeta^2 - q_\xi^2 - q_\eta^2) & 0 & \sqrt{3}(q_\eta^2 - q_\xi^2) & 0 \\ 0 & (2q_\zeta^2 - q_\xi^2 - q_\eta^2) & 0 & \sqrt{3}(q_\eta^2 - q_\xi^2) \\ \sqrt{3}(q_\eta^2 - q_\xi^2) & 0 & (2q_\zeta^2 - q_\xi^2 - q_\eta^2) & 0 \\ 0 & \sqrt{3}(q_\eta^2 - q_\xi^2) & 0 & -(2q_\zeta^2 - q_\xi^2 - q_\eta^2) \end{pmatrix} \begin{matrix} :|-\frac{3}{2}\rangle \\ :|-\frac{1}{2}\rangle \\ :|\frac{1}{2}\rangle \\ :|\frac{3}{2}\rangle \end{matrix}$$

For $X = T_2$ two linearly independent invariants exist which we label by the indices a and b. $I_a(T_2)$ is hermitian and is shown below. The other invariant $I_b(T_2)$ is anti-hermitian and will be associated with that part of the Hamiltonian which changes sign under time conjugation. Such a Hamiltonian is rather rare and the invariant $I_b(T_2)$ will not be given here.

$$I_a(T_2) = (3\sqrt{2})^{-1} \times \begin{pmatrix} 0 & -q_\zeta(q_\xi - iq_\eta) & -iq_\xi q_\eta & 0 \\ -q_\zeta(q_\xi + iq_\eta) & 0 & 0 & -iq_\xi q_\eta \\ iq_\xi q_\eta & 0 & 0 & q_\zeta(q_\xi - iq_\eta) \\ 0 & iq_\xi q_\eta & q_\zeta(q_\xi + iq_\eta) & 0 \end{pmatrix} \begin{matrix} :|-\frac{3}{2}\rangle \\ :|-\frac{1}{2}\rangle \\ :|\frac{1}{2}\rangle \\ :|\frac{3}{2}\rangle \end{matrix}$$

(B) *Cubic invariants*

(1) $\Gamma = E$　　$\Lambda = E$

$$I(A_1) = \tfrac{1}{4}\sqrt{2}(3q_\epsilon^2 q_\theta - q_\theta^3)\mathbf{I}$$

$$I(E) = \frac{\sqrt{6}}{8} (q_\theta^2 + q_\epsilon^2) \begin{pmatrix} -q_\theta & q_\epsilon \\ q_\epsilon & q_\theta \end{pmatrix} \begin{matrix} :|E_\theta\rangle \\ :|E_\epsilon\rangle \end{matrix}$$

(2) $\Gamma = T_2$ (or T_1) $\Lambda = E$

$$I(A_1) = \frac{1}{2\sqrt{3}}(3q_\epsilon^2 q_\theta - q_\theta^3)\mathbf{I}$$

$$I(E) = \frac{\sqrt{2}}{8}(q_\theta^2 + q_\epsilon^2)\begin{pmatrix} q_\theta - \sqrt{3}q_\epsilon & 0 & 0 \\ 0 & q_\theta + \sqrt{3}q_\epsilon & 0 \\ 0 & 0 & -2q_\theta \end{pmatrix}\begin{matrix} :|\xi\rangle \\ :|\eta\rangle \\ :|\zeta\rangle \end{matrix}$$

(3) $\Gamma = T_2$ (or T_1) $\Lambda = T_2$

$$I(A_1) = -\sqrt{2}\,q_\xi q_\eta q_\zeta\mathbf{I}$$

For $Y = T_2$ there are two linearly independent invariants, which are denoted by $I(T_2, 1)$ and $I(T_2, 2)$

$$I(T_2, 1) = (6\sqrt{5})^{-1}\begin{pmatrix} 0 & q_\zeta(3q_\xi^2 + 3q_\eta^2 - 2q_\zeta^2) & q_\eta(3q_\zeta^2 + 3q_\xi^2 - 2q_\eta^2) \\ q_\zeta(3q_\xi^2 + 3q_\eta^2 - 2q_\zeta^2) & 0 & q_\xi(3q_\eta^2 + 3q_\zeta^2 - 2q_\xi^2) \\ q_\eta(3q_\zeta^2 + 3q_\xi^2 - 2q_\eta^2) & q_\xi(3q_\eta^2 + 3q_\zeta^2 - 2q_\xi^2) & 0 \end{pmatrix}\begin{matrix} :|\xi\rangle \\ :|\eta\rangle \\ :|\zeta\rangle \end{matrix}$$

$$I(T_2, 2) = -(30)^{-\frac{1}{2}}(q_\xi^2 + q_\eta^2 + q_\zeta^2)\begin{pmatrix} 0 & q_\zeta & q_\eta \\ q_\zeta & 0 & q_\xi \\ q_\eta & q_\xi & 0 \end{pmatrix}\begin{matrix} :|\xi\rangle \\ :|\eta\rangle \\ :|\zeta\rangle \end{matrix}$$

(4) $\Gamma = G_{\frac{3}{2}}$ $\Lambda = E$

$$I(A_1) = \frac{1}{4}(3q_\theta q_\epsilon^2 - q_\theta^3)\mathbf{I}$$

The invariant $I(A_2)$ is anti-hermitian and is not listed here.

$$I(E) = -\frac{\sqrt{3}}{8}(q_\theta^2 + q_\epsilon^2)\begin{pmatrix} q_\theta & 0 & q_\epsilon & 0 \\ 0 & -q_\theta & 0 & q_\epsilon \\ q_\epsilon & 0 & -q_\theta & 0 \\ 0 & q_\epsilon & 0 & q_\theta \end{pmatrix}\begin{matrix} :|-\tfrac{3}{2}\rangle \\ :|-\tfrac{1}{2}\rangle \\ :|\tfrac{1}{2}\rangle \\ :|\tfrac{3}{2}\rangle \end{matrix}$$

(5) $\Gamma = G_{\frac{3}{2}}$ $\Lambda = T_2$

$$I(A_1) = -(3/2)^{\frac{1}{2}}q_\xi q_\eta q_\zeta\mathbf{I}$$

For $Y = T_1$ two linearly independent invariants exist. They are both anti-hermitian and are not presented here.

For $Y = T_2$ the symbol n in equation (A IV.9) takes two values 1 and 2. For each of these there exist two linearly independent invariants $I_a(T_2, n)$ and $I_b(T_2, n)$ of which the first is hermitian and the second anti-hermitian. These invariants may be written in terms of the functions F, equation (A IV.11) [having removed the indices 1, 2 and 3 from χ in equation (A IV.10)].

$$n = 1 \qquad\qquad\qquad n = 2$$

$$F(^{T2}_\xi, n) = (30)^{-\frac{1}{2}}[3q_\zeta^2 + 3q_\eta^2 - 2q_\xi^2]q_\xi, \qquad 5^{-\frac{1}{2}}q^2 q_\xi$$

$$F(^{T2}_\eta, n) = (30)^{-\frac{1}{2}}[3q_\xi^2 + 3q_\zeta^2 - 2q_\eta^2]q_\eta, \qquad 5^{-\frac{1}{2}}q^2 q_\eta$$

$$F(^{T2}_\zeta, n) = (30)^{-\frac{1}{2}}[3q_\eta^2 + 3q_\xi^2 - 2q_\zeta^2]q_\zeta, \qquad 5^{-\frac{1}{2}}q^2 q_\zeta$$

In the following invariant the functions F will be written as F_ξ, F_η and F_ζ for brevity. $F_\xi \equiv F(^{T2}_\xi, n)$, etc.

$$I_a(T_2, n) = \frac{1}{\sqrt{12}} \begin{pmatrix} 0 & F_\eta - iF_\xi & iF_\zeta & 0 \\ F_\eta + iF_\xi & 0 & 0 & iF_\zeta \\ -iF_\zeta & 0 & 0 & -F_\eta + iF_\xi \\ 0 & -iF_\zeta & -F_\eta - iF_\xi & 0 \end{pmatrix} \begin{matrix} :|-\tfrac{3}{2}\rangle \\ :|-\tfrac{1}{2}\rangle \\ :|\tfrac{1}{2}\rangle \\ :|\tfrac{3}{2}\rangle \end{matrix}$$

$$n = 1, 2$$

A IV.4 References

1. B. Halperin and R. Englman (Unpublished).
2. U. Fano and G. Racah, *Irreducible Tensorial Sets*. (New York: Academic Press, 1959).
3. J. S. Griffith, *The Irreducible Tensor Method for Molecular Symmetry Groups*. (London: Prentice Hall, 1962).
4. G. Racah, *Phys. Rev.*, **61**, 186 (1942).
5. G. F. Koster, J. O. Dimmock, R. G. Wheeler and H. Statz, *Properties of the 32 Point Groups*. (Cambridge: MIT Press, 1963).

Appendix V

Matrices of vibronic coupling within $t_2^n e^m$ (for the ground states of the strong crystal field case and including spin–orbit coupling)

In a $t_2^n e^m$ configuration in cubic symmetry spin–orbit coupling has first order matrix elements only within the t_2-states. It is therefore frequently a good approximation to disregard the spin–orbit coupling on the e-electrons. For the t_2-states one makes use of the t_2–p equivalence to define, within t_2^n, an effective total orbital angular momentum L' and an effective total angular momentum J'. The matrices for vibronic coupling which are obtained within t_2^n are shown below (Tables A V.1–2). To derive similar matrices for the more general configurations $t_2^n e^m$ would be rather laborious, since one would have to obtain the cubic field terms of the configurations and thereby lose the advantages of using t_2–p equivalence. One would also need W-coefficients within the double group of O (or O_h or T_d), which are not available. However, in the strong crystal field limit of an octahedral environment it turns out (Table 7.1) that either t^n or e^m (or both) gives rise to an orbital singlet in the ground state so that the orbital parts of e^m and t^n are not coupled. Therefore either the pure $E \otimes \varepsilon$

vibronic coupling case (Chapter 3) applies or the matrices below, which are based on the t_2–p equivalence, can be used. These matrices must be supplemented by the diagonal matrices of the spin–orbit interaction. The use of the matrices is immediate for t_2^n or $t_2^n e^4$ when $n = 1, 2$. $n = 3$ has an orbital singlet for the ground state. For $n > 3$ one can utilize the conjugate nature of the t_2^n and t_2^{6-n} states. For $t_2^n e^2$ two further steps are required before turning to the matrices. Let us as an example take Fe^{2+}, a $(3d)^6$ configuration whose ground state term $^5T_2$ arises from $e^2(^3A_2)$ $t_2^4(^3T_1)$. Clearly the spins of the two subshells e^2 and t_2^4 are aligned by Hund's rule to give a resultant total spin $S = 2$. The orbital state in e^2 being a singlet, we can represent the spins and the orbital states of the two subshells e^2 (subshell I) and t_2^4 (II) by $S_I = 1, L_I = 0$ and $S_{II} = 1, L_{II} = 1$ respectively. The total quasi-orbital angular momentum will be $L' = 1$ and the total angular momentum will take the values $J' = 3, 2, 1$. $J' = 1$ is lowest, being separated form $J' = 2$ in MgO by 105 cm^{-1}. Suppose that we wish to obtain the vibronic couplings in this $J' = 1$ ground manifold. As a first step we recouple the J', J'_z states by means of W-coefficients in the manner

$$|e^2(^3A_2), S_I = 1, L_I = 0), t_2^4(^3T_1, S_{II} = 1, L_{II} = 1), S_I + S_{II} = S = 2,$$
$$L' = 1, J' = 1, J'_z\rangle$$

$$= [S]^{\frac{1}{2}}\sum_{J_{II}}(-1)^{L'+S'_I+S'_{II}+J'}[J_{II}]^{\frac{1}{2}} \times W\begin{pmatrix} L' & S_{II} & J_{II} \\ S_I & J' & S \end{pmatrix} |S_I, (L_{II}S_{II})J_{II},$$
$$J' = 1, J'_z\rangle$$

$$= \sqrt{5}\left\{\frac{\sqrt{5}}{30}\middle| S_I, J_{II} = 2, J' = 1, J'_z\rangle + \frac{\sqrt{3}}{6}\middle| S_I, J_{II} = 1, J' = 1, J'_z\rangle \right.$$
$$\left. + \tfrac{1}{3}\middle| S_I, J_{II} = 0, J' = 1, J'_z\rangle\right\}$$

In the second step we use vector coupling or V-coefficients to express the states of the right hand side in terms of the states of the two subshells,

E.g. $|S_I = 1, J_{II} = 1, J' = 1, J'_z = 0\rangle =$
$$\{-|S_I = 1, S_{Iz} = 1\rangle|J_{II} = 1, J_{IIz} = -1\rangle +$$
$$|S_I = 1, S_{Iz} = -1\rangle|J_{II} = 1, J_{IIz} = 1\rangle\}/\sqrt{2}$$

The matrices of Tables A V.1–2 show the matrix elements of the vibronic coupling within the states of $J_{II} = 2, 1, 0$. The eigenstates of S_I in which (we recall) the orbital part is a singlet, have trivial vibronic matrix elements. The matrix elements in the $J' = 1$ triplet are therefore found. These are actually proportional to the corresponding vibronic matrix elements in the one-electron t_2 manifold (a result which follows from the Wigner–Eckart theorem). However, the magnitude of the many-electron matrix elements is reduced, e.g. for coupling to the ε-modes by as much as a factor 10.

(In the following matrices the t_2^n-subshell symbol II, used in the preceding example, is deleted.)

Table A V.1 Vibronic matrices in the $^2t_2$-manifold.
$$(S = \tfrac{1}{2},\, L' = 1,\, J' = \tfrac{3}{2}, \tfrac{1}{2}.)$$

Ratios of the reduced matrix elements (r.m.e.) in $^2t_2$ to the r.m.e. in t_2.

J'	$\tfrac{3}{2}$	$\tfrac{1}{2}$
$\tfrac{3}{2}$	$\sqrt{\tfrac{2}{3}}$	$-\sqrt{\tfrac{2}{3}}$
$\tfrac{1}{2}$	$\sqrt{\tfrac{2}{3}}$	0

Table A V.1 (continued)

J'	J'_z	$\frac{3}{2}$				$\frac{1}{2}$	
		$-\frac{3}{2}$	$-\frac{1}{2}$	$\frac{1}{2}$	$\frac{3}{2}$	$-\frac{1}{2}$	$\frac{1}{2}$
$\frac{3}{2}$	$-\frac{3}{2}$	$L_\epsilon q_\theta$	$L'_\tau(q_\eta + iq_\xi)$	$L_\epsilon q_\epsilon - iL'_\tau q_\zeta$	0	$-\dfrac{L'_\tau}{\sqrt{2}}(q_\eta + iq_\xi)$	$-\sqrt{2}(L_\epsilon q_\epsilon - iL'_\tau q_\xi)$
	$-\frac{1}{2}$	$-L_\epsilon q_\theta$	0		$L_\epsilon q_\epsilon - iL'_\tau q_\zeta$	$\sqrt{2}L_\epsilon q_\theta$	$\dfrac{3}{\sqrt{2}}L'_\tau(q_\eta + iq_\xi)$
	$\frac{1}{2}$		$-L_\epsilon q_\theta$		$-L'_\tau(q_\eta + iq_\xi)$	$-\dfrac{\sqrt{3}}{2}L'_\tau(q_\eta - iq_\xi)$	$-\sqrt{2}L_\epsilon q_\theta$
	$\frac{3}{2}$				$L_\epsilon q_\theta$	$\sqrt{2}(L_\epsilon q_\epsilon + iL'_\tau q_\xi)$	$-\dfrac{L'_\tau}{\sqrt{2}}(q_\eta - iq_\xi)$
$\frac{1}{2}$	$-\frac{1}{2}$					0	0
	$\frac{1}{2}$						0

Matrix ($\times 2\sqrt{3}$) for vibronic coupling to ϵ and τ_2-modes. The coefficients L_ϵ and L'_τ denote the coefficient L_ϵ and $(-\sqrt{2/3})L_\tau$ of equation (3.53).

Table AV.2 Vibronic matrices in the t_2^2 subshell configuration
$$(S = 1, L' = 1, J' = 2, 1, 0.)$$

Ratios of the r.m.e. in t_2^2 to the r.m.e. in the one-electron orbital t_2-manifold

J'	2	1	0
2	$-\frac{1}{2}\sqrt{\frac{7}{3}}$	$\sqrt{\frac{3}{4}}$	$-\frac{1}{\sqrt{3}}$
1	$-\sqrt{\frac{3}{4}}$	$\frac{1}{2}$	0
0	$-\frac{1}{\sqrt{3}}$	0	0

Note: the vibronic matrices for the spin triplet states of the sp configuration are numerically equal to the matrices of this table, (Reference 80 in Chapter 3) and differ from it by an overall change of sign.

J' → M ↓	2					1			0
	−2	−1	0	1	2	−1	0	1	0
2, −2	$-L_\epsilon q_\theta$	$-\frac{\sqrt{3}}{2}L'_\tau(q_\eta + iq_\xi)$	$-\frac{1}{\sqrt{2}}(L_\epsilon q_\epsilon - iL'_\tau q_\zeta)$	0	0	$\frac{\sqrt{3}}{2}L'_\tau(q_\eta + iq_\xi)$	0	0	$-L_\epsilon q_\epsilon + iL'_\tau q_\zeta$
2, −1		$\frac{1}{2}L_\epsilon q_\theta$	$-\frac{L'_\tau}{2\sqrt{2}}(q_\eta + iq_\xi)$	$-\sqrt{\frac{3}{2}}(L_\epsilon q_\epsilon - iL'_\tau q_\zeta)$	0	$-\frac{3}{2}L_\epsilon q_\theta$	$\frac{\sqrt{3}}{2\sqrt{2}}L'_\tau(q_\eta - iq_\xi)$	$-\frac{\sqrt{3}}{2}(L_\epsilon q_\epsilon - iL'_\tau q_\zeta)$	$L'_\tau(q_\eta + iq_\xi)$
2, 0			$L_\epsilon q_\theta$	$\frac{L'_\tau}{2\sqrt{2}}(q_\eta + iq_\xi)$	$-\frac{1}{\sqrt{2}}(L_\epsilon q_\epsilon + iL'_\tau q_\zeta)$	$\frac{3L'_\tau}{2\sqrt{2}}(q_\eta - iq_\xi)$	0	$-\frac{3L'_\tau}{2\sqrt{2}}(q_\eta + iq_\xi)$	$-\sqrt{2}L_\epsilon q_\theta$
2, 1				$\frac{1}{2}L_\epsilon q_\theta$	$\sqrt{\frac{3}{2}}L'_\tau(q_\eta + iq_\xi)$	$-\frac{\sqrt{3}}{2}(L_\epsilon q_\epsilon + iL'_\tau q_\zeta)$	$-\frac{\sqrt{3}}{2\sqrt{2}}L'_\tau(q_\eta - iq_\xi)$	$\frac{3}{2}L_\epsilon q_\theta$	$-L'_\tau(q_\eta - iq_\xi)$
2, 2					$-L_\epsilon q_\theta$	0	$-\frac{\sqrt{3}}{2}(L_\epsilon q_\epsilon + iL'_\tau q_\zeta)$	$\frac{\sqrt{3}}{2}L'_\tau(q_\eta - iq_\xi)$	$-(L_\epsilon q_\epsilon + iL'_\tau q_\zeta)$
1, −1						$\frac{1}{2}L_\epsilon q_\theta$	$\frac{\sqrt{3}}{2}L'_\tau(q_\eta + iq_\xi)$	$\frac{\sqrt{3}}{2}(L_\epsilon q_\epsilon + iL'_\tau q_\zeta)$	0
1, 0							$-L_\epsilon q_\theta$	$-\frac{\sqrt{3}}{2}L'_\tau(q_\eta + iq_\xi)$	0
1, 1								$\frac{1}{2}L_\epsilon q_\theta$	0
0, 0									0

Matrix ($\times -2\sqrt{3}$) for vibronic coupling to ϵ and τ_2-modes. L_ϵ and L'_τ denote the coefficient L_ϵ and $(-\sqrt{2/3})\,L_\tau$ of equation (3.53)

Appendix VI
The semi-classical approximation for the line-shape

This method presupposes that while the total intensity (of say, absorption) is determined by the electronic transition operator, the distribution of the intensity, i.e. the band shape, depends on the motion of the system in configuration space. This latter dependence operates in two ways. First, the intensity of transition from any chosen point in configuration space is proportional to the probability of the system being at that point. Secondly, since the system has no time to move in configuration space during the fast electronic transition, the transition takes place vertically and the energy of the photon absorbed is the difference between the energies of the two electronic states of the chosen value of the configuration coordinate. We shall take for this difference the difference between the potentials, one of which is the J-T potential. In some earlier works (e.g. Reference 1) the energy of the initial state at low temperatures was taken as a constant, independent of the coordinate, namely the energy of zero point motion. The alternative used here appears to agree better with experiment[2] and depends on the assumption that the electronic wave-functions may be obtained in either the ground or the excited state, by treating the vibrational coordinate as classical position coordinates or adiabatic parameters.

In the mathematical formulation of the method one writes the transition probability as a product of an electronic factor and a line shape function $I(\omega)$ (normalized to unity) which depends on the vibrational motion only. This factorization (called the Condon approximation) is a rather bold step in the present circumstances, since the J-T-E negates the factorization of the wave-function into electronic and vibrational parts. However, disregarding this blemish we can write the line shape function as

$$I(\omega) = \int \cdots \int \prod_n dQ_n P_i(\{Q\}) \delta[(\hbar\omega - E_f(\{Q\}) - E_i(\{Q\}))]$$

(A VI.1)

where $\prod_n dQ_n P_i(\{Q\})$ represents the probability that when the system is in the initial electronic state i it will be found in the volume element $\prod_n dQ_n$ near the point $\{Q\}$. In the product n runs over all vibrational coordinates of the set $\{Q\}$. $E_i(\{Q\})$ and $E_f(\{Q\})$ are the energies in the initial and final states: at least one of them is a multi-valued function of $\{Q\}$, being the solution of the static J-T problem.

For the probability P two forms are in vogue. The low temperature form (used in the present context in Reference 3) is based on a quantum mechanical probability, the square of the ground state wave-function of the set $\{Q\}$ of harmonic oscillators, i.e.

$$P(\{Q\}) = \left[\prod_n \frac{M\omega_n}{\hbar\sqrt{\pi}}\right] \exp\left[-\sum_n (M\omega_n/\hbar)Q_n^2\right]$$

The high temperature form[4] assumes a Boltzmann-distribution within the energy continuum of the initial electronic state

$$P(\{Q\}) = e^{-E_i(\{Q\})/kT} / \int \cdots \int \prod_n dQ_n\, e^{-E_i(\{Q\})/kT}$$

The natural interpolation between the two limits is[1, 5-6]

$$P \propto \exp -\sum_n Q_n^2 \Big/ \left[\frac{\hbar}{M\omega_n} \coth\left(\frac{\hbar\omega_n}{2kT}\right)\right]$$

This is the form we shall favour, more as a kind of shorthand for both low and high temperature forms than because of its intrinsic justification. Occasionally we shall also abbreviate our notation, writing

$$\frac{\hbar\omega_n}{2k} \coth \frac{\hbar\omega_n}{2kT} = T_n,$$

an effective temperature characteristic of the n mode.

In practice, and especially in works which aim at analytical formulae, one is forced to work with a limited number of modes. The modes chosen in Chapter 3 are the J-T active normal vibrations of the quasi-molecular system.

The semi-classical approximation for $T \otimes \tau_2$, as shown in Figure 3.28, was compared with the quantum mechanical result, depicted in Figure 3.30, in §3.3.7. For $E \otimes \varepsilon$ attention was drawn in the caption of Figure 3.16 that the high-energy band (or peak) differs in shape and strength (is weaker) than the low-energy band, whereas the semi-classical treatment yields bands of equal strength.[7]

A VI.1 References

1. C. C. Klick, D. A. Patterson and R. S. Knox, *Phys. Rev.*, **133**, A1717 (1964).
2. P. R. Moran, *Phys. Rev.*, **137**, A1016 (1965).
3. M. C. M. O'Brien, *Proc. roy. Soc. A*, **281**, 323 (1964).
4. Y. Toyozawa and M. Inoue, *J. Phys. Soc. Japan*, **21**, 1663 (1966).
5. F. E. Williams and M. H. Hebb, *Phys. Rev.*, **84**, 1181 (1951).
6. M. Lax, *J. Chem. Phys.*, **20**, 1752 (1952).
7. T. A. Fulton and D. B. Fitchen, *Phys. Rev.*, **179**, 846 (1969).

Appendix VII

Selection rules for electric dipole transitions

Tables A VII.1–4 give the polarizations (i.e. the direction of the electric field component in the electromagnetic wave) of electric dipole transitions between electronic or vibronic states. The absence of any entry indicates the forbiddenness of the transition. The tables cover cubic groups, some axial groups and the linear groups $C_{\infty v}$ and $D_{\infty h}$. The first line for each group and for each representation shows which transitions are allowed between an initial state (a row) and a final state (a column). This part of the table can be used either for electronic states without vibrational assistance or for vibronic states belonging to the given representations. An entry like $M_x + M_y$, as opposed to $M_{x,y}$, indicates that the polarizations M_x and M_y are inequivalent. Then we denote by e.g. $\varepsilon : M_x$ the polarization M_x of a transition between electronic states which is made allowed by weak vibronic interaction acting in the first-order in the vibrational amplitude ε (or, for linear molecules, acting in the second order) in a degenerate electronic state. We then show by e.g. $\varepsilon^0 : M_x$ the polarization M_x of transitions between electronic states which are allowed if the system in its initial state assumes the vibronically stabilized distortion denoted by ε^0. Similarly ε^{00}, $\varepsilon^{0'}$, ε^1, etc., denote other frozen-in distortions of the molecular system occurring along the vibrational mode ε.

The electronic state which is stabilized in the static distortion is unambiguously specified in the footnote. It should be noted that the tables give the selection rule for the transition to the full manifold of the final electronic state, i.e. regardless of any possible splitting of the final state due to the distortion.

For axial molecules z is the direction of main axis of rotation. If a distortion occurs, the convention is used that y retains more elements of symmetry (e.g. an axis of rotation, a plane of reflection) than x.

The words 'add (g, u)-rule' mean in these tables the following: affix the subscripts g to all vibrational modes and u to the representation of the dipole operator, and use the table given with the additional restriction that only g ↔ u transitions are allowed.

Table A VII.1 Octahedral and tetrahedral groups: electric dipole transition selection rules

		A_1	A_2	E	T_1	T_2	$E_{1/2}$	$E_{5/2}$	$G_{3/2}$
A_1					$M_{x,y,z}$				
A_2						$M_{x,y,z}$			
E	ε:				$M_{x,y,z}$	$M_{x,y,z}$			
	ε^0:				$M_{x,y,z}$	$M_{x,y,z}$			
	ε^{00}:		$M_{z,y}$		$M_{x,y}+M_z$	$M_{x,y}$			
					$M_{x,y}$	$M_{x,y}+M_z$			
T_1	ε:	$M_{x,y,z}$		$M_{x,y,z}$	$M_{x,y,z}$	$M_{x,y,z}$			
	ε^0:	M_z		M_z	$M_{x,y}$	$M_{x,y}$			
	ε^{00}:	M_z	M_y	M_y	$M_{x,y}$	$M_{x,y}$			
	ε^1:	M_y	$M_{y'}$	$M_{y'}$	$M_{x'}+M_z$	$M_{x}+M_z$			
	ε^2:	$M_{y'}$	$M_{y'}$	$M_{x',y'}$	$M_{x'}+M_z$	$M_{x'}+M_z$			
	τ_2:	$M_{x,y,z}$	$M_{x,y,z}$	$M_{x,y,z}$	$M_{x,y,z}$	$M_{x,y,z}$			
	τ_2^0:	$M_{z'}$		$M_{x',y'}$	$M_{x,y}+M_z$	$M_{x',y}+M_{z'}$			
	τ_2^1:	$M_{y'}$	$M_{x'}+M_{z'}$	$M_{x'}+M_{y'}+M_{z'}$	$M_{x}+M_{y'}+M_{z'}$	$M_{x}+M_{y'}+M_{z'}$			
	τ_2^2:	$M_{x'}+M_{z'}$	$M_{y'}$	$M_{x'}+M_{y'}+M_{z'}$	$M_{x'}+M_{y'}+M_{z'}$	$M_{x'}+M_{y'}+M_{z'}$			

Table A.   (continued)

T_2

ε:		$M_{x,y,z}$	$M_{x,y,z}$	$M_{x,y,z}$	$M_{x,y,z}$	$M_{x,y,z}$
ε^0:	$M_{x,y,z}$	$M_{x,y,z}$	$M_{x,y,z}$	$M_{x,y,z}$	$M_{x,y,z}$	$M_{x,y,z}$
ε^{00}:	M_z	M_z	M_z	M_z	M_z	$M_{x,y}$
ε^1:	M_y	M_y	M_y	M_z	M_z	$M_{x,y}$
ε^2:	$M_{y'}$	$M_{y'}$	$M_{y'}$	M_y	$M_x + M_z$	$M_x + M_z$
τ_2:	$M_{x,y,z}$	$M_{x,y,z}$	$M_{x,y,z}$	$M_{x,y,z}$	$M_{x,y,z}$	$M_{x,y,z}$
$\tau_2{}^0$:	$M_{y'}$	$M_{x'}$	$M_{x'}$	$M_{x',y'}$	$M_{x',y'}$	$M_{x',y'}$
$\tau_2{}^1$:	$M_{x'} + M_{z'}$	$M_{x'} + M_{y'} + M_{z'}$	$M_{x'} + M_{y'} + M_{z'}$	$M_{x'} + M_{y'} + M_{z'}$	$M_x + M_{y'} + M_{z'}$	$M_x + M_{y'} + M_{z'}$
$\tau_2{}^2$:	$M_{y'}$	$M_x + M_{y'} + M_{z'}$	$M_{x'} + M_{y'} + M_{z'}$	$M_{x'} + M_{y'} + M_{z'}$	$M_x + M_{y'} + M_{z'}$	$M_x + M_{y'} + M_{z'}$

(c)

$E_{1/2}$

$M_{x,y,z}$	$M_{x,y,z}$

$E_{5/2}$

$M_{x,y,z}$	$M_{x,y,z}$

$G_{3/2}$

ε:	$M_{x,y,z}$	$M_{x,y,z}$	$M_{x,y,z}$
ε^0:	$M_{x,y,z}$	$M_{x,y,z}$	$M_{x,y,z}$
ε^{00}:	$M_{x,y} + M_z$	$M_{x,y}$	$M_{x,y} + M_z$
τ_2:	$M_{x,y}$	$M_{x,y} + M_z$	$M_{x,y} + M_z$
$\tau_2{}^0$:	$M_{x,y,z}$	$M_{x,y,z}$	$M_{x,y,z}$
$\tau_2{}^{00}$:	$M_{x',y'}$	$M_{x',y'} + M_{z'}$	$M_{x',y'} + M_{z'}$
	$M_{x',y'} + M_{z'}$	$M_{x',y'} + M_{z'}$	$M_{x',y'} + M_{z'}$

(c)

(c)

Table A VII.1—(continued)

Group O: $M_{x,y,z} = T_1$

ε^0 is a distortion in the ε-mode such that the component (E_θ) of the doublet E symmetric under $C_2(y')$ is stabilized. The broken symmetry is D_4. The axis y' is oriented at 45° from the cubic axes.

ε^{00} is a distortion in the ε-mode such that the component (E_ϵ) of the doublet E antisymmetric under $C_2(y')$ is stabilized. The broken symmetry is D_4.

Both ε^0 and ε^{00} stabilize the z-component of the triplets T_1, T_2.

$\varepsilon^{0'}$ is a distortion in the ε-mode such that the Kramers doublet stabilized maintains its sign upon a $\pi/2$ rotation about the z-axis. The broken symmetry is D_4.

$\varepsilon^{00'}$ is a distortion in the ε-mode such that the Kramers doublet stabilized changes its sign upon a $\pi/2$ rotation about the z-axis. The reduced symmetry is D_4.

ε^1 and ε^2 are distortions additional to either ε^0 or ε^{00}, such that the excited state doublet arising from either T_1 or T_2 splits and one of the states achieves relative stabilization. The nature of the distortion and the stabilized state are for ε^1 as for $\beta_1{}^0$ under D_4 (see there) and for ε^2 as for $\beta_2{}^0$ under D_4. $\tau_2{}^0$ is a distortion in the τ_2-mode stabilizing the z'-component (A_2 or A_1 in D_3) of the triplet T_1 or T_2. The trigonally oriented coordinate axes are denoted by x', y', z'.

$\tau_2{}^{0'}$ is a distortion in the τ_2-mode such that the $|S_z| = \frac{3}{2}$ spinor-like components of $G_{3/2}$ are stabilized. The symmetry is D_3.

$\tau_2{}^{00'}$ is a distortion in the τ_2-mode such that the $|S_z| = \frac{1}{2}$ spinor-like components of $G_{3/2}$ are stabilized. The symmetry is D_3.

$\tau_2{}^1$ and $\tau_2{}^1$ are distortions additional to $\tau_2{}^0$, such that the excited state doublet arising from either T_1 or T_2 splits and one of its components achieves relative stabilization. The broken symmetry is C_2. The nature of the distortion and the stabilized state are for $\tau_2{}^1$ as described for ε^0 under D_3 and for $\tau_2{}^2$ as for ε^{00} under D_3.

Group O_h: Add (g, u) rule

Group T_d: $M_{x,y,z} = T_2$

Use this table but interchange the entries of the following columns: $1 \leftrightarrow 2$, $4 \leftrightarrow 5$, $6 \leftrightarrow 7$. The meanings of the distortional types are (analogous to, but) different from those in O, since the group operations differ. Further tables have been prepared (for different groups and transition operators) which yield additional information.

(*) Only these transitions are allowed for higher order vibronic coupling.

Table A VII.2 Tetragonal groups:
electric dipole transition selection rules

		A_1	A_2	B_1	B_2	E	
A_1			M_z			$M_{x,y}$	
A_2		M_z				$M_{x,y}$	
B_1					M_z	$M_{x,y}$	
B_2				M_z		$M_{x,y}$	
E		$M_{x,y}$	$M_{x,y}$	$M_{x,y}$	$M_{x,y}$	M_z	
	β_1:	$M_{x,y}$	$M_{x,y}$	$M_{x,y}$	$M_{x,y}$	M_z	(*)
	β_2:	$M_{x,y}$	$M_{x,y}$	$M_{x,y}$	$M_{x,y}$	M_z	(*)
	β_1^0:	M_y	M_x	M_y	M_x	M_z	
	β_2^0:	$M_{y'}$	$M_{x'}$	$M_{y'}$	$M_{x'}$	M_z	

Group D_4: $M_z = A_2$ $M_{x,y} = E$

β_1^0 is a distortion in the β_1-mode, such that the component of E symmetric under $C_2(y)$ is stabilized. The broken symmetry is D_2. The selection rules for the case when the other component is stabilized follow by interchanging M_x and M_y.

β_2^0 is a distortion in the β_2-mode such that the component of E symmetric under $C_2(y')$ is stabilized. The broken symmetry is again D_2, however the two-fold axes $C_2(x')$, $C_2(y')$ in this symmetry are rotated by 45° from the axes of the previous case. The selection rules for the case when the other component of E is stabilized follow upon interchange of $M_{x'}$ and $M_{y'}$.

Group D_{4h}: Add (g, u) rule.

Group C_{4v}: $M_z = A_1$, $M_{x,y} = E$

In the first four rows interchange the entries of the following columns: 1–2, 3–4. In the caption for D_4 replace the symmetry elements $C_2(y)$, $C_2(y')$ by $\sigma(yz)$, $\sigma(y'z)$ and D_2 by C_{2v}.

(*)   In higher order vibronic coupling only these transitions are allowed.

Table A VII.3 Trigonal groups:
electric dipole transition selection rules

		A_1	A_2	E
A_1			M_z	$M_{x,y}$
A_2		M_z		$M_{x,y}$
E		$M_{x,y}$	$M_{x,y}$	$(M_{x,y} + M_z)$
	ε:	$(M_{x,y} + M_z)$	$(M_{x,y} + M_z)$	$(M_{x,y} + M_z)$
	ε^0:	M_y	$(M_x + M_z)$	$(M_x + M_y + M_z)$
	ε^{00}:	$(M_x + M_z)$	M_y	$(M_x + M_y + M_z)$

Group D_3: $M_z = A_2$, $M_{x,y} = E$

ε^0 is a distortion in the ε-mode such that the state (E_θ) symmetric under $C_2(y)$ is stabilized.

ε^{00} is a distortion in the ε-mode such that the state (E_ϵ) antisymmetric under $C_2(y)$ is stabilized.

Broken symmetry in either case is C_2, with two fold axis along y. In higher order vibronic coupling all transitions are allowed.

Group D_{3d}: Add (g, u) rule.

Group C_{3v}: $M_z = A_1$, $M_{x,y} = E$

Use this table after interchanging the entries of M_z between the first and second columns. In the caption for D_3 replace the symmetry element $C_2(y)$ by $\sigma(yz)$ and C_2 by C_s.

Table A VII.4 Continuous axial groups: electric dipole transition selection rules

$C_{\infty v}$		Σ^+	Σ^-	Π	Δ	Φ	Λ
Σ^+		M_z		$M_{x,y}$			
Σ^-			M_z	$M_{x,y}$			
Π	$(\pi)^2$:	$M_{x,y}$	$M_{x,y}$	M_z	$M_{x,y}$	$M_{x,y}$	
	π^0:	$M_{x,y}$	$M_{x,y}$	M_z	$M_{x,y}$	$M_{x,y}$	
	π^{00}:	$(M_y + M_z)$	M_x	$(M_x + M_y + M_z)$	$(M_x + M_y + M_z)$	$(M_x + M_y + M_z)$	
		M_x	$(M_y + M_z)$	$(M_x + M_y + M_z)$	$(M_x + M_y + M_z)$	$(M_x + M_y + M_z)$	
Δ	$(\pi)^2$:	M_z	M_z	M_z	M_z	M_z	
	π^0:	$(M_y + M_z)$	M_x	$(M_x + M_y + M_z)$	M_z	M_z	
	π^{00}:	M_x	$(M_y + M_z)$	$(M_x + M_y + M_z)$	$(M_x + M_y + M_z)$	$(M_x + M_y + M_z)$	
					$(M_x + M_y + M_z)$	$(M_x + M_y + M_z)$	
Φ	$(\pi)^2$:	$M_{x,y}$	$M_{x,y}$	M_z	M_z	M_z	
	π^0:	$(M_y + M_z)$	M_x	$(M_x + M_y + M_z)$	M_z	M_z	
	π^{00}:	M_x	$(M_y + M_z)$	$(M_x + M_y + M_z)$	$(M_x + M_y + M_z)$	$(M_x + M_y + M_z)$	
					$(M_x + M_y + M_z)$	$(M_x + M_y + M_z)$	
Λ							M_z

Table A VII.4—(continued)

Group $C_{\infty v}$: $M_z = \Sigma^+$, $M_{x,y} = \Pi$

$(\pi)^2$: gives the selection rules for vibronic coupling second order in the π-vibrational mode.

π^0 is a distortion in the y-component of the π-mode such that the state symmetric in $\sigma(yz)$ is stabilized. The symmetry is reduced to C_s.

π^{00} is a distortion in the y-component of the π-mode such that the state antisymmetric in $\sigma(yz)$ is stabilized. The symmetry is reduced to C_s.

In higher order vibronic coupling, transitions are allowed for M_z between two doubly degenerate representations whose rotational quantum numbers (Λ) have the same parity, and for $M_{x,y}$ if they have opposing parity.

Λ stands for a general two dimensional representation whose quantum number Λ satisfies $|\Lambda| > 4$.

Group $D_{\infty h}$: $M_z = \Sigma_u^+$, $M_{x,y} = \Pi_u$

For selection rules in zero, second or higher order vibronic coupling add (g, u)-rule to this table.

Appendix VIII
Critical phenomena in spinels

Part of the data was listed in *Phys. Rev.*, **B2**, 75 (1970) by B. Halperin and the author.

Table A VIII.1 Mn^{3+} (d^4, ^{5}E ground state) at B-sites

(ΔH = latent heat of transition in kcal/mole. ΔS = entropy change in cal/mole/deg)

Compound	Probable formula	Methods of measurement	T_t in °K	c/a at R.T.	c/a near T_t	Thermal data	Critical concentration at R.T.
Mn$_3$O$_4$	$Mn^{2+}[Mn_2^{3+}]O_4$	X-ray [1,2,3,4] Electrical conductivity [2] Thermal measurements [1,2,5] Infrared spectroscopy [6]	1443	1·15-1·16	1·13 [3]	$\Delta H = 4\cdot5$-5 [1,5] $\Delta S = 3\cdot1$-$3\cdot5$ [1,5]	
Mn$_3$O$_4$—MgAl$_2$O$_4$ [7]	Varying	X-ray		1·11-1·16			0·58
Mg$_z$Mn$_{3-2z}$O$_4$	$Mg_z^{2+}Mn_{1-z}^{2+}[Mn_2^{3+}]O_4$ [8] $0 \leqslant z < 1$ $Mg^{2+}[Mn_2^{3+}]O_4$ $z = 1, T < 1070$ [1] $Mg_{0\cdot45}^{2+}Mn_{0\cdot45}^{3+}[Mg_{0\cdot45}^{2+}M_{1\cdot55}^{3+}]O_4$ $z = 1, T = T_t$ [1] $Mg_{0\cdot51}^{2+}Mn_{0\cdot49}^{3+}[Mg_{0\cdot49}^{2+}Mn_{1\cdot51}^{3+}]O_4$ $z = 1$, when quenched from 1520 °K [9]	X-ray [1,8,9,10] Electrical conductivity [8] Thermal measurements [1]	1123-1443	1·15-1·16 1·147 $z = 1$ [10]	$\leqslant 1\cdot16$ [1]	$\Delta H = 3\cdot5$ $\Delta S = 3\cdot0$ [1]	
MgMn$_2$O$_4$—MgAl$_2$O$_4$ [7]	Varying	X-ray		1·10-1·15			0·59
MgGa$_{2-z}$Mn$_z$O$_4$ [11]	Varying	X-ray and neutron scatt.	>1123	~1·02	~1·15		~0·5
MgCr$_{2-z}$Mn$_z$O$_4$ [11]	Varying	X-ray and neutron scatt.	>1123	~1·02	~1·15		~0·48
Zn$_z$Mn$_{3-2z}$O$_4$	$Zn_z^{2+}Mn_{1-z}^{2+}[Mn_2^{3+}]O_4$ [1,8] $0 \leqslant z \leqslant 1$ $Zn^{2+}[Zn_{z-1}^{2+}Mn_{z-1}^{4+}Mn_{4-2z}^{3+}]O_4$ [8] $1 \leqslant z < 1\cdot2$	X-ray [1,8,6] Electrical conductivity [8] Thermal measurements [8] Infrared spectroscopy [6]	1120-1470 [8] 1323 $z = 1$ [1]	1·11-1·16 1·02	1·15 [1] $z = 1$	$\Delta H = 5$ [1] $z = 1$ $\Delta S = 3\cdot8$ [1] $z = 1$	

Compound	Cation distribution	Experimental method	Temperature (°C)	c/a		x
$Zn_2Ge_{1-z}Mn_2O_4$ [4]	$Zn_{0.6}^{2+}Ge_{0.4}^{4+}[Mn_{0.8}^{2+}\ Mn_{1.2}^{3+}]O_4$ $z = 0.6$	X-ray Magnetic data		1·15		0·6
$Zn_zGe_{1-z}Co_{2-2z}Mn_{2z}O_4$ [4] $0 \leqslant z \leqslant 1$	$Zn_z^{2+}Ge_{1-z}^{4+}[Co_{2-2z}^{2+}Mn_{2z}^{3+}]O_4$	X-ray Magnetic data	700–1300	$\leqslant 1.144$		0·65
$Fe_{3-z}Mn_zO_4$	$Mn_z^{2+}Fe_{1-z}^{3+}[Mn_{2-t}^{3+}Fe_{1-z+t}^{2+}Fe_{1-z+2t}^{3+}]O_4$ [12,13] $Mn_{0.8}^{2+}Fe_{0.2}^{2+}[Fe_{1.8}^{3+}Mn_{0.2}^{3+}]O_4$ [14] $z = 1$	X-ray [3,14,15] Mössbauer effect [13] NMR [16]	$\geqslant 670$	1·15	1·06–1·13	~0·60 [3]
$Co_{3-z}Mn_zO_4$	$Co^{2+}[Co_{2-z}^{3+}Mn_z^{3+}]O_4$ $0 \leqslant z < 2$ [4] $Co_{3-z}^{2+}Mn_{z-2}^{2+}[Mn_2^{3+}]O_4$ $2 \leqslant z \leqslant 3$ [4]	X-ray [1,2,4] Electrical conductivity [1] Thermal measurements [4] Magnetic data [1,2]	600–1443 [2] 1173 $z = 2$ [1]	1·03–1·16 [2] $z = 2$ [1]	$\Delta H = 3.5$ [1] $z = 2$ $\Delta S = 3.0$ [1] $z = 2$	0·56 [2]
$Cu_zCo_{1-z}Mn_2O_4$ [10]	$Co_{1-z}^{2+}Cu_z^{2+}[Mn_2^{3+}]O_4$	X-ray	1173 $z = 0$	1·123 $z = 0$		$z = 0.53$
$Cu_zCd_{1-z}Mn_2O_4$ [10]	$Cd_{1-z}^{2+}Cu_z^{2+}[Mn_2^{3+}]O_4$	X-ray		1·200 $z = 0$		$z = 0.48$
$Cu_zMg_{1-z}Mn_2O_4$ [10]	$Mg_{1-z}^{2+}Cu_z^{2+}[Mn_2^{3+}]O_4$	X-ray	1123 $z = 0$	1·147 $z = 0$		$z = 0.28$
$Cu_zZn_{1-z}Mn_2O_4$ [10]	$Zn_{1-z}^{2+}Cu_z^{2+}[Mn_2^{3+}]O_4$	X-ray	1323 $z = 0$	1·142 $z = 0$		$z = 0.85$

Table A VIII.2 Cu^{2+} (d^9, ^{2}E ground state) at B-sites (ΔH and ΔS as in Table A VIII.1)

Compound	Probable formula	Methods of measurement	T_t in °K	c/a at R.T.	c/a near T_t	Thermal data	Critical concentration
$CuFe_2O_4$	$Fe^{3+}[Cu^{2+}Fe^{3+}]O_4$ [17] $Fe_{1-t}^{3+}Cu_t^{2+}$ $[Cu_{1-t}^{2+}Fe_{1+t}^{3+}]O_4$ [18]	X-ray [17,18,19,20] Thermal measurements [21] Mössbauer effect [19] Neutron scattering [17]	663 [21] After quenching: 363 [19], 633 [18]	1·03–1·06 [17,18,19]	1·01 [19] 1·03 [19]	$\Delta H = 0·25$ [21]	0·38 [18]
$CuFe_{2-z}Cr_zO_4$ $z < 1$	$Fe_{1-t}^{3+}Cu_t^{2+}$ $[Cu_{1-t}^{2+}Fe_{1+t-z}^{3+}Cr_z^{3+}]O_4$ [20]	X-ray [18,20] Magnetic data [20]	220–630 [18,22]	1·04–1·06 [22]	1·02–1·04 [22]		0·38 [18]
$Ge_{0·2}Cu_{1·2}Fe_{1·6}O_4$ [23]		X-ray Mössbauer effect	783	1·075	1·05		

Table A VIII.3 Cu^{2+} and Mn^{3+} at B-sites

Compound	Probable formula	Methods of measurement	T_t in °K	c/a at R.T.	Critical composition at R.T.
$Cu_zMn_{3-z}O_4$	$Mn^{2+}[Cu_z^{2+}Mn_z^{4+}Mn_{2-2z}^{3+}]O_4$ $0 \leqslant z \leqslant 0.2$ [8] $Mn_{1-z}^{2+}Cu_z^{2+}[Mn_2^{3+}]O_4$ [10]	X-ray[8,10] Electrical conductivity[8] Thermal measurements[10]	1320–1443	1·161	$z = 0.58$
$(ZnMn_2O_4)_{1-z}$ $(Cu_2GeO_4)_z$ $0 \leqslant z \leqslant 0.6$ [24]	$Zn_{1-z}Cu_z[Cu_zGe_zMn_{2-2z}]O_4$	X-ray Neutron scattering	>300	1·07–1·15	$z = 0.4$
$(ZnMn_2O_4)_{1-z}$ $(Zn\,CuGe\,O_4)_z$ $0 \leqslant z \leqslant 0.6$ [24]		X-ray Neutron scattering	>300	1·04–1·15	$z = 0.55$

Table A VIII.4 Fe^{2+} (d^6, ^5E ground state) at A-sites

Compound	Methods of measurement	Transition temperature in °K	c/a
$FeCr_2O_4$	Mössbauer effect[25] Infrared[26]	135	0·978 at 100 °K
$Fe_{3-x}Cr_xO_4$	X-ray[28,29]	100–170 depending on x	>1 for $1 < x < 1\cdot3$,[28] $1\cdot4 < x < 1\cdot5$[27]
$\{Fe^{3+}_{1-y}Fe^{2+}_{y}[Fe^{2+}_{1-x+y}Fe^{3+}_{1-x+y}Cr^{3+}_{x}]O_4\}$[27]	Magnetic data[29]		<1 for $1\cdot5 < x < 2$[28,29]
$FeCr_2S_4$	Mössbauer effect[30]		= 1 down to $4\cdot2$ °K[30]
FeV_2O_4	Mössbauer effect[25]	127[27]–140[25]	0·985 at 110 °K[25]
$Fe_xCo_{1-x}Cr_2O_4$	X-ray[28]	>80	<1 for $0\cdot7 < x < 1$
$Fe_xNiCr_{2-x}O_4$	X-ray[28]	Varying with x	>1 for $x < 0\cdot2$ at 86 °K; <1 for $x > 0\cdot28$ at 86 °K orthorhombic in between
$Fe_xNi_{1-x}Cr_2O_4$	X-ray[31]	Varying with x	>1 for $x < 0\cdot8$ at 175 °K; <1 for $x > 0\cdot8$ at 175 °K with also an orthorhombic region[31,32,33]

Table A VIII.5 Ni^{2+} (d^8, $^3T_1$ ground state) at A-sites

Compound	Methods of measurement	Transition temperature in °K	c/a
$NiCr_2O_4$	X-ray[18,26,32,34] Infrared[26]	275,[35] 294,[33] 310 [18,26,34]	1·04 at 0 °K [35]

Table A VIII.6 Cu^{2+} (d^9, $^2T_2$ ground state) at A-sites

Compound	Methods of measurement	Transition temperature in °K	c/a
$CuCr_2O_4$	X-ray[20] Magnetic data[20]	860	0·91 at 0 °K
$CuMn_2O_4$	X-ray[19]		~1
$CuCs_2Cl_4$	X-ray[36] Infrared[37]		<1 at R.T.
$CuCo_{1-x}Cr_2O_4$	X-ray[28]		>1 for $x > 0\cdot48$ at R.T.
$CuFe_{2-x}Cr_xO_4$	X-ray[20,38]	Varying with x	Varies from $c/a < 1$ to >1 between $x = 0$ and $=2$, with $c/a = 1$ near 1
$\{(Cu_tFe_{1-t})[Cu_{1-t}Fe_{1+t-z}Cr_z]O_4\}$[29]	(See Table A VIII.2)		
$Cu_xFe_yCr_2O_4$	X-ray[39] Mössbauer effect[39]	90	1·04 at 0 °K
$Cu_xZn_{1-x}Cr_2O_4$	X-ray[31]		<1, $c/a = 0\cdot95$ at R.T. for $x = 0\cdot5$ [32]
$Cu_xNi_{1-x}Cr_2O_4$	X-ray[28,31]	Varying with x	>1 for $x < 0\cdot12$ at R.T. <1 for $x > 0\cdot14$ at R.T. orthorhombic for intermediate x's

A VIII.1 References

1. K. S. Irani, A. P. B. Sinha and A. B. Biswas, *J. Phys. Chem. Solids*, **23,** 711 (1962).
2. I. Aoki, *J. Phys. Soc. Japan*, **17,** 53 (1962).
3. H. F. McMurdie, B. M. Sullivan and F. A. Mauer, *J. Research NBS*, **45,** 35 (1950).
4. D. G. Wickham and W. J. Croft, *J. Phys. Chem. Solids*, **7,** 351 (1958).
5. J. C. Southard and G. E. Moore, *J. Am. Chem. Soc.*, **64,** 1769 (1942).
6. K. Siratori and Y. Aiyama, *J. Phys. Soc. Japan*, **20,** 1962 (1965).
7. K. S. Irani, A. P. B. Sinha and A. B. Biswas, *J. Phys. Chem. Solids*, **17,** 101 (1960).
8. M. Rosenberg, P. Nicolau, R. Manaila and P. Pausescu, *J. Phys. Chem. Solids*, **24,** 1419 (1963).
9. R. Manaila and P. Pausescu, *Phys. Stat. Sol.*, **9,** 385 (1967).
10. S. T. Kshirsagar and A. B. Biswas, *J. Phys. Chem. Solids*, **28,** 1493 (1967).
11. M. Grenot and M. Huber, *J. Phys. Chem. Solids*, **28,** 2441 (1967).
12. G. I. Finch, A. P. B. Sinha and K. P. Sinha, *Proc. roy. Soc. A*, **242,** 28 (1957).
13. M. Tanaka, T. Mizoguchi and Y. Aiyama, *J. Phys. Soc. Japan*, **18,** 1091 (1963).
14. L. Cervinka, S. Krupicka and V. Synecek, *J. Phys. Chem. Solids*, **20,** 167 (1961).
15. L. Cervinka and D. Vetterkind, *J. Phys. Chem. Solids*, **29,** 171 (1968).
16. T. Kubo, A. Hirai and H. Abe, *J. Phys. Soc. Japan*, **26,** 1094 (1969).
17. E. Prince and R. G. Treuting, *Acta Cryst.*, **9,** 1025 (1956).
18. H. Ohnishi, T. Teranishi and S. Miyahara, *J. Phys. Soc. Japan*, **14,** 106 (1959).
19. T. Yamadaya, T. Mitui, T. Okada, N. Shikazono and Y. Hamaguchi, *J. Phys. Soc. Japan*, **17,** 1897 (1962).
20. S. Miyahara and H. Ohnishi, *J. Phys. Soc. Japan*, **12,** 1296 (1956).
21. T. Inoue and S. Iida, *J. Phys. Soc. Japan*, **13,** 656 (1958).
22. H. Ohnishi and T. Teranishi, *J. Phys. Soc. Japan*, **16,** 35 (1961).
23. M. Tanaka, T. Mizoguchi and Y. Aiyama, *J. Phys. Soc. Japan*, **18,** 1089 (1963).
24. M. Robbins and P. K. Baltzer, *J. Appl. Phys.*, **36,** 1039 (1965).
25. M. Tanaka, T. Tokoro and Y. Aiyama, *J. Phys. Soc. Japan*, **21,** 262 (1966).
26. K. Siratori, *J. Phys. Soc. Japan*, **23,** 948 (1967).
27. J. B. Goodenough, *J. Phys. Chem. Solids*, **25,** 151 (1964).
28. R. J. Arnott, A. Wold and D. B. Rogers, *J. Phys. Chem. Solids*, **25,** 161 (1964).
29. M. H. Francombe, *J. Phys. Chem. Solids*, **3,** 37 (1957).
30. M. Eibschütz, S. Shtrikman and Y. Tenenbaum, *Phys. Letters*, **24A,** 563 (1967).
31. Y. Kino and S. Miyahara, *J. Phys. Soc. Japan*, **21,** 2732 (1966).
32. J. Kanamori, M. Katooka and Y. Itoh, *J. Appl. Phys.*, **39,** 688 (1968).
33. P. Pausescu and R. Manaila, *Kristalografiya*, **13,** 627 (1968). [English transl. *Soviet Phys.–Cristallography*, **13,** 533 (1969).]
34. F. K. Lotgering, *Philips Res. Rept.*, **11,** 195 (1956).

35. S. Tsushima, *J. Phys. Soc. Japan*, **17,** Suppl. BI, 189 (1962).
36. L. Helmholz and R. F. Kruh, *J. Am. Chem. Soc.*, **74,** 1175 (1952).
37. I. R. Beattie, T. R. Gibson and G. A. Ozin, *J. Chem. Soc.*, **A1969,** 534 (1969).
38. S. Miyahara, *J. Phys. Soc. Japan*, **17,** Suppl. BI, 181 (1962).
39. T. Yamadaya, T. Mitui, T. Okada, N. Shikazono and Y. Hamaguchi, *J. Phys. Soc. Japan*, **17,** 1897 (1966).

Appendix IX
Parameters for vibronic systems

The following parameters can be found in the table: the Jahn–Teller (or pseudo Jahn–Teller) stabilization energy E_{JT} (or E_{PJT}) due to linear coupling; the ratio of E_{JT} to the effective vibrational quantum $\hbar\omega$, the barrier 2β for the rotational motion in the q_θ, q_ε plane; 3Γ the tunnelling energy between the minima of the potential in that plane, the linear coupling coefficient L, or $L(\Gamma_1, \Gamma_2)$ for the P-J-T-E; $2W$ or $3W$, the P-J-T splitting between the electronic states (which is positive if a symmetric or a singlet state is lower). For linear molecules the angle of distortion is listed. Other parameters are explained in the table.

The energies are given in cm^{-1}.

The data were collected or extracted from experiments or calculation without much sieving, except that results superseded by more reliable data were omitted. So were results which were based on evidently faulty reasoning. On the other hand, the uncertainties or error limits which would be reasonably attached to the numbers are not shown in the table. These uncertainties are frequently large (and at times the numbers may therefore be regarded as 'the least improbable values').

Ag^{2+}	in KCl, O_h
	^2E_g (ground state), $E \otimes \varepsilon$, $E_{JT}/\hbar\omega \gg 1$, $2\beta = 4000$
	[Reference 1] Optical absorption
Anthracene	as dimer
	$(A + B) \otimes \beta$, $E_{JT}/\hbar\omega = 2$, $2W = \hbar\omega = 0\cdot4$
	(in sandwich dimer), $2\cdot7$ (in stable dimer)
	[References 2, 3] Optical absorption

Benzene

C_6H_6, D_{6h}
$^3B_{1u}$ and $^3E_{1u}$ (excited states). $(E + A) \otimes \varepsilon$,
$2W = 6886$,[4] $\hbar\omega = 1600$
$L(E, A)/\hbar\omega = 2.1$,[5] 2.5–2.8 [6,7]
Optical absorption and calculated

$^1B_{1u}$ and $^1E_{1u}$ (ex. states), $(E + A) \otimes \varepsilon$,
$2W = 3200$,[8] $\hbar\omega = 1600$ [8]
$L(E, A)/\hbar\omega = 2.9$,[8] $E_{PJT} = 2500$,[9]
$2\beta = 500$ [9]
Calculated

Benzene Ion

$C_6H_6^+$ or $C_6H_6^-$, D_{6h}
$^2E_{1g}$ or $^2E_{2u}$ (gr. st.), $E \otimes (\varepsilon_1 + \varepsilon_2 + \ldots)$,
$E_{JT} = 490$,[10] 560,[11] 380,[12] 700,[13] 430,[14]
$2\beta = 48$,[10] 82,[11] 0,[12,14] -73,[13]
$L < 0$
$E_{JT}(\varepsilon_1) = 560$, $\hbar\omega_1 = 1600$;
$E_{JT}(\varepsilon_2) = 245$, $\hbar\omega_2 = 611$ [15]
Calculated

Bi

in GaP, T_d
T_1 (gr. st.), $T \otimes (\varepsilon + \tau_2)$, $(E_{JT}(\tau_2)/\hbar\omega_\tau)$
$\qquad\qquad\qquad\qquad > (E_{JT}(\varepsilon)/\hbar\omega_\varepsilon) \simeq 1 - 3$
[Reference 16] Luminescence with stress

$G_{3/2}$ (exciton from bound hole and electron pair),
$G \otimes (\varepsilon + \tau_2)$, $E_{JT}(\varepsilon)/\hbar\omega = 3$
[Reference 17] Luminescence

CCl_4

T_d
T_2 (gr. st.), $T \otimes (\varepsilon + \tau_2)$, $(E_{JT}(\tau)/\hbar\omega_\tau) \gg$
$\qquad\qquad\qquad\qquad\qquad (E_{JT}(\varepsilon)/\hbar\omega_\varepsilon)$
[References 18, 19] Raman spectroscopy

CH_4^+

T_d
T_2 (gr. st.), $T \otimes (\varepsilon + \tau_2)$, $E_{JT}(\varepsilon)/\hbar\omega_\varepsilon = 0.095$,[20]
$\qquad\qquad\qquad\qquad\qquad 9.4$,[20a] $\hbar\omega_\varepsilon = 1530$
8.75,[20b] $\hbar\omega_\varepsilon = 1300$[20b]
$E_{JT}(\tau)/\hbar\omega_\tau = 2.8$,[20] 6.0,[20a] 7.55,[20b] $\hbar\omega_\tau = 1300$
Calculated
(For $C_3H_4^+$ and $C_3H_6^+$ see p. 328.]

CF$_2$ D$_{\infty h}$
$^1A_1$ (gr. st. in C$_{2v}$), $\widehat{FCF} = 104\cdot9°$

$^1B_1$ (ex. st. in C$_{2v}$), $\widehat{FCF} = 134\cdot8°$
[References 21, 22] Optical absorption

CF$_4^+$ T$_d$
T$_2$ (gr. st.), T $\otimes$ ($\varepsilon + \tau_2$),
$E_{JT}(\varepsilon)/\hbar\omega_\varepsilon = 5 \times 10^{-3}$, $\hbar\omega_\varepsilon = 640$
$E_{JT}(\tau)/\hbar\omega_\tau = 2\cdot1 \times 10^{-2}$, $\hbar\omega_\tau = 1400$
[Reference 20] Calculated

CO$_2^-$ D$_{\infty h}$
$^2A_1$ (gr. st. in C$_{2v}$), $\widehat{OCO} = 134°$
[Reference 23] ESR

CO^{2+} in MgO, O$_h$
$^4T_{2g}$ (ex. st.), T $\otimes$ ε, $E_{JT}/\hbar\omega = 1\cdot1$
[Reference 23a] Luminescence and optical
absorption

in CdS, T$_d$
$^4T_1^a$ (ex. st.), T $\otimes$ ε, $E_{JT}/\hbar\omega = 1\cdot3$
[References 24, 25] Optical absorption

CO^{3+} in (CoF$_6$)$^{3-}$, O$_h$
^5E_g (ex. st.), E $\otimes$ ε, $E_{JT}/\hbar\omega = 13$(K$_3$CoF$_6$),
10(Na$_3$CoF$_6$), 10(K$_2$NaCoF$_6$), 16(Li$_3$CoF$_6$),
58(Ba$_3$(CoF$_6$)$_2$)
[Reference 26] Optical absorption

^5E_g (ex. st.), E $\otimes$ ε, $E_{JT} = 1800$
[Reference 27] Calculated

Coronene$^-$ C$_{24}$H$_{12}^-$, D$_{6h}$
$^2E_{2u}$ (gr. st.), E $\otimes$ ε, $E_{JT} = 100$
[Reference 28] Calculated

Cr^{2+} in Al$_2$O$_3$, $\sim$O$_h$
^5E_g (gr. st.), E $\otimes$ ε, $E_{JT}/\hbar\omega = 3\cdot2$, $\hbar\omega = 730$,
$3\Gamma = 1\cdot74$
[Reference 29] Acoustic paramagnetic resonance
and thermal conduction

in MgO, O_h

^5E_g (gr. st.), $E \otimes \varepsilon$, $E_{JT}/\hbar\omega = 3\cdot5\text{--}4\cdot5$,

$\hbar\omega = 300$, $3\Gamma = 7\cdot5$,[31] 16*

A.p.r., thermal conduction and calculation.

[*L. J. Challis, A. M. de Goer, K. Guckelsberger and G. A. Slack (to be published)]

in $(CrF_6)^{4-}$, O_h

^5E_g (gr. st.), $E \otimes \varepsilon$, $E_{JT} = 6520$

[Reference 27] Calculated

Cr^{3+} in water, O_h

$^5T_{2g}$ (ex. st.), $T \otimes (\varepsilon + \tau_2)$, $E_{JT}(\varepsilon) = 1770$,

$E_{JT}(\tau) = 45$

[Reference 32] Calculated

Cu^{2+} in water, O_h

^2E_g (gr. st.), $E \otimes \varepsilon$, $E_{JT}/\hbar\omega = 25$, $\hbar\omega = 200$,

$2\beta = 1200$, $3\Gamma = 10^{-5}$

[Reference 33] Calculated

in $La_2Mg_3(NO_3)_{12}\cdot24H_2O$, O_h

^2E_g (gr. st.), $E \otimes \varepsilon$, $E_{JT} = 1100$, $2\beta = 900$,

$3\Gamma = 10^{-1}$

[References 34, 35] ESR

in $CuSiF_6\cdot6H_2O$, O_h

^2E_g (gr. st.), $E \otimes \varepsilon$, $E_{JT} = 400$

[Reference 36] Magnetic susceptibility and anisotropy

in $Ca(OH)_2$, O_h

^2E_g (gr. st.), $E \otimes \varepsilon$, $2\beta = 1300$

[Reference 37] ESR

in CaO, O_h

^2E_g (gr. st.), $E \otimes \varepsilon$, $E_{JT} = 500$, $3\Gamma = 50$

[References 38, 39, 40] ESR (above 4 °K)

in CaO, O_h

^2E_g (gr. st.), $E \otimes \varepsilon$, $2\beta < 0$, $3\Gamma = 0\cdot006$

[Reference 41] ESR at 1·2 °K

in MgO, O_h
^2E_g (gr. st.), $E \otimes \varepsilon$, $E_{JT} = 500$
[References 39, 42] ESR

in LiCl, O_h
^2E_g (gr. st.), $E \otimes \varepsilon$, $2\beta < 55$
[Reference 43] ESR

in NaCl, O_h
^2E_g (gr. st.), $E \otimes \varepsilon$, $E_{JT}/\hbar\omega = 6$, $\hbar\omega = 200$
[Reference 44] ESR with strain

in $(CuF_6)^{4-}$, O_h
^2E_g (gr. st.), $E \otimes \varepsilon$, $E_{JT} = 1350$, $2\beta = 400$,
$L > 0$
[Reference 27] Calculated

in $(CuCl_6)^{4-}$, O_h
^2E_g (gr. st.), $E \otimes \varepsilon$, $E_{JT} = 2025$, $2\beta = 39$
[Reference 45] Calculated

in $(CuCl_4)^{2-}$, T_d
$^2T_2$ (gr. st.), $T \otimes (\varepsilon + \tau_2)$, $E_{JT}(\varepsilon) = 2300^{46}$,
$3200,^{47}$ $\hbar\omega_\varepsilon = 85$ ^{46}
$E_{JT}(\tau) = 2120$ ^{47}
Calculated

in NH_4Cl, D_{4h}
$^2A_{1g}$ (gr. st.) and $^2B_{1g}$ (ex. st.) in
$CuCl_4(NH_4^+\text{-vacancy})_2$ complex
$(A + B) \otimes \varepsilon$, $2W = 4000$, $L(A, B) = 3000$
[Reference 48] ESR

in NH_4Cl, C_{4v}
$^2A_1$ (gr. st.) and $^2B_1$ (ex. st.) in
$CuCl_4 \cdot NH_4^+(NH_4^-\text{-vacancy})$ complex
$(A + B) \otimes \varepsilon$, $2W = 4000$, $L(A, B) = 4000$
[Reference 48] ESR

Cyclobutadiene C_4H_4, D_{4h}
$^1B_{1g}$ (gr. st.) and ^1A_g (ex. st.), $(A + B) \otimes \varepsilon$,
$2W = 22\,000$
$L^2(A, B)/(2W . \hbar\omega) = 2\cdot7$
[Reference 49] Calculated

Cycloheptatrienyl

C_7H_7, D_{7h} in naphthalene
$^2E'$ (gr. st.), $E \otimes \varepsilon$, $E_{JT} = 700$
[Reference 50] Calculated

Cyclooctatetraene
anion

$C_8H_8^-$, D_{8h}
^2E_g (gr. st.), $E \otimes \varepsilon$, $E_{JT}/\hbar\omega = 1/2$, $\hbar\omega = 1300$
[Reference 51] Calculated

Cyclopentadienyl

C_5H_5, D_{5h}
$^2E'$, $E \otimes \varepsilon$, $E_{JT} = 730$,[10] 560,[11] 500,[12] 3240,[13]
$2\beta = 9$,[10] 4,[11] 84,[13] 50 [52]
Calculated and ESR[52]

Ethylene

trimers, barrelene[a] and triquinacene[b]
singlet and doublet excited states
$(E + A) \otimes \varepsilon$, $3W/\hbar\omega = -21^{[a]}$, $+0{\cdot}3^{[b]}$
$\hbar\omega = 1600$, $L(E, A)/\hbar\omega = 5{\cdot}7$
[Reference 8] Calculated

Eu^{2+}

in CaF_2 and SrF_2, O_h
^2E (ex. st.), $E \otimes \varepsilon$, $3\Gamma = 15{\cdot}3(CaF_2)$, $6{\cdot}5(SrF_2)$
[References 53, 54] Optical absorption and ESR

F-centres

in cesium halides, O_h
$G_{3/2u}$ (ex. st.), $G_{3/2} \otimes \varepsilon$, $E_{JT}(\varepsilon) = 20(CsF)$,
$64(CsCl)$, $89(CsBr)$
[Reference 55] Optical absorption [See also p. 328]

F^+

in CaO, O_h
$^2T_{1u}$ (ex. st.), $T \otimes (\varepsilon + \tau_2)$, $E_{JT}(\varepsilon)/\hbar\omega_\varepsilon =$
$E_{JT}(\tau)/\hbar\omega_\tau = 2{\cdot}7\text{--}3{\cdot}3$, $\hbar\omega_\varepsilon = \hbar\omega_\tau = 280$,
$K(T_1) = 0{\cdot}02$
[References 56, 57] Optical absorption with stress

Fe^{2+}

in MgO, O_h
T_{2g} (gr. st.), $T \otimes (\varepsilon + \tau_2)$, $E_{JT}(\varepsilon) = E_{JT}(\tau) = 100$
[Reference 58] Optical and infrared spectroscopy,
ESR and NMR

in water, O_h
$^5T_{2g}$ (gr. st.), $T \otimes (\varepsilon + \tau_2)$, $E_{JT}(\varepsilon) = 530$,
$$E_{JT}(\tau) = 1160,$$
^5E_g (ex. st.), $E \otimes \varepsilon$, $E_{JT} = 4100$
[Reference 47] Calculated

in silver halides, O_h
^5E_g (ex. st.), $E \otimes \varepsilon$, $E_{JT}/\hbar\omega = 13$ (AgBr, AgCl)
$\hbar\omega = 150$ (AgBr), 170 (AgCl)
[Reference 59] Optical absorption

in other monohalides, O_h
^5E_g (ex. st.), $E \otimes \varepsilon$, $E_{JT}/\hbar\omega = 17$ (CdBr),
18 (CdCl, FeCl), 12 (MgCl), 15 (MnCl),
$\hbar\omega = 66$ (CdBr), 90 (CdCl, FeCl), 107 (MgCl),
97 (MnCl)
[Reference 60] Optical absorption

in other ionic compounds, O_h
^5E_g (ex. st.), $E \otimes \varepsilon$, $E_{JT}/\hbar\omega = 8$ (MgO),[61]
2 (KFeF$_3$),[61] 6 (FeF$_2$ at R.T.),[62]
2·2 (FeF$_2$ at 810 °K),[63] $\hbar\omega = 230$ (MgO),[61]
770 (KFeF$_3$),[61] 750 (FeF$_2$)[63]
Optical absorption

in tetrahedral coordination, T_d
$^5T_2$ (ex. st.), $T \otimes \varepsilon$, $E_{JT}/\hbar\omega = 1·8$,[64] 0·6[65] (CdTe),
1·8 (MgAl$_2$O$_4$),[64]
$\hbar\omega = 140$ (CdTe),[64] 495 (MgAl$_2$O$_4$)[64]
Optical absorption [See also p. 328]

in haemoglobin, C_4
^2E, $E \otimes \varepsilon$, $E_{JT}/\hbar\omega = 0·8$, $\hbar\omega = 265$
[Reference 66] ESR

Ga$^+$ in KCl, O_h
$^1T_{1u}$ (ex. st.), $T \otimes \tau_2$, $E_{JT} = 2600$
[Reference 67] Optical absorption

Hydride molecules   XH$_2$, $D_{\infty h}$

$^2A_{1u}$ (gr. st. in D_{2h}), $\widehat{HAlH} = 119°$ [68]

$\widehat{HBH} = 131°$ [69]

$^1A_{1u}$ (gr. st.), $\widehat{HCH} = 102·4°$, $^1B_1$ (ex. st.),

$\widehat{HCH} < 150°$ [70]

$^2B_{1u}$ (gr. st.), $\widehat{HNH} = 103°$ [71,71a]

$\widehat{HPH} = 91°$ [72,73]

$^2A_{1u}$ (ex. st.), $\widehat{HNH}$ = 144° [71,71a]

$$\widehat{HPH} = 123° \text{ [72,73]}$$

Optical absorption and theory[71a]

$\widehat{ABH}$, $C_{\infty v}$

States arising from $^1\Delta$ (in $C_{\infty v}$):

$^1A'$ (gr. st. in C_s) $\widehat{HNO}$ = 108·6°, [74]

$\widehat{HPO}$ = 104·7° [75]

$^1A''$ (gr. st.), $\widehat{HCP}$ = 128, [68] $\widehat{HCN}$ = 125° [76]

$^1A''$ (ex. st.), $\widehat{HNO}$ = 116·3° [74]

State arising from $^2\pi$

$^2A'$ (gr. st. in C_s), $\widehat{HCO}$ = 119·5° [77]

Optical absorption

H_3 D_{3h}
$^2E'$, $E \otimes \varepsilon$, E_{JT} = 18 000
[Reference 78] Calculated

In^+ in CsBr, O_h
$^1T_{1u}$ (ex. st.), $T \otimes \tau_2$, $E_{JT}/\hbar\omega = 2$, $\hbar\omega = 100$
[Reference 79] Optical absorption

in KCl, O_h
$^1T_{1u}$ (ex. st.), $T \otimes \tau_2$, E_{JT} = 3150,
[Reference 67] Optical absorption

Mn^{2+} in RbMnF$_3$, O_h
$^4T_{1g}(G)$ (ex. st.), $T \otimes \varepsilon$, $E_{JT}/\hbar\omega = 0.7$
[Reference 80] Magnon spectroscopy

Mn^{3+} in water, O_h
^5E_g (gr. st.), $E \otimes \varepsilon$, $|2\beta| = 3000$, $3\Gamma = 3 \times 10^{-3}$,
$\hbar\omega = 300$
[Reference 27] Calculated

in $(MnF_6)_2^{3-}$, O_h
^5E_g (gr. st.), $E \otimes \varepsilon$, E_{JT} = 3485,[81] 4000 [27,32]
Reflectance in K_2MnF_5 [81] and calculated[27,32]

	in YAlG, O_h
	^5E_g (gr. st.), $E \otimes \varepsilon$, $2\beta = 220$, $\hbar\omega = 90$
	[Reference 82] Acoustic attenuation
Mo^{5+}	in $MoCl_5$, D_{3h}
	$^2E''$ (gr. st.), $E \otimes \varepsilon$, $E_{JT} > 600$
	[Reference 83] Optical absorption and infrared spectroscopy
N	in diamond, T_d
	T_2 (gr. st.), $T \otimes \tau$, $E_{JT} = 5600$,[84,85] $40,000$[85a]
	[References 84, 85] ESR [85a] Calculated
N_2^-	in NaN_3, D_{3d}
	Doublet (stabilized by P-J-T-E), $E \otimes \varepsilon$,
	$E_{JT} = 250$, $3\Gamma < 15$
	[Reference 86] ESR
NO_2	$D_{\infty h}$
	arising from $^2\pi_u$ (gr. st.), $^2A_{1u}$ (gr. st. in D_{2h}),
	$\widehat{ONO} = 134 \cdot 2°$
	[Reference 87] Optical absorption
NH_3^+	D_{3h} and C_{3v}
	E'' or E excited states, $E \otimes \varepsilon$,
	$E_{JT}/\hbar\omega = 0 \cdot 02$ (E'' in D_{3h}), $5 \cdot 7$ (E in C_{3v})
	[Reference 20] Calculated
Ni^{3+}	in Al_2O_3, $\sim O_h$
	^2E_g (gr. st.), $E \otimes \varepsilon$, $E_{JT}/\hbar\omega = 9 \cdot 5$, $2\beta = 90$,
	$3\Gamma = 1$, $\hbar\omega = 80$
	[References 39, 82, 88] ESR and acoustic attenuation
	in $SrTiO_3$, O_h
	^2E_g (gr. st.), $E \otimes \varepsilon$, $E_{JT}/\hbar\omega = 80$, $3\Gamma \gg 23$,
	$\hbar\omega = 200$
	[Reference 89] ESR
	in monoxides, O_h
	^2E_g (gr. st.), $E \otimes \varepsilon$, $E_{JT} \leqslant 2500$ (CaO),
	1300 (MgO)
	[Reference 90] ESR

Ni⁻

in Ge, T_d
^2E (gr. st. stabilized by PJTE), $E \otimes \varepsilon$, $E_{JT} = 1800$, $2\beta = 1500$
[Reference 91] ESR

Os^{6+}

in OsF_6
$E + T_2$ (gr. states, nearly degenerate),
$(E + T) \otimes \varepsilon$, $E_{JT}(\varepsilon)/\hbar\omega_\varepsilon = 0.03 - 0.04$,
$\hbar\omega_\varepsilon = 668$,
$T \otimes \tau_2$, $E_{JT}(\tau)/\hbar\omega_\tau = 0.071$, $\hbar\omega_\tau = 276$
[Reference 92] Infrared spectroscopy

Pb^+

in chlorides, O_h
$^1T_{1u}$ (ex. st.), $T \otimes \tau_2$, $E_{JT} = 6700$ (NaCl),
4900 (KCl), 7800 (RbCl)
[Reference 67] Optical absorption

R-centre

in KCl, C_{3v}
E (gr. st.), $E \otimes \varepsilon$, $E_{JT}/\hbar\omega = 0.5$
[Reference 93] Stress induced dichroism

R⁻-centre

in MgO, C_{3v}
^2E (ex. st.), $E \otimes \varepsilon$, $E_{JT}/\hbar\omega < 0.05$, $\hbar\omega = 350$
[Reference 94] Optical absorption

R′-centre

in LiF, C_{3v}
^3E (ex. st.), $E \otimes \varepsilon$, $E_{JT}/\hbar\omega = 1.5-3$
[Reference 95] Optical absorption

Re^{6+}

in ReF_6, O_h
$G_{3/2g}$ (gr. st.), $G_{3/2} \otimes (\varepsilon + \tau_2)$, $E_{JT}(\varepsilon)/\hbar\omega = 0.063,$[92]
$0.07,$[96,97] $\hbar\omega_\varepsilon = 670$, $E_{JT}(\tau)/\hbar\omega_\tau = 0.078,$[92]
$0.2,$[96] $\hbar\omega_\tau = 295$ [97]
Infrared spectroscopy and calculated

Re^{4+}

in K_2ReCl_6, O_h
$G_{3/2g}(^2T_{2g})$ (ex. st.), $G_{3/2} \otimes \varepsilon$, $E_{JT}/\hbar\omega = 2.7$
[Reference 98] Optical absorption

Ru^{6+}

in RuF_6, O_h
T_{2g} (gr. st.), $T \otimes \tau_2$, $E_{JT}/\hbar\omega = 0.11$, $\hbar\omega_\tau = 283$
[Reference 92] Infrared spectroscopy

Sc^{2+}

in fluorites, O_h
^2E_g (gr. st.), $E \otimes \varepsilon$, $E_{JT}/\hbar\omega = 10$ (CaF_2, SrF_2),
$3\Gamma = 10$ (CaF_2), 8 (SrF_2)
[References 99, 100] ESR

Sm^{2+}

in fluorites, O_h
^5E_g(f^5d) (ex. st.), $E \otimes \varepsilon$, $E_{JT}/\hbar\omega = 3\cdot2$ (CaF_2),
$4\cdot6$ (SrF_2), $\hbar\omega = 250$ (CaF_2), 210 (SrF_2)
[Reference 101] Optical absorption

Sn^{2+}

in rubidium halides, O_h
$^1T_{1u}$ (ex. st.), $T \otimes \tau_2$, $E_{JT} = 11,700$ (RbCl),
$10,000$ (RbBr)
[Reference 67] Optical absorption

Tc^{6+}

in TcF_6, O_h
$G_{3/2g}$ (gr. st.), $G_{3/2} \otimes (\varepsilon + \tau_2)$,
$E_{JT}(\varepsilon)/\hbar\omega_\varepsilon = 0\cdot69$, $\hbar\omega_\varepsilon = 640$,
$E_{JT}(\tau)/\hbar\omega_\tau = 0\cdot071$, $\hbar\omega_\tau = 297$
[Reference 92] Infrared spectroscopy

Ti^{3+}

in water, O_h
$^2T_{2g}$ (gr. st.), $T \otimes (\varepsilon + \tau_2)$, $E_{JT}(\varepsilon) = 100$,[32]
790,[47] $E_{JT}(\tau) = 180$,[32] 2100 [47]
Calculated

in corundum, $\sim O_h$
$^2T_{2g}$ (gr. st.), $T \otimes \varepsilon$, $E_{JT}/\hbar\omega = 1$,[102]
^2E_g (ex. st.), $E \otimes \varepsilon$, $E_{JT}/\hbar\omega = 20$,[103] $\hbar\omega = 200$
ESR and optical absorption

in $(TiF_6)^{3-}$, O_h
^2E_g (ex. st.), $E \otimes \varepsilon$, $E_{JT} = 7000$,[27]
$E_{JT}/\hbar\omega = 25$ (NaK_2TiF_6, $(NH_4)_3TiF_6$),[104]
$\hbar\omega = 300$
[Reference 27] Calculated
[Reference 104] Optical absorption

in silver halides, O_h
^2E_g (ex. st.), $E \otimes \varepsilon$, $E_{JT}/\hbar\omega = 8$ (AgCl), 34 (AgBr),
$\hbar\omega = 370$ (AgCl), 190 (AgBr)
[Reference 59] Optical absorption

                                        in CaF_2, O_h
$^2T_{2g}$ (ex. st.), $T \otimes \tau_2$, $E_{JT}/\hbar\omega = 1.4$, $\hbar\omega = 250$
[Reference 105] Optical absorption

Tl$^+$ in alkali halides, O_h
$^1T_{1u}$ (ex. st., C-band), $T \otimes \tau_2$, $E_{JT} = 4500$ (NaCl),
3000 (KCl)[67]
$^3T_{1u}$ (ex. st., A-band), $T \otimes (\varepsilon + \tau_2)$,
$E_{JT}(\varepsilon)/\hbar\omega_\varepsilon = 2.6$ (KCl), 0.3 (KBr), 2.6 (KI) [106]
$E_{JT}(\tau)/\hbar\omega_\tau = 1.1$ (KCl), 0.3 (KBr), 0.2 (KI),[106]
$\hbar\omega_\tau = \hbar\omega_\varepsilon = 144$ (KCl), 116 (KBr), 96 (KI),
$^3T_{2u}$ (ex. st., B-band), $T \otimes (\varepsilon + \tau_2)$,
$E_{JT}(\varepsilon)/\hbar\omega_\varepsilon = 0.2$ (KI), $E_{JT}(\tau)/\hbar\omega_\tau = 0.5$ (KI)[106]
Optical absorption[67] with stress[106]

Triphenylene D_{6h}
$^2E_{1g}$ (gr. st.), $E \otimes \varepsilon$, $E_{JT} = 100$
[Reference 28] Calculated

Vacancy   in diamond, T_d
V$^+$: $^2T_2$ (gr. st.), $T \otimes \varepsilon$, $E_{JT} = 570$
V^0: ^1E (gr. st.), $E \otimes \varepsilon$, $E_{JT} = 400$
      $^3T_1$ (ex. st.), $T \otimes \tau_2$, $E_{JT} = 340$
V$^-$: $^2T_1$ (ex. st.), $T \otimes \tau_2$, $E_{JT} = 480$
[Reference 107] Calculated

Vacancy and ion   in silicon, C_{3v}
pair   Doublet (gr. st.), $E \otimes \varepsilon$, $E_{JT} = 10,000$ (Al),
4600 (As), $11,000$ (P), 5500 (Sb), $|2\beta| = 480$ (As),
500 (P), 560 (Sb)
[References 108, 109, 110] ESR and ENDOR
[See also p. 329]

V^{2+}   in KMgF$_3$, O_h
$^4T_{2g}$ (ex. st.), $T \otimes \varepsilon$, $E_{JT}/\hbar\omega = 0.87$, $\hbar\omega = 150$
[Reference 111] Optical absorption

                                          in CaF_2, O_h
$^4T_{1g}$ (gr. st.), $T \otimes (\varepsilon + \tau_2)$, $E_{JT}(\varepsilon) = 6$,
$E_{JT}(\tau) = 140$, $\hbar\omega_D = 330$
[Reference 112] Calculated

V^{3+}

in water, O_h
$^3T_{1g}$ (gr. st.), $T \otimes (\varepsilon + \tau_2)$, $E_{JT}(\varepsilon) = 165$,
$E_{JT}(\tau) = 145$
[Reference 32] Calculated

in Al_2O_3, $\sim O_h$
$^3T_{2g}$ (ex. st.), $T \otimes \varepsilon$, $E_{JT}/\hbar\omega = 2\cdot5$, $\hbar\omega = 200$
[Reference 113] Optical absorption

in MgO, O_h
$^3T_{2g}$ (ex. st.), $T \otimes \varepsilon$, $E_{JT}/\hbar\omega = 0\cdot6$, $\hbar\omega = 500$
[Reference 114] A.p.r. and calculated

V^{4+}

in Al_2O_3, $\sim O_h$
$^2T_{2g}$ (gr. st.), $T \otimes \varepsilon$, $E_{JT}/\hbar\omega = 1\cdot6$, $\hbar\omega = 200$
[Reference 115] ESR

in VCl_4, T_d
^2E (gr. st.), $E \otimes \varepsilon$, $E_{JT} = 73$,[116] 1600 [117]
$\hbar\omega = 118$,[116] 112 [117]
Electron diffraction[116] and calculated[117-9]

Additions

Allene cation $C_3H_4^+$, D_{2d}
E (gr. st.), $E \otimes \varepsilon$, $E_{JT} = 4680$
[Reference 120] Calculation

Cr^{2+} in tetrahedral coordination (CdS, CdTe, ZnS, ZnSe, ZnTe), T_d
$^5T_2$ (gr. st.), $T \otimes \varepsilon$, $E_{JT}/\hbar\omega = 7\cdot5$, $\hbar\omega = 50-90$
[Reference 121] I.R. and optical spectroscopy

Cyclopropane $C_3H_6^+$, D_{3h}
cation $^2E'$ (gr. st.), $E \otimes \varepsilon$, $E_{JT} = 3260,^{120}\ 3630^{122}$
$\hbar\omega = 1051,^{122-3}\ 480^{124}$
[References 120, 122] calculation, [124] photoelectron spectroscopy

F centres in alkali halides, O_h
$^2T_{1u}$ (ex. st.), $T \otimes \varepsilon$, $E_{JT}/\hbar\omega = 1\cdot2^{125}$ (KCl),
$1\cdot5^{126-8}$ (KF),
$^2T_{1u}$ and $^2A_{1u}$ (ex. states), $(T + A) \otimes \tau_2$,
R'(P-J-T ratio, equation (1.1)) $= 2\cdot4-3$ (KCl),
$1\cdot9-3$ (KF), $\geqslant 1\cdot1$ (NaF)^{126}
$2W = 140$ (KCl), 130 (KF), 100–130 (NaF)$^{125-9}$
Stark effect and MCD

F' in CaO, O_h
$^3T_{1u}$ (ex. st.), $T \otimes \varepsilon$, $E_{JT}/\hbar\omega = 1$
[Reference 57] ESR and MCD

Fe^{2+} in CdTe, T_d
^5E (gr. st.), $E \otimes \varepsilon$, $E_{JT}/\hbar\omega = 0\cdot11$, $\hbar\omega = 38$
[Reference 130] I.R. spectroscopy
in ZnS, T_d
$^5T_2$ (ex. st.), $T \otimes (\varepsilon_1 + \varepsilon_2)$, $(E_{JT}/\hbar\omega)_{1,2} = 2, 0\cdot3$
$\hbar\omega_{1,2} = 100, 300$
[Reference 131] I.R. spectroscopy and luminescence

La^{2+} in $SrCl_2$, O_h
^2E (gr. st.), $E \otimes \varepsilon$, $E_{JT}/\hbar\omega = 1$, $\hbar\omega = 300$
[Reference 132] ESR

Mn^{3+} in acetylacetonate, $\sim O_h$
^5E (gr. st.), $E \otimes \varepsilon$, $E_{JT}/\hbar\omega = 17$, $\hbar\omega = 500$
$3\Gamma = 10$
[Reference 133] calculation, optical absorption, ESR

NCO $C_{\infty v}$
$^2\Pi$ (gr. st.), $\Pi \otimes \pi$, ε (Renner–Teller parameter) $= -0.16$
$\hbar\omega = 540$
[References 134–5] ESR and calculation

U^{4+} in ThO$_2$, O_h
T_{2u} (gr. st.), $T \otimes \varepsilon$, $E_{JT}/\hbar\omega = 0.35$
[References 136–7] Magnetic susceptibility

Vacancy–impurity in SiC, C_{3v}
pair Doublet (ex. st.), $E \otimes \varepsilon$, $|2\beta| = 260$
$\hbar\omega = 530$
[Reference 138] Luminescence

A IX.1 References

1. C. J. Delbecq, W. Hayes, M. C. M. O'Brien and P. H. Yuster, *Proc. roy. Soc. A*, **271**, 243 (1963).
2. E. A. Chandross, J. Ferguson and E. G. MacRae, *J. Chem. Phys.*, **45**, 3546 (1966).
3. M. Garcia-Sucre, F. Geny and R. Lefebvre, *J. Chem. Phys.*, **49**, 458 (1968).
4. S. D. Colson and E. R. Bernstein, *J. Chem. Phys.*, **43**, 2661 (1965).
5. H. C. Longuet-Higgins and L. Salem, *Proc. roy. Soc. A*, **251**, 172 (1959).
6. C. A. Coulson and A. Golebiewski, *Proc. Phys. Soc.*, **78**, 1310 (1961).
7. J. H. van der Waals, A. M. D. Berghuis and M. S. de Groot, *Mol. Phys.*, **13**, 301 (1967).
8. M. H. Perrin and M. Gouterman, *J. Chem. Phys.*, **46**, 1019 (1967).
9. A. D. Liehr, *Z. Naturforsch*, **16a**, 642 (1961).
10. A. D. Liehr, *Zeits. Phys. Chem.*, **9**, 338 (1956).
11. L. C. Snyder, *J. Chem. Phys.*, **33**, 619 (1960).
12. W. D. Hobey and A. D. MacLachlan, *J. Chem. Phys.*, **33**, 1965 (1960).
13. C. A. Coulson and A. Golebiewski, *Mol. Phys.*, **5**, 71 (1962).
14. W. D. Hobey, *J. Chem. Phys.*, **43**, 2187 (1965).
15. B. Sharf and J. Jortner, *Chem. Phys. Letters*, **2**, 68 (1968).
16. A. Onton and T. N. Morgan, *Phys. Rev.*, **B1**, 2592 (1970).
17. T. N. Morgan, *J. Luminescence*, **1**, 420 (1970).
18. W. Holzer and H. Moser, *J. Mol. Spectrosc.*, **13**, 430 (1964).
19. E. M. Verlan, *Optika i Spektroskopiya*, **24**, 378 (1967), [English transl. *Optics and Spectry.*, **24**, 197 (1968)].
20. C. A. Coulson and H. L. Strauss, *Proc. roy. Soc. A*, **269**, 443 (1962).
20a. J. Arents and L. C. Allen, *J. Chem. Phys.*, **53**, 73 (1970).
20b. R. N. Dixon, *Mol. Phys.*, **20**, 113 (1971).
21. P. Venkateswarlu, *Phys. Rev.*, **77**, 676 (1950).
22. C. W. Matthews, *J. Chem. Phys.*, **45**, 1068 (1966).
23. D. W. Ovenall and D. H. Whiffen, *Mol. Phys.*, **4**, 135 (1961).
23a. J. E. Ralph and M. G. Townsend, *J. Chem. Phys.*, **48**, 149 (1968).
24. H. Weakliem, *J. Chem. Phys.*, **36**, 2117 (1962).
25. M. D. Sturge, *Solid State Phys.*, **20**, 91 (1967).
26. F. A. Cotton and M. D. Meyers, *J. Am. Chem. Soc.*, **82**, 5023 (1960).
27. A. D. Liehr and C. J. Ballhausen, *Ann. Phys. (N.Y.)*, **3**, 304 (1958).
28. L. C. Snyder, *Bull. Am. Phy. Soc.*, **6**, 165 (1961).
29. C. A. Bates and J. M. Dixon, *J. Phys. C: Solid State Phys.*, **2**, 2209 (1969).
30. C. A. Bates, J. M. Dixon, J. R. Fletcher and K. W. H. Stevens, *J. Phys. Soc., C: Solid State Phys.*, **1**, 859 (1968).
31. J. R. Fletcher and K. W. H. Stevens, *J. Phys. C: Solid State Phys.*, **2**, 444 (1969).
32. J. H. Van Vleck, *J. Chem. Phys.*, **7**, 61, 72 (1939).
33. U. Öpik and M. H. L. Pryce, *Proc. roy. Soc. A*, **238**, 425 (1957).
34. D. P. Breen, D. C. Krupka and F. I. B. Williams, *Phys. Rev.*, **179**, 241 (1969).
35. F. I. B. Williams, D. C. Krupka and D. P. Breen, *Phys. Rev.*, **179**, 255 (1969).

36. B. D. Bhattacharya and S. K. Datta, *Ind. J. Phys.*, **41**, 181 (1968).

37. R. G. Wilson, F. Holuj and N. E. Hedgecock, *Phys. Rev.*, **B1**, 3609 (1970).

38. W. Low and J. T. Suss, *Phys. Letters*, **7**, 310 (1963).

39. K. A. Müller, 'Jahn–Teller effects in magnetic resonance' in *Magnetic Resonance and Relaxation*, Ed. Blinc (Amsterdam: North Holland, 1967), p. 192.

40. R. Englman, *Phys. Letters*, **31A**, 473 (1970).

41. R. E. Coffman, D. L. Lyle and D. R. Mattison, *J. Phys. Chem.*, **72**, 139 (1968).

42. U. T. Höchli, K. A. Müller and P. Wysling, *Phys. Letters*, **15**, 1 (1965).

43. J. R. Pilbrow, R. W. H. Stevenson, *Phys. Stat. Sol.*, **34**, 293 (1969).

44. R. H. Borcherts, A. Kanzaki and H. Abe, *Phys. Rev.*, **B2**, 23 (1970).

45. L. L. Lohr, Jr., *Inorg. Chem.*, **6**, 1890 (1967).

46. G. Felsenfeld, *Proc. roy. Soc. A*, **236**, 500 (1950).

47. I. B. Bersuker, *Teor. i Eksper. Khim.*, **1**, 5 (1965).

48. J. R. Pilbrow and J. M. Spaeth, *Phys. Stat. Sol.*, **20**, 225, 237 (1967).

49. R. Buenker and S. D. Peyerimhoff, *J. Chem. Phys.*, **48**, 354 (1968).

50. H. J. Silverstone, D. E. Wood and H. M. McConnell, *J. Chem. Phys.*, **41** 2311 (1962).

51. A. D. MacLachlan and L. C. Snyder, *J. Chem. Phys.*, **36**, 1159 (1962).

52. G. R. Liebling and H. M. McConnell, *J. Chem. Phys.*, **42**, 3931 (1965).

53. A. A. Kaplyanskii and A. K. Przhevuskii, *Opt. i Spektroskopiya*, **19**, 597 (1965) [English transl.: *Opt. Spectry.*, **19**, 331 (1965)].

54. L. L. Chase, *Phys. Rev. Letters*, **23**, 275 (1969).

55. P. R. Moran, *Phys. Rev.*, **137**, A1016 (1965).

56. A. E. Hughes, *J. Phys. C: Solid State Phys.*, **3**, 627 (1970).

57. Y. Merle d'Aubigné and A. Roussel, *Phys. Rev.*, **B3**, 1421 (1971).

58. F. S. Ham, W. M. Schwarz and M. C. M. O'Brien, *Phys. Rev.*, **185**, 548 (1969).

59. H. D. Koswig, U. Retter and W. Ulrici, *Phys. Stat. Sol.*, **24**, 605 (1967).

60. T. E. Freeman and G. D. Jones, *Phys. Rev.*, **182**, 411 (1969).

61. G. D. Jones, *Phys. Rev.*, **155**, 259 (1967).

62. W. E. Hatfield and T. S. Piper, *Inorg. Chem.*, **3**, 1295 (1964).

63. J. P. Young, *Inorg. Chem.*, **8**, 825 (1969).

64. G. A. Slack, F. S. Ham and R. M. Chrenko, *Phys. Rev.*, **152**, 376 (1966).

65. A. S. Marfunin, A. N. Platonov and V. E. Fedorov, *Fiz. Tverd. Tela*, **9**, 3616 (1967) [English transl.: *Soviet Phys.–Solid State*, **9**, 2847 (1968)].

66. H. Kamimura and S. Mizuhashi, *J. Appl. Phys.*, **39**, 684 (1968).

67. Y. Toyozawa and M. Inoue, *J. Phys. Soc. Japan*, **21**, 1663 (1966).

68. G. Herzberg, *Electronic Spectra of Polyatomic Molecules* (Princeton, Van Nostrand, 1966).

69. G. Herzberg and J. W. C. Johns, *Proc. roy. Soc. A*, **298**, 142 (1967).

70. G. Herzberg and J. W. C. Johns, *Proc. roy. Soc. A*, **295**, 107 (1966).

71. K. Dressler and D. A. Ramsay, *Phil. Trans. roy. Soc. A*, **251**, 553 (1959).

71a. R. N. Dixon, *Mol. Phys.*, **9**, 357 (1965).

72. R. N. Dixon, G. Duxbury and D. A. Ramsay, *Proc. roy. Soc. A*, **296**, 137 (1967).

73. P. C. Jordan, *J. Chem. Phys.*, **41**, 1442 (1969).

74. F. W. Dalby, *Can. J. Phys.*, **36**, 1336 (1958).

75. M. Lam Tanh and M. Peyron, *J. Chim. Phys.*, **60**, 1289 (1963); **61**, 1531 (1964).

76. G. Herzberg and K. K. Innes, *Can. J. Phys.*, **35**, 842 (1957).

77. G. Herzberg and D. A. Ramsay, *Proc. roy. Soc. A*, **233**, 34 (1955).

78. R. N. Porter, R. M. Stevens and M. Karplus, *J. Chem. Phys.*, **49**, 5163 (1968).

79. A. Fukuda and S. Makishima, *Phys. Letters*, **24A**, 267 (1967).

80. M. Y. Chen and D. S. McClure, *Bull. Am. Phys. Soc.*, **14**, 79 (1969).

81. T. S. Davis, J. P. Fackler and M. J. Weeks, *Inorg. Chem.*, **7**, 1994 (1968).

82. E. M. Gyorgy, M. D. Sturge, D. B. Fraser and R. C. LeCraw, *Phys. Rev. Letters*, **15**, 19 (1965).

83. R. F. W. Bader and Kun Po Huang, *J. Chem. Phys.*, **43**, 3760 (1965).

84. L. A. Shulman, I. M. Zaritskii and G. A. Podzyarei, *Fiz. Tverd. Tela*, **8**, 2307 (1966) [English transl.: *Soviet Phys.–Solid State*, **8**, 1842 (1967)].

85. J. H. N. Loubser and W. P. Ryneveld, *Brit. J. Appl. Phys.*, **18**, 1029 (1967).

85a. R. P. Messner and G. D. Watkins, *Phys. Rev. Letters*, **25**, 656 (1970).

86. E. Gelerinter and R. H. Silsbee, *J. Chem. Phys.*, **45**, 1703 (1966).

87. A. E. Douglas and K. P. Huber, *Can. J. Phys.*, **43**, 74 (1965).

88. F. S. Ham, in *Electron Paramagnetic Resonance*, ed. Geschwind (New York, Plenum Press, 1971).

89. J. C. Slonczewski, K. A. Müller and W. Berlinger, *Phys. Rev. B.*, **1**, 3545 (1970).

90. U. T. Höchli, K. A. Müller and P. Wysling, *Phys. Letters*, **15**, 5 (1965).

91. G. W. Ludwig and H. H. Woodbury, *Phys. Rev.*, **113**, 1014 (1959).

92. B. Weinstock and G. L. Goodman, *Adv. Chem. Phys.*, **9**, 169 (1963).

93. R. H. Silsbee, *Phys. Rev.*, **138**, A180 (1965).

94. I. K. Ludlow, *J. Phys. C: Solid State Phys.*, **1**, 1194 (1968).

95. J. A. Davis and D. B. Fitchen, *Sol. State Commun.*, **6**, 505 (1968).

96. M. S. Child and A. C. Roach, *Mol. Phys.*, **9**, 281 (1965).

97. A. A. Kiseljov, *J. Phys. B*, **2**, 270 (1969).

98. J. C. Eisenstein, *J. Chem. Phys.*, **34**, 1628 (1961).

99. U. T. Höchli and T. L. Estle, *Phys. Rev. Letters*, **18**, 128 (1967).

100. U. T. Höchli, *Phys. Rev.*, **162**, 262 (1967).

101. P. P. Sorokin, M. J. Stevenson, J. R. Lankard and G. D. Pettit, *Phys. Rev.*, **127**, 503 (1962).

102. R. M. Macfarlane, J. Y. Wong and M. D. Sturge, *Phys. Rev.*, **166**, 250 (1968).

103. D. S. McClure, *J. Chem. Phys.*, **36**, 2757 (1962).

104. H. D. Bedon, S. M. Horner and S. Y. Tyree, Jr., *Inorg. Chem.*, **3**, 647 (1964).

105. W. Low and A. Rosenthal, *Phys. Letters*, **26A**, 143 (1968).

106. D. Bimberg, W. Dultz and W. Gebhardt, *Phys. Stat. Solidi*, **31**, 661 (1969).

107. J. Friedel, M. Lannoo and G. Leman, *Phys. Rev.*, **164**, 1056 (1967).

108. E. Elkin and G. D. Watkins, *Phys. Rev.*, **174**, 881 (1968).

109. G. D. Watkins, *Phys. Rev.*, **155**, 802 (1955).

110. G. D. Watkins and J. W. Corbett, *Phys. Rev.*, **134**, A1359 (1964).

111. M. D. Sturge, *Phys. Rev.*, **B1**, 1005 (1970).

112. L. K. Aminov and B. Z. Malkin, *Fiz. Tverd. Tela*, **9**, 1316 (1967) [English transl.: *Soviet Phys.–Solid State*, **9**, 1030 (1967)].

113. W. C. Scott and M. D. Sturge, *Phys. Rev.*, **146**, 262 (1966).

114. T. Ray, *Solid St. Commun.*, **9**, 911 (1971).

115. R. M. McFarlane, J. Y. Wong and M. D. Sturge, *Phys. Rev.*, **166**, 250 (1967).

116. Y. Morino and M. Uehara, *J. Chem. Phys.*, **45**, 4543 (1966).
117. C. A. Coulson and B. M. Deb, *Mol. Phys.*, **16**, 545 (1969).
118. C. A. L. Becker and J. P. Dahl, *Theoret. Chim. Acta*, **19**, 135 (1970).
119. D. A. Copeland and C. J. Ballhausen, *Theoret. Chim. Acta*, **20**, 317 (1971).
120. E. Haselbach, *Chem. Phys. Letters*, **7**, 427 (1970).
121. J. T. Vallin, G. A. Slack, S. Roberts and A. E. Hughes, *Phys. Rev.*, **B2**, 4313 (1970).
122. C. Rowland, *Chem. Phys. Letters*, **9**, 169 (1971).
123. S. J. Cyvin, *Spectrochim. Acta*, **16**, 1022 (1960).
124. H. Basch, M. B. Robin, N. A. Kuebler, C. Baker and D. W. Turner, *J. Chem. Phys.*, **51**, 52 (1969).
125. M. P. Fontana, *Phys. Rev.*, **B2**, 4304 (1970).
126. L. F. Stiles, Jr., M. P. Fontana and D. B. Fitchen, *Phys. Rev.*, **B2**, 2077 (1970).
127. M. P. Fontana and D. B. Fitchen, *Phys. Rev. Letters*, **23**, 1497 (1969).
128. T. Iida, K. Kurata and S. Muramatsu in Proceedings of International Conference of Colour Centres at Reading, September 1971.
129. L. D. Bogan and D. B. Fitchen, *Phys. Rev.*, **B1**, 4122 (1970).
130. J. T. Vallin, *Phys. Rev.*, **B2**, 2390 (1971).
131. F. S. Ham and G. A. Slack, *Phys. Rev.*, **B4**, 777 (1971).
132. J. R. Herrington, T. L. Estle and L. A. Boatner, *Phys. Rev.*, **B3**, 2933 (1971).
133. B. D. Bhattacharyya, *Phys. Stat. Sol.*, **43**, 495 (1971).
134. R. N. Dixon, *Phil. Trans. Roy. Soc. A*, **252**, 165 (1960).
135. A. Carrington, A. R. Fabris, B. J. Howard and J. D. Lucas, *Mol. Phys.*, **20**, 961 (1971).
136. K. Sasaki and Y. Obata, *J. Phys. Soc. Japan*, **28**, 1157 (1970).
137. F. S. Ham, *J. Physique (Suppl.)*, **32**, C1–953 (1971).
138. W. J. Choyke and L. Patrick, *Phys. Rev.*, **B4**, 1843 (1971).

Author Index

Subject Index